D1356621

Biodiversity *in* Agroecosystems

Advances in Agroecology

Series Editor: Clive A. Edwards

Agroecosystem Sustainability: Developing Practical Strategies,
Stephen R. Gliessman

Agroforestry in Sustainable Agricultural Systems,
Louise E. Buck, James P. Lassoie, and Erick C.M. Fernandes

Biodiversity in Agroecosystems,
Wanda Williams Collins and Calvin O. Qualset

Interactions Between Agroecosystems and Rural Communities
Cornelia Flora

Landscape Ecology in Agroecosystems Management
Lech Ryszkowski

Soil Ecology in Sustainable Agricultural Systems,
Lijbert Brussaard and Ronald Ferrera-Cerrato

Soil Tillage in Agroecosystems
Adel El Titi

Structure and Function in Agroecosystem Design and Management
Masae Shiyomi and Hiroshi Koizumi

BIODIVERSITY IN AGROECOSYSTEMS

Edited by

Wanda W. Collins and Calvin O. Qualset

Library of Congress Cataloging-in-Publication Data

Biodiversity in agroecosystems / edited by Wanda Williams Collins, Calvin O. Qualset.
p. cm.— (Advances in agroecology)
Includes bibliographical references and index.
ISBN 1-56670-290-9 (alk. paper)
1. Agriculatural ecology. 2. Biological diversity. 3. Agricultural systems.
I. Collins, Wanda Williams. II. Qualset, Calvin O. III. Series.
S589.7.B575 1998
630—dc21 98-13056
CIP

Direct all inquiries to CRC Press LLC, 2000 N.W. Corporate Blvd., Boca Raton, Florida 33431.

Lewis Publishers is an imprint of CRC Press LLC

International Standard Book Number 1-56670-290-9
Library of Congress Card Number 98-13056
Printed in the United States of America 4 5 6 7 8 9 0
Printed on acid-free paper

Preface

Diversity in biological resources (biodiversity) evolved to fill the innumerable environmental niches of the Earth. It is manifest in all except the most extreme global environments. Humans have exploited this diversity and shaped it over millennia to meet basic needs for survival. Thus, biodiversity has been an integral and essential feature in the evolution of agriculture. Agrobiodiversity, that part of the full spectrum of diversity on which humans directly depend for food and fiber, represents plants, animals, and microbes that are greatly modified by the domestication process to the extent that many of them cannot survive in the environments of their progenitors.

Ancient farmers knew well the necessity of having different crops or animal breeds and different varieties of each crop or breed in case of failure of one or the other for various reasons. They also explored and utilized the rich diversity of flavors, textures, and aromas available in their plant and animal species.

The management of agrobiodiversity is temporally and spatially integrated into agroecosystems. How agroecosystem management affects ecosystems in general is a subject of intensive research, debate, and controversy. Both intensive and extensive agricultural systems are essential to the sustainability of the global human population. In turn, these systems are dependent upon the existence of a rich reservoir of agrobiodiversity and its use in ways that complement and enhance biodiversity in its natural state. As is evident throughout this book, there are examples of severe negative impacts of agricultural activities on biodiversity, mainly through degradation of natural habitats. Biodiversity losses are becoming recognized and, through research and practice, are slowly being mitigated. At the same time, there are examples of agriculture practices that enhance biodiversity and one of the great challenges for the next century will be to discover additional ways for achieving complementarity of food, fiber, and energy production and biodiversity conservation.

Today's farmers, using both traditional and high-technology methods, also understand that the continued success of their agriculture depends upon conserving, maintaining, and using the diversity that is so under threat now. It is also important that conservationists and policy makers have a thorough understanding of agricultural processes so they can discover practical ways to facilitate both agricultural production and natural resource conservation.

This book is designed to emphasize just how important biodiversity is to agriculture. The various chapters point out the positive effects of agriculture on helping to conserve and protect the diversity on which it depends. The effects of agricultural growth, genetic uniformity, and dependence on high levels of external inputs are also highlighted. The book is intended to present the reader with a broad view of the interplay of biodiversity and agriculture with chapters addressing soil microbes, insects, plants, animals, rangelands, and agro-forestry. Other chapters discuss the efforts to conserve, maintain, and effectively use plant and animal diversity most important to agriculture. Still other chapters delve into the value of genetic diversity to agriculture and the interaction with surrounding habitats and species. Finally, the integration of new biotechnologies into traditional and industrial agricultural systems can have great impact on the quality of biodiversity and help shape strategies for its conservation.

We believe that sharing this knowledge and these experiences by some of the world's most knowledgeable experts will provide readers with a broad appreciation and heightened awareness of the stakes involved in the future preservation of these natural resources.

Wanda W. Collins
Calvin O. Qualset

The Editors

Wanda W. Collins, Ph.D., currently is the Deputy Director General for Research at the International Potato Center in Lima, Peru. She is the author of more than 40 journal articles on genetics and plant breeding, several book chapters, and numerous extension and popular publications.

Dr. Collins is a graduate of North Carolina State University where she received her Ph.D. in both Plant Pathology and Genetics in 1976. She joined the school's Department of Horticultural Science that same year and became a full professor in 1986. She has been doing research on vegetable breeding and genetics (sweetpotato and potato) since 1976 and traveled to Peru to research the genetic resources of the *Ipomoea* species (wild relatives of the sweetpotato). Between 1996 and 1998, Dr. Collins also spent over 2 years on study leave at the World Bank in Washington, D.C. as Agricultural Research Advisor, Environmentally Sustainable Development Vice Presidency.

She is one of the originators and the first chair of the Sweetpotato Crop Genetics Committee which advises the U.S. Germplasm System on sweetpotato germplasm. Her many memberships and contributions to the field of horticulture include Chair of the Board of Trustees for the International Plant Genetic Resources Institute, member of the Executive Board for the American Society of Horticultural Science, and as the current President of the American Society of Horticultural Science.

Calvin O. Qualset, Ph.D., is Director of the Genetic Resources Conservation Program of the Division of Agriculture and Natural Resources at the University of California, Davis. Before retiring as a professor in 1994, he held the positions of Chair of the Department of Agronomy and Range Science and Assistant Dean for Plant Sciences and Pest Management at UC Davis. He also served as Assistant Professor at the University of Tennessee, Knoxville from 1964 to 1967 before moving to California.

He is active in developing biological conservation strategies and conducts research on plant genetic resource conservation and genetics. He serves on the Board of Trustees of the American Type Culture Collection as a representative of the Genetics Society of America. He coordinates the International Triticeae Mapping Initiative and is Principal Investigator on McKnight Foundation and BARD-funded projects in Mexico and Israel on *in situ* conservation of crop plants and genetic mapping of wheat. He has served on numerous reviews of international centers for the Consultative Group for International Agricultural Research, of university departments, and of current issues in agriculture and biology for the National Science Foundation and National Research Council.

Dr. Qualset held Fulbright Fellowships to Australia and Yugoslavia and served as president of the American Society of Agronomy and of the Crop Science Society of America. He has received citations for work in plant breeding from the U.S. National Council of Commercial Plant Breeders and the Mexican National Institute of Agriculture and Forestry. He has authored more than 190 scientific papers and developed numerous varieties of wheat, oat, and triticale.

Contributors

Miguel A. Altieri
Center for Biological Control
ESPM-Division of Insect Biology
201 Wellman – 3012
University of California at Berkeley
Berkeley, CA 94720-3112
(agroeco3@nature.Berkeley.edu)

Stuart S. Bamforth
Department of Ecology, Evolution, and Organismal Biology
Tulane University
New Orleans, LA 70118-5698

Mary E. Barbercheck
Department of Entomology
North Carolina State University
Raleigh, NC 27695-7634
(mary_barbercheck@ncsu.edu)

Gary W. Barrett
Institute of Ecology
University of Georgia
Athens, GA 30602-2202
(gbarrett@sparrow.ecology.uga.edu)

Terry A. Barrett
Institute of Ecology
University of Georgia
Athens, GA 30606-2202
(barrett@negia.net)

Harvey W. Blackburn
Research Leader
USDA Sheep Experiment Station
H.C. 62, Box 2010
Dubois, ID 83423
(usdaspur@micron.net)

Lara M. Coburn
Department of Wildlife and Fisheries Science
Texas A&M State University
210 Nagle Hall
College Station, TX 77843-2258
(lcoburn@wfscgate.tamu.edu)

Joel I. Cohen
Project Manager, Intermediary Biotechnology Service
International Service for National Agricultural Research (ISNAR)
The Hague, The Netherlands
(j.cohen@cgnet.com)

Wanda W. Collins (Editor)
Deputy Director General for Research
International Potato Center
Apartado 1558
Lima 12, Peru
(w.collins@cgnet.com)

Cornelis de Haan
World Bank
1818 H Street N.W.
Washington, D.C. 20433
(cdehaan@worldbank.org)

Michael I. Goldstein
Department of Wildlife and Fisheries Science
Texas A&M State University
210 Nagle Hall
College Station, TX 77843-2258
(mgoldstein@wfscgate.tamu.edu)

Douglas Gollin
Department of Economics
Fernald House, Williams College
Williamstown, MA 01267 and
Affiliate Scientist, Economics Program
Centro Internacional de Mejoramiento de Maiz y Trigo (CIMMYT)
Lisboa 27, Apdo. Postal 6-641,
CP 06600 Mexico, D.F.
(*dgollin@williams.edu*)

Keith Hammond
Animal Production and Health Division
Food and Agriculture Organization of the United Nations
00100 Rome, Italy
(keith.hammond@fao.org)

Geoffrey C. Hawtin
Director General
International Plant Genetic Resources Institute (IPGRI)
Via delle Sette Chiese 142
Rome, Italy
(g.hawtin@cgnet.com)

Ann C. Kennedy
USDA-ARS
Pullman, WA 99164-6421
(akennedy@mail.wsu.edu)

Thomas E. Lacher, Jr.
Department of Wildlife and Fisheries Science
Texas A&M State University
210 Nagle Hall
College Station, TX 77843-2258
(tlacher@wfscgate.tamu.edu)

Roger R. B. Leakey
Institute of Terrestrial Ecology, Bush Estate
Penicuik, Midlothian EH26 0QB
Scotland, U.K.
(r.leakey@ite.ac.uk)

Helen W. Leitch
International Livestock Research Institute
Box 30709
Nairobi, Kenya
(h.leitch@cgnet.com)

Virginia D. Nazarea
Department of Anthropology
University of Georgia
Athens, GA 30602
(*vnazarea@uga.cc.uga.edu*)

Deborah A. Neher
Department of Biology
University of Toledo
Toledo, OH 43606
(dneher@uoft02.utoledo.edu)

Clara I. Nicholls
Department of Entomology
University of California at Davis
Davis, CA 95616
(cinicholls@ucdavis.edu)

John David Peles
Savannah River Ecology Laboratory
Drawer E
Aiken, SC 29806
(peles@srel.edu)

Keith S. Pike
Irrigated Agriculture Research and Extension Center
Washington State University
Prosser, WA 99350
(kpike@tricity.wsu.edu)

Jon K. Piper
Department of Biology
Bethel College
North Newton, KS 67117
(smilax@bethelks.edu)

Calvin O. Qualset (Editor)
Director, Genetic Resources Conservation Program
University of California at Davis
Davis, CA 95616
(*coqualset@ucdavis.edu*)

Robert E. Rhoades
Department of Anthropology
University of Georgia
Athens, GA 30602
(*rrhoades@uga.cc.uga.edu*)

R. Douglas Slack
Department of Wildlife and Fisheries Science
Texas A&M State University
210 Nagle Hall
College Station, TX 77843
(dslack@wfscgate.tamu.edu)

Melinda Smale
Economics Program
Centro Internacional de Mejoramiento
de Maiz y Trigo (CIMMYT)
Lisboa 27, Apdo. Postal 6-641
CP 06600 Mexico, D.F
(*msmale@cimmyt.mx*)

Petr Starý
Institute of Entomology
Czech Academy of Sciences
Branisovska 31,
370 05 Ceské Budejovice
Czech Republic

Neil E. West
Department of Rangeland Resources
Utah State University
Logan, UT 84322
(new369@cc.usu.edu)

Contents

CHAPTER 1

Microbial Diversity in Agroecosystem Quality

Ann C. Kennedy

CONTENTS

INTRODUCTION

Agroecosystems comprise 30% of Earth's surface (Altieri, 1991). Microbial diversity is one factor that controls agroecosystem productivity and quality. Biological diversity studies generally have been centered around plants, animals, and insect

1-56670-290-9/99/$0.00+$.50

Table 1 Justification for Investigations of Microbes and Microbial Diversity

Microbes are an important source of knowledge about life strategies and limitations.
Microbes are key to understanding evolutionary history.
The untapped diversity of microorganisms is a resource for new genes and organisms of value to biotechnology.
Microbes and microbial diversity patterns can be used as indicators of environmental changes.
Microbial communities are key to understanding biological interactions.
Microbes play a key role in conservation and restoration biology of higher organisms.
Microbes are critical to sustainability.

Source: Microbial Diversity Research Priorities, American Society of Microbiology, Washington, D.C., 1994. With permission.

species with little attention being given to microorganisms. Without microorganisms and their biochemical processes, life on Earth would not be possible (Price, 1988). Microbial diversity can directly influence plant productivity and diversity by influencing plant growth and development, plant competition, and nutrient and water uptake. Microbes provide sources of genetic material; in addition, they can be used as indicators of environmental change (American Society of Microbiology, 1994; Table 1).

Microorganisms are key to understanding biological interactions. They play a role in conservation and restoration of agroecosystems and are essential to quality considerations. Soil organisms constitute a large, dynamic nutrient source in all ecosystems and play a major role in residue decomposition and nutrient cycling (Smith and Paul, 1990). In addition, microorganisms are responsible for chemical changes in the soil, such as the accumulation of soil organic matter, dinitrogen fixation, and other changes in soil properties that may affect plant growth. When a disturbed system begins to recover, it is not only due to plant recolonization, but is also dependent upon the action of microbes in promoting favorable soil conditions and fertility.

We are unaware of the true extent or dimension of the diversity of soil microbes, although molecular investigations suggest that populations in soil are greater than we are able to understand with cultural techniques (Holben and Tiedje, 1988; Torsvik et al., 1990). Microbial diversity is critical to ecosystem functioning as a result of the myriad of processes for which microbes are responsible, such as decomposition and nutrient cycling, soil aggregation, and pathogenicity. It is necessary for us to increase our knowledge of biotic and functional diversity to understand better the desired makeup of microbial communities and the optimum management practices for an agroecosystem.

Anthropogenic influences may affect ecosystem functioning and diversity. Some of the most dramatic examples of ecosystem disturbances are occurring as a result of soil erosion from agroecosystem perturbations. These disturbances may be leading to progressive declines in biological, chemical, and physical stability in ecosystems. Continued disturbance of these areas may cause massive changes in global carbon cycling, resulting from significant organic carbon loss from the terrestrial environment. Microorganisms are highly sensitive to disturbance, such as those introduced by agriculture (Elliott and Lynch, 1994), and they may function as early warning indicators of changes in quality. Fluxes in microbial diversity and functional diversity may contribute greatly to the understanding of soil quality and the development of sustainable agroecosystems (di Castri and Younes, 1990; Hawksworth, 1991a; Thomas and Kevan, 1993). Soil organisms are useful in classifying disturbed or contaminated systems, since diversity can be affected by minute changes in the ecosystem. The use of microorganisms and their functioning for examination of environmental stress and declining biological diversity needs to be exploited for the benefit of agroecosystems (Office of Technology Assessment, 1987). The purpose of this chapter is to explore the issue of microbial diversity in agroecosystems and its influence on the quality of that agroecosystem.

Taxonomic Diversity

The number of known microbial species is estimated to be over 110,000, but only a fraction of these species is identified and even fewer are being studied or are in culture collections (Hawksworth, 1991b). By one estimate, only 3 to 10% of Earth's microbial species have been identified, leaving a vast portion of that biota unknown and therefore unstudied (Hawksworth, 1991a). According to other estimates, the number of microbial species may be as many as 1 million organisms (American Society for Microbiology, 1994). Even with conservative estimates, the microbial populations and their diversity require more investigation. Microbial communities are highly diverse and thought to exhibit even greater diversity than that seen in higher orders of organisms (Torsvik et al., 1990; Ward et al., 1992). Obviously, we are aware of only a miniscule portion of the total potential of the system. Within the soil microbial population, there is a wealth of genetic information waiting to be discovered.

Functional Diversity

Functional diversity and taxonomic diversity are often two vastly different measurements. Functional diversity includes the magnitude and capacity of soil inhabitants that are involved in key roles. These processes are selected to represent biologically meaningful processes, such as carbon or nitrogen cycling, decomposition of various compounds, and other transformations (Zak et al., 1994). Taxonomic diversity is determined by culturability and isolation of species, and may represent 20% or less of the microbes present and active in soil. Speciation relies on characterization of known phenotypic or genetic characterizations that may not be present in soil and possibly have no bearing on soil processes (Lee and Pankhurst, 1992).

Microbial diversity indexes have been used to describe the status of microbial communities and their response to natural or human disturbances. Microbial diversity indexes can function as bioindicators of the stability of a community and can be used to describe the ecological dynamics of a community and the impact of stress on that community (Mills and Wassel, 1980; Atlas, 1984). Chemically stressed or heavy metal–stressed soils were found to decrease in microbial diversity depending on the type of chemical applied (Atlas, 1984; Reber, 1992). A factor limiting the greater use of these indexes is the absence of detailed information on the microbial species composition of soil environments (Torsvik et al., 1990).

The ability of an ecosystem to buffer the effects of extreme disturbances may depend in part on the diversity of the system (Elliott and Lynch, 1994). It may be important to monitor diversity as an indicator of change or in response to a stress. The extinction rate of species within a system may be an important indicator of the status of the system and critical in determining the level of diversity necessary to maintain an agroecosystem. The actual numbers of species and species composition may not be as important as the flux of species within a system and the functioning of those individuals within that system.

Diversity indexes can be used to indicate the effect of disturbance; however, greater diversity may not always be desirable. Greater diversity should not be equated with a more stable system; rather, the changes in diversity with management may be more informative of the status of a soil microbial community. Diversity issues in understanding quality of soil may not be enlightening unless the functions of the system are taken into account. Basic shifts in large groups in an ecosystem may indicate a change, but may not be able to address functioning of that altered system. Microbial communities and their processes need to be examined, not only in relation to the individuals that make up the community, but also with regard to the effect of perturbations or environmental stresses on those communities.

SOIL MICROORGANISMS IN AGROECOSYSTEMS

The soil is full of microorganisms such as bacteria, actinomycetes, fungi, algae, viruses, and protozoa. Microbes are classified in categories based on their utilization of carbon and energy and their nutrient requirements (i.e., phototrophs, chemotrophs, autotrophs, heterotrophs, or lithotrophs). This provides limited information, since many situations overlap and refinement is needed (Alexander, 1977; Walker, 1992; Zak et al., 1994). Bacteria and fungi, as primary decomposers in the cycling of nutrients, occupy a critical position in the soil food web. Of all nutrient cycling, 90 to 95% passes through these two groups of organisms to higher trophic levels. Thus, the function and possibly the diversity of bacteria and fungi will be a large factor in determining the quality of agroecosystems (Lynch, 1983).

Bacteria and actinomycetes are the most numerous of microbial groups in soils, but because of their small size, 1 to 10 μm, they only account for 50% of the total biomass in soil (Alexander, 1977). Bacteria are found in soil at populations of 10^4 to 10^9 cells g^{-1} soil. As a group they are diverse metabolically and use many different sources of energy and carbon. Bacteria play an important role in the breakdown of organic material and nutrient cycling. Most natural and xenobiotic compounds can be broken down by soil microflora with few compounds becoming recalcitrant (Dorn et al., 1974). Some bacteria have the potential for nitrogen fixation (Sprent, 1979) or methane production or utilization (Jones, 1991). Denitrification and sulfate reduction involve a variety of facultative and obligate anaerobic bacteria (Tiedje et al., 1984). Nitrification and sulfur oxidation, on the other hand, are the result of the activity of a limited number of genera of aerobic autotrophic bacteria (Belser and Schmidt, 1978; Bock et al., 1989).

Fungi are less diverse than bacteria and less numerous in soil at 10^4 to 10^6 propagules g^{-1} soil. Fungi however, are responsible for up to 70% of the biomass (Lynch, 1983). Fungi are found in soil, in associations with plant roots, or as saprophytes on detrital material (Swift and Boddy, 1984). Fungi can withstand adverse soil conditions better than other microorganisms, and they survive at lower water potentials than bacteria (Papendick and Campbell, 1975). Hyphal strands allow fungi to overcome the environmental constraints of low moisture and depleted nutrients by translocation of water and nutrients. Also, several fungi excrete organic acids that solubilize otherwise unavailable nutrients (Sollins et al., 1981). Fungi are active decomposers of cellulose, lignin, and other organic materials. The products of decomposition are then released and used by other organisms, especially bacteria. Many fungi are plant pathogens, yet some form beneficial relationships with plant roots (e.g., mycorrhizae).

Algae are most numerous in surface soil and found at populations of 10^2 to 10^6 cells g^{-1} soil. In some agricultural systems, algae contribute to nitrogen cycling by nitrogen fixation or to soil stabilization (Metting and Rayburn, 1983). Protozoa inhabit soil at populations between 10^3 and 10^5 cells g^{-1} soil. These organisms are major predators of bacteria and thus regulate bacterial populations (Opperman et al., 1989).

Rhizosphere

Plants may be an important selective force for the diversity of rhizosphere populations of bacteria and fungi through their influence on soil nutrients. Rhizosphere soil can be defined as that volume of soil adjacent to and influenced by plant roots (Metting, 1993). It is a region of intense microbial activity because of its proximity to plant root exudates, making rhizosphere microbial communities distinct from those of bulk soil (Curl and Truelove, 1986; Whipps and Lynch, 1986). Microbial activity is stimulated in this area because of the nutrients provided by the root or germinating seed (Rouatt and Katznelson, 1961). Microbial populations and functioning in an agroecosystem are influenced by the root and the soil environment, including mineral and organic material. The aboveground plant community can influence microbial spatial heterogeneity in soil. Those microorganisms that respond to root exudates or related substrates will dominate the rhizosphere. Decaying root systems also function as a source of nutrients for the surrounding microorganisms (Swinnen et al., 1995). Microbial populations decrease with distance from the roots (Yeates and Darrah, 1991).

Bacteria account for the largest number of inhabitants of the rhizosphere. Gram-negative, non-spore-forming rods with simple nutritional requirements are stimulated more by roots than are coccoid forms and Gram-positive, spore-forming rods (Curl and Truelove, 1986). The composition of the plant community may influence the diversity of the microbial community due to the variability in chemical composition of the exudates (Christensen, 1989).

The development of rhizosphere microbial communities also is influenced by different plant species (Rovira, 1956), plant phenology (Smith, 1969), and environmental factors influencing plant growth (Rovira, 1959; Vancura, 1967; Martin and Kemp, 1980). Bacterial and fungal abundance in the rhizosphere is influenced by the nutrient status of both plant and soil. The percent mycorrhizal cover on roots of *Plantago lanceolata* was positively correlated with leaf nitrogen and phosphorus, while the percent cover of bacteria and other fungi was negatively correlated with phosphorus (Newman et al., 1981). Nitrogen fertilization increased numbers of fungi and Gram-negative bacteria in a rice rhizosphere (Emmimath and Rangaswami, 1971). It may be difficult to separate the effects of soil nutrients on rhizosphere populations from effects involved with increased or altered root exudation of organic compounds. Grasses grown in monoculture can modify nitrogen availability (Wedin and Tilman, 1990), and it has been hypothesized that such changes in soil nitrogen availability influenced by plant species affects the composition of vesicular arbuscular mycorrhizae fungal communities (Johnson et al., 1992b).

MICROBIAL IMPACTS ON AGROECOSYSTEMS

Microbes impact agroecosystems through a large list of functions for which they are responsible (Table 2). Soil humus formation, cycling of nutrients, and building soil tilth and structure (Lynch, 1983; Wood, 1991) are distributed among a large number of different genera and species. Microorganisms are responsible for many transformations in soil related to plant nutrition and health. The majority of soil microbes are beneficial to plant growth, but they need to be managed effectively (Lynch, 1983). Potential harmful effects from soil microorganisms include plant disease, production of plant-suppressive compounds, and loss of plant-available nutrients. Specific microorganisms can be manipulated to produce beneficial effects for agriculture (Lynch, 1983), for example, rhizobia to increase plant available nitrogen (Sprent, 1979), mycorrhizal associations to enhance nutrient uptake (Mohammad et al., 1995), or biological control of plant pests to reduce chemical inputs (Cook and Baker, 1983; Kennedy et al., 1991).

Beneficial soil bacteria can enhance plant performance by an increase in mineral solubilization (Okon, 1982), dinitrogen fixation (Albrecht et al., 1981), the production of hormones (Brown, 1972), and the suppression of harmful pathogens (Chang and Kommendahl, 1968; Cook and Baker, 1983). The symbiotic relationship between bacteria and legumes is one of the most widely studied and applied plant–microbial

Table 2 Several Functions of Soil Microbes

Decomposition of organic residues with release of nutrients
Formation of beneficial soil humus by decomposing organic residues and through synthesis of new compounds
Release of plant nutrients from insoluble inorganic forms
Improved plant nutrition through mycorrhizal relationships which are symbiotic relationships between fungi and plant roots
Transformation of atmospheric dinitrogen to plant-available N
Improvement of soil aggregation, aeration, and water infiltration
Antagonistic action against insects, plant pathogens, and weeds (biological control)

interactions (Sprent, 1979). The bacterium *Rhizobium* forms nodules on the roots of the legume plant, takes dinitrogen from the air, and transforms it to plant-available nitrogen (i.e., NH_4^+, NO_3^-). The plant provides nodules and photosynthate for the bacteria, while the bacteria give the plant the nitrogen it needs. Inoculation of legumes with dinitrogen-fixing *Rhizobium* can add appreciable amounts of nitrogen to the soil. The distribution and diversity of specific strains of dinitrogen-fixing bacteria vary with environmental conditions (Turco and Bezdicek, 1987; Hirsch et al., 1993). The plant community can influence the presence or absence of specific strains of *Rhizobium*, and thus impact the diversity of this group of microorganisms (Strain et al., 1994). The interaction involving mycorrhizal fungi and rhizobia may further affect the host plant by increasing nitrogen and phosphorus nutrition (Allen, 1992). These interactions are specific (Molina et al., 1992), further illustrating the complexity of the plant–microbe interaction and the changes in diversity of various microbial groups that can affect plant growth or impact other soil features.

Plant-suppressive bacteria reduce seed germination and delay plant development by the production of phytotoxic substances (Woltz, 1978; Suslow and Schroth, 1982; Alstrom, 1987; Schippers et al., 1987). Pathogenic fungi greatly reduce the survival, growth, and reproduction of plants (Shipton, 1977; Bruehl, 1987; Burdon, 1987), while beneficial mycorrhizal fungi can enhance plant growth by increasing nutrient (Fitter, 1977; Hall, 1978; Rovira, 1978; Ocampo, 1986) and water uptake (Tinker, 1976).

Mycorrhizal fungi have been found in a vast majority of plant communities (Allen and Allen, 1990) growing in association with 90% of terrestrial plants examined (Harley and Smith, 1983). Mycorrhizae are nonpathogenic fungi that form symbiotic associations with plant roots. The diversity of this group of organisms is vast, yet is not well studied. These associations are highly specific and indicate the diversity of this group of microorganisms (Molina et al., 1992). Mycorrhizal associations have been shown to be of greatest importance in stressed environments, phosphorus-deficient soils, eroded sites, and acidic or reclaimed lands (Harley and Smith, 1983; Barea, 1991). This association may be key in plant productivity and nutrient cycling (Barea, 1991; Allen, 1992). Mycorrhizal associations are enhanced by crop rotation and management practices favoring minimum disturbance.

Plant Growth

Interactions between plants and rhizosphere microbes may play a critical role in the outcome of plant competition. Competitive interaction among plants may also be important for the development of rhizosphere soil communities. Microbes affect nutrient uptake (Tinker, 1976; Okon, 1982). Microorganisms may play important roles affecting plant competition acting as mutualistic or pathogenic plant associates (Allen and Allen, 1990). Rhizosphere microorganisms may affect plant growth directly (Woltz, 1978; Suslow and Schroth, 1982; Gaskins et al., 1985; Alstrom, 1987; Schippers et al., 1987) or indirectly by their effects on each other and the microscale alteration of soil nutrients. Plant performance may also be affected by competitive interactions between adjacent plants (Goldberg and Fleetwood, 1987; Gurevitch et al., 1990; Goldberg and Landa, 1991).

Another example of microbial diversity influencing plant growth is the investigation of deleterious rhizobacteria, which were discovered in the early 1980s. Investigations of deleterious rhizobacteria have led to changes in management practices for many crops (Schippers et al., 1987) and may lead to biological control of weed species (Kremer et al., 1990; Kennedy et al., 1991). Deleterious rhizobacteria that specifically inhibit various grass weeds, but do not affect the crop, have been isolated from soil (Kennedy et al., 1991; 1992; Kennedy, 1994) and inhibit plant growth by the production of plant-suppressive compounds (Tranel et al., 1993). These bacteria are excellent biological control agents, in part, because they are aggressive colonizers of the roots and residue, often constituting up to 95% of the total pseudomonads on the plant root (Stroo et al., 1988; Kennedy et al., 1992). Biocontrol is critical to sustainable agriculture systems, but a greater understanding of soil microbes and their ecology is needed before biological control can be implemented. There is a wealth of genetic material contained within the soil that may have potential in biotechnology programs; thus, diversity investigations will benefit more than just one area of science (Malik and Claus, 1987; Bull et al., 1992).

Nutrient Cycling

Soil microorganisms constitute a large dynamic source and sink of nutrients in all ecosystems and play a major role in plant litter decomposition and nutrient cycling (Smith and Paul, 1990; Collins et al., 1992; Cambardella and Elliott, 1992). Microbes break down complex compounds in organic residue to simpler, smaller compounds, as well as recalcitrant compounds. This recycling into compounds of various complexity provides substrate for other microbes, further sustaining the diversity of soil microbes. Organic matter, in various forms of decay, improves soil physical properties, increases water-holding capacity, increases nutrient availability, and acts as a cementing agent for holding soil particles together. Organic matter can be maintained by incorporation of crop residues, crop rotation, and addition of animal and green manures when possible. Addition of organic matter aids in ensuring a productive soil and stimulates plant growth by providing food for microorganisms.

Soil Structure

Microbes play a major role in the formation of soil structure (Lynch and Bragg, 1985; Tisdall, 1991). Fungi and actinomycetes produce hyphal threads that bind soil particles together. Extracellular polysaccharides produced by bacteria and fungi bind soil particles together, assisting in building soil structure. Humic materials from microbial action form organic matter/clay complexes. This action reduces erosion, allows for good water infiltration, and maintains adequate aeration of the soil. Soil aggregation can be increased by the addition of residues resulting in additional microbial activity (Gilmour et al., 1948). The carbon and nitrogen pools delimit microbial biomass and decomposition rates and polysaccharide production (Knapp et al., 1983). The ability of fungi and bacteria to influence aggregation varies with species and is substrate dependent (Aspiras et al., 1971). Limited nitrogen in soil solution reduces biomass while increasing polysaccharide production, which can lead to increased aggregation (Elliott and Lynch, 1984). Soil stability may be managed by addition of different amendments to stimulate microbial activity (Jordahl and Karlen, 1993).

AGROECOSYSTEM IMPACTS ON MICROORGANISMS

Microbial diversity considerations need to be included in soil quality investigations of agricultural lands (Kennedy and Papendick, 1995). Soil quality is critical to the functioning of any ecosystem (Papendick and Parr, 1992). The quality of a soil can greatly impact land use, sustainability, and productivity. Soil quality, the inherent characteristic of a soil, comprises physical, chemical, and biological properties. These properties may be altered through intensive management practices. Reduction in aboveground plant diversity that occurs with severe disturbance, such as tillage, overgrazing, and pollutants, may decrease microbial diversity (Gochenauer, 1981; Boddy et al., 1988; Christensen, 1989).

Cropping Systems

Microbial communities in agroecosystems change with management history. In continuous cropping systems, cycling of pathogens and antagonists of pathogens and changes in crop yields can often be seen due to alteration in the disease pressure over time. An example of this is the decline of a disease of wheat called take-all decline that is the result of a change in the soil microbial community. The microbial community shifts to favor growth of antagonists of the pathogen *Gaeumannomyces graminis* var. *tritici* which results in decline of the disease (Cook, 1981). An example of the use of soil microbial diversity is the emerging area of biological control of plant pathogens (Cook and Baker, 1983). Microbes have the potential to be used in biological control, which is the suppression of one pest by using its natural predator or antagonist. Biocontrol can be used to control insects, pathogens, and weeds by either lowering the populations of the pest or by reducing the impact of the pest (DeBach, 1964). Microbes function as a direct delivery system for the natural pesticide they produce. Bacteria and fungi that produce different types of antibiotics can be used to control many plant pathogens (Cook and Baker, 1983).

Crop rotation is a key component in sustainable systems because it enhances beneficial microbes, interrupts the cycle of pathogens, and reduces weed populations. Legumes in rotation supply symbiotically fixed nitrogen to the system, aid in maintaining proper water status, and reduce pathogen load. Studies have shown the positive effects of crop rotation on crop growth, attributing this to changes in the microbial community composition (Shipton, 1977; Cook, 1981; Johnson et al., 1992a). Continuous monocropping led to changes in the soil community which increased the pathogen load and reduced barley growth when compared with grains in multiple-crop rotation (Olsson and Gerhardson, 1992). The populations and aggressiveness of pathogens can be altered with crop rotation, illustrating the changes in microbial diversity and function due to management (El Nashaar and Stack, 1989). In a long-term study, *Cochliobolus sativus*, a pathogen of spring wheat, was found in higher numbers and individual isolates exhibited greater aggressiveness or ability to cause severe disease in continuous wheat rotation, when compared with wheat in a 3-year rotation.

Tillage

In a study of the diversity of prairie and cultivated soils, diversity indexes were greater in disturbed or cultivated systems when compared with grassland (Kennedy and Smith, 1995). The increase in diversity with disturbance indicates a change in the microbial community to one that exhibited a greater range of substrate utilization and stress resistance. Soil microorganisms may affect plant growth and may influence plant competition. In turn, plants may act as a selective force for rhizosphere microbial populations through their influences on soil nutrients.

The ecology of root–microbe interactions after minimum tillage practices is vastly different from that after extensive plowing to prepare the seedbed. The changes in the physical and chemical properties of the soil resulting from tillage greatly alter the matrix-supporting growth of the microbial population. Within a given soil, there is considerable variation in the composition of the microbial community and diversity with depth in the profile. In no-till agricultural systems, microbial activities differed drastically with depth, with the greatest microbial activity occurring near the no-till surface; in the tilled system, activities were more evenly distributed throughout the plow layer (Doran, 1980). The composition of the microbial community influenced the rate of residue decomposition and nutrient cycling in both no-till and conventionally tilled systems (Beare et al., 1993). Decomposition in the no-till system was dominated by fungi, while the bacterial component was found to be responsible for a greater portion of the decomposition of residue in conventionally tilled systems. These studies illustrate the alteration of the makeup of the microbial communities and possibly the diversity of basic microbial groups with changes in management systems.

POTENTIAL APPLICATION OF MICROBIAL INDICATORS

The diversity of the microbial community, as well as the functions within the community, affect the stability and resilience of the soil system. Neither a higher nor lower degree of diversity in a system can be said to be better or worse; however, changes in the activity or community structure may influence the quality of the soil. The microbial biomass has recently been used as a sensitive indicator of management-induced changes (Doran, 1987; Powlson et al., 1987; Visser and Parkinson, 1992). Researchers have found discrepancies in microbial biomass responses to perturbations such as tillage. Kennedy and Smith (1995) found increased diversity with tillage. Utilizing microbial characteristics may better forecast change in soils since they respond to perturbations more rapidly than other indicators (Kennedy and Papendick, 1995).

Soil quality measures have recently become important in system comparisons of management options. Integral to soil quality assessment are measures of the microbial community. Microbiological properties can identify changes in the overall soil quality before changes occur in certain physical and chemical parameters that eventually impact overall crop and soil viability (Visser and Parkinson, 1992). Activity levels, biomass numbers, and community shifts can reflect the stability of a system with respect to the level of nutrient cycling, the amount of carbon utilized in a system, and the overall community structure and function in a soil system. Since microbial diversity of agricultural soils is important to the maintenance of soil formation, toxin removal, and elemental cycling (Borneman et al., 1996), diversity measures will most likely become increasingly important in evaluations of soil quality. Examples of rapid shifts in community structure such as these may serve as early indicators of changes in soil quality (Turco et al., 1994).

Soil resilience is defined as the ability of the soil to recover after disturbance (Elliott and Lynch, 1994). Soil resilience can be restored and biodiversity protected by, among other things, reducing tillage and increasing crop rotations (Elliott and Lynch, 1994). Studies have shown that by reducing the impact of disturbances, such as tillage, microbial biomass levels and soil resilience will improve (Elliot and Lynch, 1994). In addition, crop rotations of at least 3 years can decrease disease-related problems (Rovira et al., 1990), which can also increase crop resilience. However, soil resilience may be different depending on the disturbance; disturbances such as tillage may affect the general biological status, while pesticides or other stressors may only affect individual functional groups (Swift, 1994). By further establishing the relationships between the microbial biodiversity and soil resilience, a greater understanding of microbial indicators can be achieved.

CONCLUSION

The diversity of microorganisms is greater than any other group of organisms on Earth, but our knowledge of the diversity and genetic wealth in these groups is limited. Microorganisms are key to the integrated functioning of nutrient cycles and decomposition, soil structure, and plant growth in agricultural systems. Research is needed to increase our understanding of the diversity and function of microbial communities in agroecosystems. In agroecosystems, microbial diversity will influence all the other levels within the ecosystem by functions such as those involved in nitrogen and carbon cycling, soil structure maintenance, and biological control. The challenge ahead is to identify the level of microbial diversity, species composition and distribution, and the resiliency of the community to withstand stress and maintain a quality ecosystem. We need to determine the extent of microbial diversity in agroecosystems and increase our knowledge of the functional roles of microbes to assess their role in agroecosystem quality and productivity.

ACKNOWLEDGMENTS

Support from the U.S. Department of Agriculture, Agricultural Research Service in cooperation with the College of Agriculture and Home Economics, Washington

State University, Pullman, WA is appreciated. Trade names and company names are included for the benefit of the reader and do not imply endorsement or preferential treatment of the product by the U.S. Department of Agriculture or Washington State University. All programs and services of the U.S. Department of Agriculture are offered on a nondiscriminatory basis without regard to race, color, national origin, religion, sex, age, marital status, or handicap.

REFERENCES

Albrecht S. L., Okon, Y., Lonnquist, J., and Burris, R. H., 1981. Nitrogen fixation by corn–*Azospirillum* associations in a temperate climate, *Crop Science,* 21:301–306.

Alexander, M., 1977. *Introduction to Soil Microbiology,* 2nd ed., Academic Press, New York.

Allen, E. B. and Allen, M. F., 1990. The mediation of competition by mycorrhizae in successional and patchy environments, in *Perspectives on Plant Competition,* J. B. Grace and D. A. Tilman, Eds., Academic Press, San Diego, CA, 367–389.

Allen, M. F., 1992. *Mycorrhizal Functioning: An Integrative Plant-Fungal Process,* Chapman & Hall, New York.

Alstrom, S., 1987. Factors associated with detrimental effects of rhizobacteria on plant growth, *Plant Soil,* 102:3–9.

Altieri, M. A., 1991. How best can we use biodiversity in agroecosystems? *Outlook Agric.,* 20:5–23.

American Society for Microbiology, 1994. *Microbial Diversity Research Priorities,* American Society for Microbiology, Washington, D.C.

Aspiras, R. B., Allen, O.N., Harris, R. F., and Chester, G, 1971. The role of microorganisms in stabilization of soil aggregates, *Soil Biol. Biochem.,* 3:347–353.

Atlas, R. M., 1984. Use of microbial diversity measurements to assess environmental stress, in *Current Perspectives in Microbial Ecology,* M. J. Klug and C. A. Reddy, Eds., American Society for Microbiology, Washington, D.C., 540–545.

Barea, J. M., 1991. Vesicular-arbuscular mycorrhizae as modifiers of soil fertility, *Adv. Soil Sci.,* 15:2–40.

Beare, M. H., Pohlad, B. R., Wright, D. H., and Coleman, D. C., 1993. Residue placement and fungicide effects on fungal communities in conventional and no-tillage soils, *Soil Sci. Soc. Am. J.,* 57:392–399.

Belser, L. and Schmidt, E., 1978. Diversity in ammonia-oxidizing population of a soil, *Appl. Environ. Microbiol.,* 36:584–588.

Bock, E., Koops, H.-P., and Harms, H., 1989. Nitrifying bacteria, in *Autotrophic Bacteria,* H. Schlegel and B. Bowein, Eds., Science Tech Publishers, Madison, WI, 81–96.

Boddy, L., Watling, R., and Lyon, A. J. E., 1988. Fungi and ecological disturbance, *Proc. R. Soc. Edinburgh,* Section B, 94.

Borneman, J., Skroch, P. W., O'Sullivan, K. M., Palus, J. A., Rumjanek, N. G., Jansen, J. L., Nienhuis, J., and Triplett, E. W., 1996. Molecular microbial diversity of an agricultural soil in Wisconsin, *Appl. Environ. Microbiol.,* 62:1935–1943.

Brown, M. E., 1972. Plant growth substances produced by microorganisms of soil and rhizosphere, *J. Appl. Bacteriol.,* 35:443–451.

Bruehl, G. W., 1987. *Soilborne Plant Pathogens,* Macmillian, New York.

Bull, A. T., Goodfellow, M., and Slater, J. H., 1992. Biodiversity as a source of innovation in biotechnology, *Annu. Rev. Microbiol.,* 46:219–252.

Burdon, J. J., 1987. *Diseases and Plant Population Biology*, Cambridge University Press, New York.

Cambardella, C. A. and Elliott, E. T., 1992. Particulate soil organic matter changes across a grassland cultivation sequence, *Soil Sci. Soc. Am. J.*, 56:777–783.

Chang, I. P. and Kommendahl, T., 1968. Biological control of seedling blight of corn by coating kernels with antagonistic microorganisms, *Phytopathology*, 58:1395–1401.

Christensen, M., 1989. A view of fungal ecology, *Mycologia*, 81:1–19.

Collins, H. P., Rasmussen, P. E., and Douglas, C. L., Jr., 1992. Crop rotation and residue management effects on soil carbon and microbial dynamics, *Soil Soc. Am. J.*, 56: 783–788.

Cook, R. J., 1981. The influence of rotation crops on take-all decline phenomena, *Phytopathology*, 71:189–192.

Cook, R. J. and Baker, K. F., 1983. *The Nature and Practice of Biological Control of Plant Pathogens*, American Phytopathological Society, St. Paul, MN.

Curl, E. and Truelove, B., 1986. *The Rhizosphere*, Springer-Verlag, Berlin.

DeBach, P., 1964. *Biological Control of Insect Pests and Weeds*, Reinhold, New York.

di Castri, F. and Younes, T., 1990. Ecosystem function of biological diversity, *Biol. Int.*, Special Issue 22:1–20.

Doran, J. W., 1980. Soil microbial and biochemical changes associated with reduced tillage, *Soil Sci. Soc. Am. J.*, 44:765–771.

Doran, J. W., 1987. Microbial biomass and mineralizable nitrogen distributions in no-tillage and plowed soils, *Biol. Fertil. Soils*, 5:68–75.

Dorn, E., Hellwig, M., Reineke, W., and Knackmuss, H. J., 1974. Isolation and characterization of a 3-chlorobenzoate degrading pseudomonad, *Arch. Microbiol.*, 99:61–70.

El Nashaar, H. M. and Stack, R. W., 1989. Effect of long-term continuous cropping of spring wheat on aggressiveness of *Cochliobolus sativus*, *Can. J. Plant Sci.*, 69:395–400.

Elliott, L. F. and Lynch, J. M., 1984. *Pseudomonads* as a factor in the growth of winter wheat (*Triticum aestivum* L.), *Soil Biol. Biochem.*, 16:68–71.

Elliott, L. F. and Lynch, J. M., 1994. Biodiversity and soil resilience, in *Soil Resilience and Sustainable Land Use*, D. J. Greenland and I. Szabolcs, Eds., CAB International, Wallingford, U.K., 353–364.

Emmimath, V. S. and Rangaswami, G., 1971. Studies of the effect of heavy doses of nitrogenous fertilizer on the soil and rhizosphere microflora of rice, *Mysore J. Agric. Sci.*, 5:39–58.

Fitter, A. H., 1977. Influence of mycorrhizal infection on competition for phosphorus and potassium by two grasses, *New Phytol.*, 69:119–125.

Gaskins, M. H., Albrecht, S. L., and Hubbell, D. H., 1985. Rhizosphere bacteria and their use to increase plant productivity: a review, *Agric. Ecosys. Environ.*, 12:99–116.

Gilmour, C. M., Allen, O. N., and Truog, E., 1948. Soil aggregation as influenced by the growth of mold species, kind in soil, and organic matter, *Soil Sci. Soc. Am. Proc.*, 13:291–296.

Gochenauer, S. E., 1981. Responses of soil faunal communities to disturbance, in *The Fungal Community: Its Organization and Role in the Ecosystem*, D. T. Wicklow and G. C. Carroll, Eds., Marcel Dekker, New York, 459–479.

Goldberg, D. and Fleetwood, L., 1987. Competitive effect and response in four annual plants, *J. Ecol.*, 75:1131–1143.

Goldberg, D. and Landa, K., 1991. Competitive effect and response: hierarchies and correlated traits in the early stages of competition, *J. Ecol.*, 79:1013–1030.

Gurevitch, J., Wilson, P., Stone, J. L., Tesse, B., and Stoutenburgh, R., 1990. Competition among old-field perennials at different soil levels of fertility and available space, *J. Ecol.*, 78:727–744.

Hall, J. R., 1978. Effects of endomycorrihzas on the competitive ability of white clover, *N. Z. J. Agric. Res.*, 21:509–515.

Harley, J. L. and Smith, S. E., 1983. *Mycorrhizal Symbiosis*, Academic Press, London.

Hawksworth, D. L., 1991a. *The Biodiversity of Microorganisms and Invertebrates: Its Role in Sustainable Agriculture*, CAB International, Redwood Press, Melksham, U.K.

Hawksworth, D. L., 1991b. The fungal dimension of biodiversity: magnitude, significance, and conservation, *Mycol. Res.*, 95:641–655.

Hirsch, P. R., Jones, M. J., McGrath, S. P., and Giller, K. E., 1993. Heavy metals from past applications of sewage sludge decrease the genetic diversity of *Rhizobium leguminosarum* biovar *trifolii* populations, *Soil Biol. Biochem.*, 25:1485–1590.

Holben, W. E. and Tiedje, J. M., 1988. Tracing tiny organisms, *Ecology*, 69:561–568.

Johnson, N. C., Copeland, P. J., Crookston, B. K., and Pfleger, F. L., 1992a. Mycorrhizae: possible explanation for yield decline with continuous corn and soybean, *Agron. J.*, 84:387–390.

Johnson, N. C., Tilman, D., and Wedin, D., 1992b. Plant and soil controls on mycorrhizal fungal communities, *Ecology*, 73:2034–2042.

Jones, J. W., 1991. Diversity and physiology of methanogens, in *Microbial Production and Consumption of Greenhouse Gases: Methane, Nitrogen Oxides, and Halomethanes*, J. E. Rogers and W. B. Whitman, Eds., American Society of Microbiology, Washington, D.C., 3–55.

Jordahl, J. L. and Karlen, D. L., 1993. Comparison of alternative farming systems. III. Soil aggregate stability, *Am. J. Alternative Agric.*, 8:27–33.

Kennedy, A. C., 1994. Biological control of annual grass weeds, in *Ecology, Management and Restoration of Intermountain Annual Rangelands*, S. B. Monsen and K. L. Johnson, Eds., Intermountain Research Station, Ogden, UT, 186–189.

Kennedy, A. C. and Papendick, R. I., 1995. Microbial characteristics of soil quality, *J. Soil Water Conserv.*, 50:243–248.

Kennedy, A. C. and Smith, K. L., 1995. Soil microbial diversity and ecosystem functioning, *Plant Soil*, 170:75–86.

Kennedy, A. C., Elliott, L. F., Young, F. L., and Douglas, C. L., 1991. Rhizobacteria suppressive to the weed downy brome, *Soil Sci. Soc. Am. J.*, 55:722–727.

Kennedy, A. C., Ogg, A. G., Jr., and Young, F. L., 1992. Biocontrol of Jointed Goatgrass. U.S. Patent 5,163,991, November 17, 1992.

Knapp, E. B., Elliott, L. F., and Campbell, G. S., 1983. Microbial respiration and growth during the decomposition of wheat straw, *Soil Biol. Biochem.*, 15:319–323.

Kremer, R. J., Begonia, M. F. T., Stanley, L., and Lanham, E. T., 1990. Characterization of rhizobacteria associated with weed seedlings, *Appl. Environ. Microbiol.*, 56:1649–1655.

Lee, K. E. and Pankhurst, C. E., 1992. Soil organisms and sustainable producitivity, *Aust. J. Soil Res.*, 30:855–892.

Lynch, J. M., 1983. *Soil Biotechnology, Microbiological Factors in Crop Productivity*, Blackwell Scientific Publications, Oxford, U.K.

Lynch, J. M. and Bragg, E., 1985. Microorganisms and soil aggregate stability, *Adv. Soil Sci.*, 2:133–171.

Malik, K. A. and Claus, D., 1987. Bacterial culture collections: their importance to biotechnology and microbiology, *Biotechnol. Genet. Eng. Rev.*, 5:137–197.

Martin, J. K. and Kemp, J. R., 1980. Carbon loss from roots of wheat cultivars, *Soil Biol. Biochem.*, 12:551–554.

Metting, B. F., 1993. *Soil Microbiology Ecology,* Marcel Dekker, New York.

Metting, B. and Rayburn, W. R., 1983. The influence of a microalgal conditioner on selected Washington soils: an empirical study, *Soil Sci. Soc. Am. J.,* 47:682–685.

Mills, A. J. and Wassel, R. A., 1980. Aspects of diversity measurement for microbial communities, *Appl. Environ. Microbiol.,* 41:578–586.

Mohammad, M., Pan, W. L., and Kennedy, A. C., 1995. Wheat responses to vesicular-arbuscular mycorrhizal fungal inoculation of soils from eroded toposequence, *Soil Sci. Soc. Am. J.,* 59:1086–1090.

Molina, R., Massicotte, H., and Trappe, J. M., 1992. Specificity phenomena in mycorrhizal symbioses: community-ecological consequences and practical implications, in *Mycorrhizal Functioning: An Integrative Plant-Fungal Process,* M. F. Allen, Ed., Chapman and Hall, New York, 357–423.

Newman, E. I., Heap, A. J., and Lawley, R. A., 1981. Abundance of mycorrhizas and root-surface microorganisms of *Plantago lanceolata* in relation to soil and vegetation: a multivariate approach, *New Phytol.,* 89:95–108.

Ocampo, J. A., 1986. Vesicular-arbuscular mycorrhizal infection of "host" and "non-host" plants, effect on the growth responses of the plants and the competition between them, *Soil Biol. Biochem.,* 18:607–610.

Office of Technology Assessment, U.S. Congress, 1987. Technologies to Maintain Biological Diversity, OTA-F330, U.S. Government Printing Office, Washington, D.C.

Okon, Y., 1982. *Azospirillum:* physiological properties, modes of association with roots and its application for the benefit of cereal and forage grass crops, *Isr. J. Bot.,* 31:214–220.

Olsson, S. and Gerhardson, B., 1992. Effects of long-term barley monoculture on plant-affecting soil microbiota, *Plant Soil,* 143:99–108.

Opperman, M. H., Wood, M., and Harris, P. J., 1989. Changes in microbial populations following the application of cattle slurry to soil at two temperatures, *Soil Biol. Biochem.,* 21:263–268.

Papendick, R. I. and Campbell, G. S., 1975. Water potential in the rhizosphere and plant and methods of measurement and experimental control, in *Biology and Control of Soil-Borne Pathogens,* G. W. Bruehl, Ed., The American Phytopathological Society, St. Paul, 39–49.

Papendick, R. I. and Parry, J. F., 1992. Soil quality — the key to a sustainable agriculture, *Am. J. Alternative Agric.,* 7:2–3.

Powlson, D. S., Brookes, P. C., and Christensen, B. T., 1987. Measurement of soil microbial biomass provides an early indication of changes in total soil organic matter, *Soil Biol. Biochem.,* 19:159–164.

Price, W. P., 1988. An overview of organismal interactions in ecosystems in evolutionary and ecological time, *Agric. Ecosyst. Environ.,* 2:369–377.

Reber, H. H., 1992. Simultaneous estimates of the diversity and the degradative capability of heavy-metal-affected soil bacterial communities, *Biol. Fertil. Soils,* 13:181–186.

Rouatt, J. W. and Katznelson, H., 1961. A study of the bacteria on the root surface and in the rhizosphere soil of crop plants, *J. Appl. Bacteriol.,* 24:164–171.

Rovira, A. D. 1956. Plant root excretions in relation to the rhizosphere effect: I. The nature of root exudates from oats and peas, *Plant Soil,* 7: 178–194.

Rovira, A. D., 1959. Root excretions in relation to the rhizosphere effect: IV. Influence of plant species, age of plant, light, temperature, and calcium nutrition on exudation, *Plant Soil,* 9:53–64.

Rovira, A. D., 1978. Microbiology of pasture soils and some effects of microorganisms on pasture plants, in *Plant Relations in Pastures,* J. R. Wilson, Ed., CSIRO, East Melbourne, Australia, 95–110.

Rovira, A. D., Elliott, L. F., and Cook, R. J., 1990. The impact of cropping systems on rhizosphere organisms affecting plant health, in *The Rhizosphere,* J. M. Lynch, Ed., John Wiley, Chichester, 389–436.

Schippers, B., Bakker, A. W., and Bakker, P. A., 1987. Interactions of deleterious and beneficial rhizosphere microorganisms and the effect of cropping practices, *Annu. Rev. Phytopathol.,* 25:339–358.

Shipton, P. J., 1977. Monoculture and soilborne plant pathogens, *Annu. Rev. Phytopathol.,* 15:387–407.

Smith, J. L. and Paul, E. A., 1990. The significance of soil microbial biomass estimations in soil, *Soil Biochem.,* 6:357–396.

Smith, W. H., 1969. Release of organic materials from the roots of tree seedlings, *For. Sci.,* 15:138–143.

Sollins, P., Cromack, K., Jr., Li, C. Y., and Fogel, R., 1981. Role of low-molecular-weight organic acids in the inorganic nutrition of fungi and higher plants, in *The Fungal Community, Its Organization and Role in the Ecosystem,* D. T. Wicklow and G. C. Caroll, Eds., Marcel Dekker, New York, 607–619.

Sprent, J. L., 1979. *The Biology of Nitrogen-Fixing Organisms,* McGraw-Hill, London, U.K.

Strain, S. R., Leung, K., Whittam, T. S., de Bruijn, F. S., and Bottomley, P. J., 1994. Genetic structure of *Rhizobium leguminosarum* bivar *trifolii* and *viciae* populations found in two Oregon soils under different plant communities, *Appl. Environ. Microbiol.,* 60:2772–2778.

Stroo, H. F., Elliott, L. F., and Papendick, R. I., 1988. Growth, survival and toxin production of root-inhibitory pseudomonads on crop residues, *Soil Biol. Biochem.,* 20:201–207.

Suslow, T. V. and Schroth, M. N., 1982. Role of deleterious rhizosphere as minor pathogens in reducing crop growth, *Phytopathology,* 72:111-115.

Swift, M. J., 1994. Maintaining the biological status of soil: a key to sustainable land management?, in *Soil Resilience and Sustainable Land Use,* D. J. Greenland and I. Szabolcs, Eds., CAB International, Wallingford, U.K., 235–248.

Swift, M. J. and Boddy, L., 1984. Animal-microbial interactions in wood decomposition, in *Invertebrate-Microbial Interactions,* J. M. Anderson, A. D. M. Rayner, and W. H. Walton, Eds., Cambridge University Press, Cambridge, U.K., 89–131.

Swinnen, J., Van Veen, J. A., and Merckx, R., 1995. Root decay and turnover of rhizodeposits in field-grown winter wheat and spring barley estimated by 14C pulse-labelling, *Soil. Biol. Biochem.,* 27:211–217.

Thomas, V. G. and Kevan, P. G., 1993. Basic principles of agroecology and sustainable agriculture, *J. Agric. Environ. Eth.,* 5:1–18.

Tiedje, J. M., Sextone, A. J., Parkin, T. B., Revsbech, N. P., and Shelton, D. R., 1984. Anaerobic processes in soil, *Plant Soil,* 76:197–212.

Tinker, P. B., 1976. Roots and water. Transport of water to plant roots in soil, *Philos. Trans. R. Soc. London, Series B, Biol. Sci.,* 273:445–461.

Tisdall, J. M., 1991. Fungal hyphae and structural stability of soil, *Aust. J. Soil Res.,* 29:729–743.

Torsvik, V., Goksoyr, J., and Daae, F. L., 1990. High diversity in DNA of soil bacteria, *Appl. Environ. Microbiol.,* 56:782–787.

Tranel, P. J., Gealy, D. R., and Kennedy, A. C., 1993. Inhibition of downy brome (*Bromus tectorum* L.) root growth by a phytotoxin from *Pseudomonas fluorescens* strain D7, *Weed Technol.,* 7:134–139.

Turco, R. F. and Bezdicek, D. F., 1987. Diversity within two serogroups of *Rhizobium leguminosarum* native to soils in the Palouse of eastern Washington, *Annu. Appl. Biol.,* 111:103–114.

Turco, R. F., Kennedy, A. C., and Jawson, M. D., 1994. Microbial indicators of soil quality, in *Defining Soil Quality for a Sustainable Environment,* American Society of Agronomy Special Publication No. 35J, W. Doran, D. C. Coleman, D. F. Bezdicek, and B. A. Stewart, Eds., American Society of Agronomy, Madison, WI, 73–90.

Vancura, V., 1967. Root exudates of plants: III. Effects of temperature and "cold shock" on the exudation of various compounds from seeds and seedlings, *Plant Soil,* 27:319–328.

Visser, S. and Parkinson, D., 1992. Soil biological criteria as indicators of soil quality: soil microorganisms, *Am. J. Alternative Agric.,* 7:33–37.

Walker, B. H., 1992. Biodiversity and ecological redundancy, *Conserv. Biol.,* 6:18–23.

Ward, D., Bateson, M., Weller, R., and Ruff-Roberts, A., 1992. Ribosomal RNA analysis of microorganisms as they occur in nature, *Adv. Microb. Ecol.,* 12:219–286.

Wedin, D. A. and Tilman, D., 1990. Special effects of nitrogen cycling: a test with perennial grasses, *Oecologia,* 84:433–441.

Whipps, J. M. and Lynch, J. M., 1986. The influence of rhizosphere on crop productivity, *Adv. Microb. Ecol.,* 9:187–244.

Woltz, S. S., 1978. Nonparasitic plant pathogens, *Annu. Rev. Phytopathol.,* 16:403–430.

Wood, M., 1991. Biological aspects of soil protection, *Soil Use Manage.,* 7:130–136.

Yeates, G. and Darrah, P. R., 1991. Microbial changes in a model rhizosphere, *Soil Biol. Biochem.,* 23:963–971.

Zak, J. C., Willig, M. R., Moorhead, D. L., and Wildman, H. G., 1994. Functional diversity of microbial communities: a quantitative approach, *Soil Biol. Biochem.,* 26:1101–1108.

CHAPTER 2

Soil Microfauna: Diversity and Applications of Protozoans in Soil

Stuart S. Bamforth

CONTENTS

INTRODUCTION

Agricultural plant production depends upon the decomposition of plant and animal residues, as well as fertilizers, into simpler compounds, many of which are transformed into microbial and animal protoplasm. These organic materials are eventually mineralized into simpler compounds, such as CO_2, ammonia, and phosphate, which are absorbed by plant roots.

ROLE OF SOIL PROTOZOA

Microarthropods and larger fauna, especially earthworms, increase the rate and amount of mineralization by comminution of organic matter and by redistribution of "hot spots" of activity through movements. However, mineralization and return

1-56670-290-9/99/$0.00+$.50

of nutrients to plants occur in the water films covering soil aggregates and filling their pores. Here, bacteria and fungi decompose organic matter and immobilize the extracted nutrients into their bodies, but grazing by the microfauna, protozoa, and nematodes regulates and modifies the size and composition of the microbial community and enhances microbial growth through microfaunal excretions. Nematodes also graze fungi (Chapter 1), but protozoa, especially amoebae, can graze bacteria in tiny pore spaces unavailable to nematodes. The degree of nutrient recycling is influenced by external factors of climate and soil management (e.g., inputs of fertilizers and biocides, compaction by farm machinery) and internally by the community of protozoa and nematodes, reflected in their biodiversity.

Most of the microfauna are located in small hot spots scattered through the soil mosaic, which is soil aggregates of 1 mm or smaller, containing bits of organic matter, detritus and the overlying litter, rhizosphere, and the "drillosphere" parts of the soil influenced by earthworm secretions and castings. The microbial-feeding microfauna constitute an essential component of the soil ecosystem; therefore, changes in their community structure can influence mineralization and soil fertility.

MEASURING PROTOZOAN BIODIVERSITY

Soil protozoa comprise four groups: the "naked" rapidly growing flagellates, amoebae, and ciliates, and the more slowly growing shelled amoebae, or testacea. The small size and flexibility of the first two groups allows them to exploit small pore spaces, and they furnish most of the protozoan numbers.

The more diverse and larger ciliates and testacea inhabit the larger pore spaces which are subject to desiccation and other stresses; consequently, both groups show a wide spectrum of species of r/K selection and degree of autochthonism (Wodarz et al., 1992). Ciliates are divided into pioneer r-selected Colpodida, competitive K-selected Polyhymenophora, and intermediate remaining taxa. Dividing the number of species of the first group by the second produces a C/P ratio, where $C/P > 1.00$ indicates a stressed soil of low productivity, and $C/P < 1.00$ a more productive soil with microarthropods and macrofauna (Foissner, 1987; Yeates et al., 1991). Among the testacea, certain species indicate soil acidity or alkalinity, and the shell conveys information about moisture fluctuations (Bonnet, 1964). Consequently, these two groups can serve as bioindicators of soil conditions.

Ideally, biodiversity studies measure both species and numbers per species. However, the small size and transparency of naked protozoa make them too difficult to find among soil particles; consequently, counting has been traditionally performed by the most probable number (MPN) technique of Singh (1946) or its modification by Darbyshire et al. (1974). There are criticisms of the method (Foissner, 1987), in response to which a second direct count method was developed by Griffiths and Ritz (1988) that separates the protozoa from soil particles by percoll phosphate gradient centrifugation and staining for fluorescent microscopy. This method is employed routinely to measure the protozoan component of the soil fauna in experimental field crop studies by the Technical University of Munich. The larger and more motile ciliates can be counted by examining a watered soil suspension drop-by-drop until at least 0.4 g of fresh soil has been examined (Foissner, 1987). The possession of a shell enables direct counting of testacea by this method, or by staining a soil suspension and mounting small samples on slides, providing a permanent record (Couteaux, 1967; Korgonova and Geltser, 1977). Combining the temporary and permanent methods provides a more complete census than either method alone.

Estimating species richness is best done by placing 10 to 50 g of sample in a petri dish and adding water until 5 to 20 ml will drain off when gently pressed with a finger. By placing several coverslips, each underlaid with a piece of lens paper on top of the sample, and examining after 1 day, a variety of flagellate species will be revealed. The culture is examined at 3 to 4 day intervals for a month to determine the succession of species of mainly ciliates and testacea (Foissner, 1987). Most amoebae will be found by streaking soil samples on bacterized water (non-nutrient) agar plates or by placing soil samples in wells cut in such plates. The amoebae migrate out from the soil particles (Bamforth, 1995a).

PROTOZOAN DIVERSITY IN AGROECOSYSTEMS

Studies on grasslands (McNaughton, 1977; Tilman, 1996) show that biodiversity stabilizes community and ecosystem processes, although individual species within the system may fluctuate considerably. Tilman (1996) found wide variations in the biomass of the 24 most abundant species of plants in an 11-year study. In a 6-month study of soil ciliates under a spruce stand, Lehle (1992) found that the proportions of the three dominant ciliate species fluctuated widely: *Cyclidium muscicola* ranged from 8 to 75% of the total populations and two colpodid species varied from 4 to 45%. The different responses in these two studies may reflect changes in the realized niches of species; thus, biodiversity furnishes a reservoir of biotic abilities contributing to ecosystem sustainability (Bamforth, 1995b). Biodiversity, like the comparison of nontillage to conventional agriculture, may not produce noticeable increases in crop production, but maintaining biodiversity can retard the deterioration that has characterized agroecosystems through 4000 years of human history. Protozoa can serve as bioindicators of ecosystem conditions, and warn of soil impoverishment.

The appeal of protozoan bioindicators is their environmental sensitivity due to their delicate cell membranes, their rapid growth rate, restricted movement in soil, ubiquity, and wide range of morphologies in ciliates and testacea, providing a multispecies approach enabling community analyses to indicate soil conditions (Foissner, 1994). Difficulties arise in the taxonomy and time needed for identification and enumeration, but, as the following applications illustrate, protozoa convey valuable information about agroecosystems because of their pivotal position in the nutrient cycling that all terrestrial ecosystems depend upon.

APPLICATIONS

Conventional agriculture creates a special ecosystem by mixing the topsoil (and compacting it) through tillage, removing plant canopies that protect the soil, adding fertilizers and biocides, and removing harvests. A more sustainable agriculture minimizes topsoil disturbance, reduces inputs, and substitutes organic for mineral fertilizers (Doran and Werner, 1990).

Plowing in conventional agriculture incorporates crop residues into the soil profile to produce homogeneous soils that favor the bacteria, protozoa, and bactivorous nematode portions of the underground food web; in contrast, minimal tillage leaves organic residues on the surface and a rich organic layer near the surface, enhancing the fungal, Collembola, and earthworm portions of the underground food web (Hendrix et al., 1986; Lee and Pankhurst, 1992). The protozoan communities differ between the two systems in the greater prominence of r-selected colpodid ciliates (reflecting less species diversity) in conventional fields (Foissner, 1992; Bamforth, unpublished data). The biomass of amoebae and flagellates, however, is greater in the surface layer of ecofarmed systems (DeRuiter et al., 1993) and is associated with increased nitrogen mineralization (DeRuiter et al., 1993). Using testacea as bioindicators, Wodarz et al. (1992) found organically farmed field and vineyard soils showed improved soil conditions over conventionally farmed counterparts.

Organic fertilizers, especially straw and animal manures, are more similar to natural organic substrates than chemical fertilizers. Microbial and protozoan activity is highest in organically enriched soils (Schnurer et al., 1985; Aescht and Foissner, 1991; 1992), and is usually accompanied by increases of most soil fauna, especially earthworms (Doran and Werner, 1990), which enhance protozoan biodiversity.

The higher protozoan activity in soils under nontillage and organic fertilizer management is enhanced by other fauna, especially earthworms, which disperse bacteria and their protozoan predators to new locations, through burrowing movements and passing ingested cysts through guts, providing new hot spots and releasing greater quantities of nutrients, which have led to increased plant yields in a few cases (Brown, 1995). Ingested active protozoa furnish a highly assimilable food source, sustaining the fauna that enhance microbial and protozoan activities (Brown, 1995). Thus, high protozoan biodiversity usually reflects earthworm abundance.

The application of biocides often influences other organisms besides those targeted. Herbicides have little effect on protozoa, although they may influence them indirectly by altering bacterial nitrogen activities and by modifying the environment in eliminating the vegetation over the soil. Insecticides and fungicides are more toxic, as shown in the study of Petz and Foissner (1990) on the effects of lindane, an insecticide, and mancozeb, a fungicide, on the soil ciliate and testacean communities of a spruce forest. The insecticide decreased both numbers and species, and altered the community structure of ciliates by increasing the abundance of several colpodids. This result shows the value of multispecies-monitoring studies, and also the value of biodiversity to an ecosystem, allowing response to changing conditions (Bamforth, 1995b). The insecticide exerted less effect on testacea, and the fungicide exerted little influence on both groups. The investigation used a randomized block design and extended the study period to the 90 days needed to ascertain if the biocide caused acute toxicity (Domsch et al., 1983). This type of study shows the precision that protozoan bioindicators can provide to assess agroecosystem conditions.

The heavy machinery used in modern farming causes soil compaction, destroying not only the worm and root channels that reduce soil porosity and the larger fauna, but also reducing pore spaces in which bacteria and their protozoan predators live. Compaction reduces testacean species diversity and eliminates large forms (Berger et al., 1985), and a number of studies relating pore space to protozoan activity (e.g., Rutherford and Juma, 1992; England et al., 1993) show less activity in smaller spaces. Griffiths and Young (1994) found the same trend and concluded that compaction influences protozoa indirectly by producing anaerobic conditions that inhibit protozoa and reduce the metabolism and reproduction of their bacterial prey.

A vital part of agricultural management is soil conservation and restoration, which can be monitored by analyzing the protozoan community to assess the degree of the comprehensive biological activity to productive farming (Yeates et al., 1991; Wodarz et al., 1992).

CONCLUSIONS

Protozoa and nematodes are pivotal organisms in agroecosystems because their predation upon bacteria increases mineralization of nutrients necessary for plant growth. Since biodiversity stabilizes community and ecosystem processes (Tilman, 1996), maintaining and increasing protozoan biodiversity will contribute to more sustainable agriculture. Ecofarming and organic fertilizer management enhance protozoan activity.

Protozoa have several unique features, such as rapid sensitivity to environmental changes and ubiquity, that favor their use as bioindicators. Protozoan biodiversity reflects the condition of an agroecosystem and can be used to monitor the effects of environmental changes.

REFERENCES

Aescht, E. and Foissner, W., 1991. Bioindikation mit mikroskopsich kleinen Bodentierren, *VDI Ber.*, 901:985–1002.

Aescht, E. and Foissner, W., 1992. Effects of mineral and organic fertilizers on the microfauna in a high altitude afforestation trial, *Biol. Fertil. Soils*, 13:17–24.

Bamforth, S. S., 1995a. Isolation and counting of protozoa, in *Methods in Applied Soil Microbiology and Biochemistry*, P. Nannipieri and K. Alef, Eds., Academic Press, New York, 174–180.

Bamforth, S. S., 1995b. Interpreting soil ciliate biodiversity, in *The Significance and Regulation of Soil Biodiversity*, H. P. Collins, G. P. Roberrtson, and M. J. Klug, Eds., Kluwer Academic, The Netherlands, 179–184.

Berger, H., Foissner, W., and Adam, H., 1985. Protozoolgische Untersuchengen an Almboden im Gasteiner Tal (Zentralalpen, Österreich). IV. Experimentelle Studien zur Wirkung der Bodenverdichtung auf die Struktur der Testaceen- und Ciliatentaxozonose, *Veröff Österr. MaB Programms*, 9:97–112.

Bonnet, L., 1964. Le peuplement thécamoebien de sols, *Rev. Écol. Biol. Sol.*, 1:123–408.

Brown, G. G., 1995. How do earthworms affect microfloral and faunal community diversity?, in *The Significance and Regulation of Soil Biodiversity,* H. P. Collins, G. P. Roberrtson, and M. J. Klug, Eds., Kluwer Academic, The Netherlands, 247–269.

Couteaux, M. M., 1967. Une technique d'observation des thécamoebiens du sol pour l'estimation de leur densité absolue, *Rev. Écol. Biol. Sol.,* 4:593–596.

Darbyshire, J. F., Wheatley, R. E., Greaves, M. P., and Inkson, R. H., 1974. A rapid micromethod for estimating bacterial and protozoan populations in soil, *Rev. Écol. Biol. Sol.,* 11:465–475.

DeRuiter, P. C., Moore, J. C., Zwart, K. B., Bouwman, L. A., Hassink, J., Bloem, J., De Vos, J. A., Marinissen, J. C. Y., Didden, W. A. M., Lebbink, G., and Brussard, L., 1993. Simulation of nitrogen mineralization in the below-ground food webs of two winter wheat fields, *J. Appl. Ecol.,* 30:95–106.

Domsch, K. H., Jagnow, G., and Anderson, T. H., 1983. An ecological concept for the assessment of side-effects of agrochemicals on soil microorganisms, *Residue Rev.,* 86:65–105.

Doran, J. W. and Werner, M. R., 1990. Management and soil biology, in *Sustainable Agriculture in Temperate Zones,* C. A. Francis and C. B. Flora, Eds., Wiley, New York, 205–225.

England, L. S., Lee, H., and Trevors, J. L., 1993. Bacterial survival in soil: effect of clays and protozoa, *Soil Biol. Biochem.,* 25: 525–531.

Foissner, W., 1987. Soil protozoa: fundamental problems, ecological significance, adaptations in ciliates and testaceans, bioindicators, and guide to the literature, *Prog. Protistol.,* 2:69–212.

Foissner, W., 1992. Comparative studies on the soil life in ecofarmed and conventionally farmed fields and grasslands of Austria, *Agric. Ecosyst. Environ.,* 40:207–218.

Foissner, W., 1994. Soil protozoa as bioindicators in ecosystems under human influence, in *Soil Protozoa,* J. F. Darbyshire, Ed., CAB International, Wallingford, 147–193.

Griffiths, B. S. and Ritz, K., 1988. A technique to extract, enumerate and measure protozoa from mineral soils, *Soil Biol. Biochem.,* 20:163–173.

Griffiths, B. S. and Young, I. M., 1994. The effects of soil structure on protozoa in a clay-loam soil, *Eur. J. Soil Sci.,* 45:285–292.

Hendrix, P. F., Parmelee, R. W., Crossley, D. A., Coleman, D. C., Odum, E. P., and Groffman, P. M., 1986. Detritus food webs in conventional and no-tillage agroecosystems, *Bioscience,* 36:374–380.

Korgonova, G. A. and Geltser, J. G., 1977. Stained smears for the study of soil Testacida (Protozoa, Rhizopoda), *Pedobiologia,* 17:222–225.

Lee, K. E. and Pankhurst, C. E., 1992. Soil organisms and sustainable productivity, *Aust. J. Soil Res.,* 30:855–892.

Lehle, E., 1992. Wimpertiere und andere Einzeller im Boden eines Fichten bestandes im Schwartzwald, *Mikrokosmos,* 81:193–198.

McNaughton, S. J., 1977. Diversity and stability of ecological communities: a comment on the role of empiricism in ecology, *Am. Nat.,* 111:515–525.

Petz, W. and Foissner, W., 1990. The effects of mancozeb and lindane on the soil microfauna of a spruce forest: a field study using a completely randomized block design, *Biol. Fertil. Soils,* 7:225–231.

Rutherford, P. M. and Juma, N. G., 1992. Influence of texture on habitable pore space and bacterial-protozoan populations in soil, *Biol. Fertil. Soils,* 12:221–227.

Schnurer, J., Clarholm, M., and Roswell, T., 1985. Microbial biomass and activity in an agricultural soil with different organic contents, *Soil Biol. Biochem.,* 17:611–618.

Singh, B. N., 1946. A method of estimating the numbers of soil protozoa, especially amoebae, based on their differential feeding of bacteria, *Annu. Appl. Biol.,* 33:112–120.

Tilman, D., 1996. Biodiversity: population versus ecosystem stability, *Ecology,* 77:350–363.

Wodarz, D., Aescht, E., and Foissner, W., 1992. A weighted coenotic index (WCI): description and application to soil animal assemblages, *Biol. Fertil. Soils,* 14:5–13.

Yeates, G. W., Bamforth, S. S., Ross, D. J., Tate, K. R., and Sparling, G. P., 1991. Recolonization of methyl bromide sterilized soils under four different field conditions, *Biol. Fertil. Soils,* 11:181–189.

CHAPTER 3

Diversity and Function of Soil Mesofauna

Deborah A. Neher and Mary E. Barbercheck

CONTENTS

INTRODUCTION

Diversity in natural communities of microbes, plants, and animals is a key factor in ecosystem structure and function. Agricultural ecosystems, however, are designed around one or several species of plants or animals. Reduction of diversity in agricultural systems, compared with that in natural ecosystems, is traditionally considered essential to increase production of food, forage, and fiber. For simplicity of

1-56670-290-9/99/$0.00+$.50

Table 1 Hierarchy of Size and Abundance of Organisms Inhabiting Soil

Class	Example(s)	Biomass (g m^{-2})	Length (mm)	Populations (m^{-2})
Microflora	Bacteria, fungi, algae, actinomycetes	1–100	Not applicable	10^6–10^{12}
Microfauna	Protozoa	1.5–6.0	0.005–0.2	10^6–10^{12}
Mesofauna	Nematodes, enchytraeids, mites, Collembola	0.01–10	0.2–10	10^2–10^7
Macrofauna	Insects	0.1–2.5	10–20	10^2–10^5
Megafauna	Earthworms	10–40	20	0–10^3

Data from Dindal (1990) and Lal (1991).

management, biological cycles are sometimes replaced by fossil fuel-based products, e.g., synthetic fertilizers. Intense management practices that include application of pesticides and frequent cultivation affect soil organisms, often altering community composition of soil fauna. Soil biological and physical properties (e.g., temperature, pH, and water-holding characteristics) and microhabitat are altered when native habitat is converted to agricultural production (Crossley et al., 1992). Changes in these soil properties may be reflected in the distribution and diversity of soil mesofauna. Organisms adapted to high levels of physical disturbance become dominant within agricultural communities, thereby reducing richness and diversity of soil fauna (Paoletti et al., 1993).

Relationships between particular groups of organisms and management practices in agriculture can be studied under specific circumstances to define expected levels of diversity. The full diversity of soil communities has not been quantified for either agricultural or native ecosystems (Lee, 1991), and, in addition, the relationship between biodiversity and ecosystem function is not understood fully (Walker, 1992). Theoretically, this knowledge could be used to establish and maintain conditions that optimize beneficial effects of these organisms. Realistically, however, ideal conditions may be difficult to attain because of constraints imposed by agricultural production practices. We do not have sufficient knowledge to determine whether or not it is necessary, possible, or desirable to duplicate in agriculture the biodiversity that may be present in natural ecosystems.

This chapter examines the diversity and some of the functions of soil mesofauna in agricultural systems (Table 1). Most research on soil biota has focused on ecosystems such as forests and grasslands that are managed less intensively than agricultural or row crop systems. Ecologists have paid more attention to the role of micro- and mesofauna in ecosystem function, whereas agricultural scientists have focused on their role in nitrogen fixation and as pests and pathogens of crops. Our understanding of the role of soil organisms in agricultural systems is increasing, but more research is needed to elucidate their significance to crop production. Mesofauna occupy all trophic levels within the soil food web and affect primary production directly by root feeding and indirectly through their contribution to decomposition and nutrient mineralization (Crossley et al., 1992). Detailed reviews of the biology of soil fauna and their relationship to soil structure and ecological function are available (Wallwork, 1976; Swift et al., 1979; Freckman, 1982; Peterson and Luxton, 1982; Pimm, 1982; Seastedt, 1984; Dindal, 1990; Beare et al., 1992).

HABITAT

Unlike soil macrofauna (e.g., earthworms, termites, ants, some insect larvae), mesofauna generally do not have the ability to reshape the soil and, therefore, are forced to use existing pore spaces, cavities, or channels for locomotion within soil. Habitable pore space (voids of sufficient size and connectivity to support mesofauna) accounts for a small portion of total pore space (Hassink et al., 1993b). Microfaunal community composition becomes increasingly dominated by smaller animals as average pore volume decreases. Within the habitable pore space, microbial and mesofaunal activity is influenced by the balance between water and air. Maximum aerobic microbial activity occurs when 60% of the pore volume is filled with water (Linn and Doran, 1984). Saturation (waterlogging) and drought are detrimental to soil faunal communities because these conditions result in anaerobiosis or dehydration, respectively.

Populations and diversity of mesofauna are greatest in soil with high porosity and organic matter, and structured horizons (Andrén and Lagerlöf, 1983). Most biological activity occurs within the top 20 cm of soil which corresponds to the "plow layer" in agricultural soils. In uncultivated soil, mesofauna are more abundant in the top 5 cm than at greater depths in the soil. The organic horizon (O) is the area of accumulation of recognizable plant materials (high C:N ratio) and animal residues (low C:N ratio). The fermentation (F or O_1) layer consists of partially decomposed, mixed plant and animal debris permeated with hyphae of fungi and actinomycetes. The humus (H or O_2) horizon contains amorphous products of decomposition with the source unrecognizable. Eventually, organic matter from these horizons becomes incorporated into the mineral soil profile. Because cultivated agricultural systems often lack a distinct organic layer on the surface, one might expect diversity of soil biota to be less than in uncultivated or no-till soils (House et al., 1984).

Plants affect soil biota directly by generating inputs of organic matter above- and belowground and indirectly by the physical effects of shading, soil protection, and water and nutrient uptake by roots. Energy and nutrients obtained by plants eventually become incorporated in detritus which provide the resource base of a complex soil food web. Plant roots also exude amino acids and sugars which serve as a food source for microorganisms (Curl and Truelove, 1986). Soil mesofauna are often aggregated spatially which is probably indicative of the distribution of favored resources, such as plant roots and organic debris (Swift et al., 1979; Goodell and Ferris, 1980; Barker and Campbell, 1981; Noe and Campbell, 1985).

BIOLOGY AND ECOLOGY OF SOIL FAUNA

Soil mesofauna are often categorized by specific feeding behaviors, often depicted as microbial feeders. However, it should be emphasized that many organisms are at least capable of feeding at other trophic groups. As a result, omnivory in soil communities may be more prevalent than assumed previously (Walter et al., 1986; Walter, 1987; Walter et al., 1988; Walter and Ikonen, 1989; Mueller et al., 1990). Our discussion will focus specifically on nematodes, Collembola (springtails), and mites because they predominate in total numbers, biomass, and species of fauna in soil (Harding and Studdart, 1974; Samways, 1992).

Soil nematodes are relatively abundant (6×10^4 to 9×10^6 per m^2), small (300 μm to 4 mm) animals with short generation times (days to a few weeks) that allow them to respond to changes in food supply (Wasilewska, 1979; Bongers, 1990). Relative to other soil microfauna, trophic or functional groups of nematodes can be identified easily, primarily by morphological structures associated with various modes of feeding (Yeates and Coleman, 1982; Freckman, 1988; Bongers, 1990). Nematodes may feed on plant roots, bacteria, fungi, algae, and/or other nematodes (Wasilewska, 1979).

Mites and collembolans can account for 95% of total soil microarthropod numbers (Harding and Studdart, 1974). Soil mites occur mainly in three suborders. The suborder Oribatida (Cryptostigmata) comprises the numerically dominant group in the organic horizons of the soil. Members of the mite suborder Mesostigmata (Gamasida) are relatively large, active mites. The mite suborder Prostigmata (Actinedida) is a large and taxonomically complex group. Soil prostigmatids have more heterogeneous feeding habits than other mite suborders (see table in Kethley, 1990, for feeding habits). Prostigmatids are mostly fungal feeders and predators.

Collembolans are abundant and distributed widely. Collembolans have relatively high metabolic, feeding, and reproductive potential. Functional classification of collembolans (Christiansen, 1964; Bödvarsson, 1970; Verhoef and Brussard, 1990) can be based on gut content or shape of the mouthparts, which are adapted to the specific feeding habit (Swift et al., 1979). Because most forms of Collembola feed on decaying vegetation and associated microflora, the distribution of mycelia and spores of saprophytic fungi may be a major factor influencing the distribution of collembolans.

Other groups of arthropods that occur commonly in soil are pseudoscorpions, symphylans, pauropods, proturans, diplurans, and the immature stages of holometabolous insects (Dindal, 1990). Ants and termites can also be very numerous; however, these macroarthropods will not be considered here (Brian, 1978).

Plant Feeders

Plant-feeding nematodes can become abundant in agricultural ecosystems (Wasilewska, 1979; Popovici, 1984; Neher and Campbell, 1996). These nematodes may affect primary productivity of plants by altering uptake of water and nutrients. These abnormalities may result from changes in root morphology and/or physiology. For many agricultural crops, a negative relationship between crop yield and populations of plant-feeding nematodes, such as *Meloidogyne, Heterodera,* and *Pratylenchus* spp., has been observed (Mai, 1985; Barker et al., 1994). However, when entire nematode communities, including free-living nematodes, are examined, a positive association has been observed between plant biomass production and total nematode populations in grassland ecosystems (Yeates and Coleman, 1982). This relationship holds for plant production measured as harvested hay and root biomass (King and Hutchinson, 1976). A negative relationship between total nematode populations and plant productivity has been observed in tropical forests (Kitazawa, 1971). The relationship between soil nematode communities and row crop yield has yet to be determined.

Microarthropods rarely harm crop plants. However, soil mesofauna may become pests when a preferred food source is absent. Some Collembola, e.g., sminthurids and onychiurids, may feed on roots. For example, root-grazing injury on sugar beet is caused by *Onychiurus* spp. (Collembola) rubbing their bristled bodies against roots (Curl et al., 1988). However, root injury decreases if specific weed species and certain kinds and amounts of organic matter are present and, thus, provide the preferred microbial food supply. Few groups of soil mites are adapted to feeding on live plant tissues in soil. Some examples occur in the Tarsonemidae (Prostigmata) and Periohmanniidae (Oribatida). Most soil mites feed on plant material only after decomposition has begun. Often, increasing vegetational diversity and the quality and quantity of organic matter in soil increases potential benefits by soil mesofauna.

Microbial Feeders

Microbial-feeding mesofauna feed on fungi (including mycorrhizae), algae, slime molds, and bacteria by removing them from clumps of decaying material or soil aggregates (Moore and de Ruiter, 1991). Generally, bacterial-feeding nematodes such as Cephalobidae and Rhabditidae (Neher and Campbell, 1996) are abundant in agricultural ecosystems (Wasilewska, 1979; Popovici, 1984). Consumption of microbes by soil mesofauna alters nutrient availability by stimulating new microbial growth and activity plus releasing nutrients immobilized previously by microbes.

In general, fungal feeding is the dominant trophic function of microarthropods. Collembolan species have preferred food sources which are maintained even after the material has passed through the digestive tracts of other animals. For example, the collembolans *Proisotoma minuta* and *O. encarpatus* feed upon the soilborne fungal plant pathogen *Rhizoctonia solani* which causes damping-off disease on cotton seedlings (Curl et al., 1988). These collembolan species prefer feeding on the fungal pathogen in soil compared with the biocontrol fungi *Laetisaria arvalis, Trichoderma harzianum,* and *Gliocladium virens* (Curl et al., 1988). Additionally, collembolan species can distinguish and graze selectively on different species of vesicular-arbuscular mycorrhizae (Thimm and Larink, 1995).

Almost all oribatid mites are microbial feeders. Examples of microbial feeding also occur in the Mesostigmata (Ameroseiidae, Uropodidae) and Prostigmatida (Tarsonemidae, Nanorchestidae, Stigmaeidae Pygmephoridae, Eupodidae, and Tydeida). Although many microarthropods are microbial feeders, recent studies indicate that other arthropods are omnivorous and shift feeding behavior as food resources change (Walter, 1987; Mueller et al., 1990).

Omnivory

Omnivores add "connectedness" to the food web by feeding on more than one food source (Coleman et al., 1983). Omnivorous nematodes, such as some Dorylamidae, make up only a small portion of the total nematodes in agricultural ecosystems (Wasilewska, 1979; Neher and Campbell, 1996). They may feed on algae, bacteria, fungi, and other nematodes. Collembolans are often microbial feeders, but may also be facultative predators of nematodes (Snider et al., 1990). Mites that feed on both microbes and decaying plant material can be found in the oribatid mite families Nothridae, Camisiidae, Liacaridae, Oribatulidae, and Galumnidae. Coprophages, which ingest dung and carrion, including dead insects, are found among the oribatid families Euphthiracaridae, Phthiracaridae, Galumnidae, and Oppiidae.

Predators

Mesofauna may be predators or serve as prey for predaceous mites and other predators, such as beetles, fly larvae, centipedes, and spiders. Predatory nematodes feed upon all the other trophic groups of nematodes (Moore and de Ruiter, 1991) and represent only a small portion of the total nematodes in agricultural ecosystems (Wasilewska, 1979). Nematode predators (e.g., members of the orders Mononchida and Tripylida) and insect-parasitic nematodes (e.g., members of the families Steinernematidae, Diplogasteridae, Mermithidae) present in the soil may affect populations of their prey (Poinar, 1979; Small, 1987; Stirling, 1991).

Soil microarthropods can be important predators on small arthropods (e.g., proturans, pauropods, enchytraeids) and their eggs, nematodes, and on each other. Predation of insect eggs in agroecosystems may constitute a major influence of controlling microarthropod populations. Brust and House (1988) found that the mite *Tyrophagus putrescentiae* is an important predator of eggs of southern corn rootworm *Diabrotica undecimpunctata howardi* in peanuts. Chaing (1970) estimated that predation by mites accounted for 20% control of corn rootworms (*Diabrotica* spp.) and 63% control following the application of manure. Mite predation on root-feeding nematodes may be significant under some conditions (Inserra and Davis, 1983; Walter, 1988). For example, one adult of the mesostigmatid mite *Lasioseius scapulatus* and its progeny consumed approximately 20,000 *Aphelenchus avenae* on agar plates in 10 days (Imbriani and Mankau, 1983). Collembolan species may also consume large numbers of nematodes (Gilmore, 1970). For example, *Entomobryoides dissimilis* consumed more than 1000 nematodes in a 24-h period. Furthermore, collembolans may consume large numbers of insect-parasitic nematodes and, thus, affect the efficacy of these nematodes used as biological control agents of soil-dwelling insect pests (Epsky et al., 1988; Gilmore and Potter, 1993).

ECOSYSTEM PROCESSES

Micro- and mesofauna contribute directly to ecosystem processes such as decomposition and nutrient cycling in complex and interactive ways (Swift et al., 1979).

Bacteria, actinomycetes, fungi, algae, and protozoa are primary decomposers of organic matter. These microorganisms are involved directly with production of humus, cycling of nutrients and energy, elemental fixation, metabolic activity in soil, and the production of complex chemical compounds that cause soil aggregation.

Microbial-grazing mesofauna affect growth and metabolic activities of microbes and alter the microbial community, thus regulating decomposition rate (Wasilewska et al., 1975; Trofymow and Coleman, 1982; Whitford et al., 1982; Yeates and Coleman, 1982; Seastedt, 1984) and nitrogen mineralization (Seastedt et al., 1988; Sohlenius et al., 1988). Nematodes feed on bacteria and fungi on decaying organic matter, but not on the organic matter itself. Nematode species with a buccal stylet (spearlike structure) feed on cell contents and juices obtained by piercing the cellular walls of plant roots or fungal mycelium. Other species have no stylets and feed on particulate food such as bacteria and small algae (Vinciguerra, 1979). Microarthropods fragment detritus and increase surface area for further microbial attack (Berg and Pawluk, 1984). For example, collembolans and mites may enhance microbial activity, accelerate decomposition, and mediate transport processes in the soil. Even when they do not transform ingested material significantly, they break it down, moisten it, and make it available for microorganisms.

There is evidence that plants benefit from increased mineralization of nitrogen by soil mesofauna. Shoot biomass and nitrogen content of plant shoots grown in the presence of protozoans and nematodes were greater when compared with plants grown without mesofauna (Verhoef and Brussard, 1990). Soil fauna are responsible for approximately 30% of nitrogen mineralization in agricultural and natural ecosystem soils. The main consumers of bacteria are protozoa and bacterial-feeding nematodes which account for 83% of nitrogen mineralization contributed by soil fauna (Elliott et al., 1988). Nematodes also excrete nitrogenous wastes, mostly as ammonium ions (Anderson et al., 1983; Ingham et al., 1985; Hunt et al., 1987). Collembola excrete nitrate in concentrations 40 times more than their food source (Teuben and Verhoef, 1992). Furthermore, large collembolan species increase mineralization by selective feeding on fungi, whereas smaller species aid in the formation of humus by nonselective scavenging and mixing of the mineral and organic fractions of soil (van Amelsvoort et al., 1988). Microfauna constitute a reservoir of nutrients. When microfauna die, nutrients immobilized in their tissues are mineralized and subsequently become available to plants.

Soil fauna transport bacteria, fungi, and protozoa (in gut or on cuticle) across regions of soil and, thus, enhance microbial colonization of organic matter (Seastedt, 1984; Moore et al., 1988). For example, Collembola and sciarid fly larvae transmit root-infecting fungi and fungal parasites (Anas and Reeleder, 1988; Whipps and Budge, 1993). Microarthropods are surrounded by and, therefore, may disseminate propagules of insect-parasitic fungi including *Beauveria* spp., *Metarhizium* spp., *Paecilomyces* spp., and *Verticillium* spp. and facultative pathogens of insects in the genera *Aspergillus* and *Fusarium* spp. Under laboratory conditions, Collembola and mites transport spores of the insect-parasite *M. anisopliae* (Zimmerman and Bode, 1983). The impact of insect-parasitic fungi on natural populations of microarthropods is unknown.

VALUE OF DIVERSITY

Diversity in form and function of biotic communities results in the formation of spatial and temporal heterogeneity of organisms that contributes to the overall function of the ecosystem. Individual taxa may have multiple functions, and several taxa may appear to have similar functions. However, function may not necessarily be redundant, because taxa performing the same function are often isolated spatially, temporally, or by microhabitat preference (Beare et al., 1995). Biodiversity allows organisms to avoid intense competition for food or space, decrease invasion and disruption, and maintain constancy of function through fluctuating environmental conditions.

Various measures of diversity are available to describe soil invertebrate communities including abundance, biomass, density, species richness, species evenness, maturity indexes, trophic/guild structure, and food web structure. Indexes of diversity, which include elements of richness (number of taxa) and evenness (relative abundances), can be applied at scales ranging from alleles and species to regions and landscapes. Diversity indexes do not reveal the taxonomic composition of the community. For example, a community composed entirely of exotic species could have the same index value as a community composed entirely of endemic species. Therefore, a diversity index, by itself, does not predict ecosystem health or productivity.

Debates concerning relationships between biodiversity and ecosystem stability became popular in the 1960s and 1970s. MacArthur (1955) was the first to argue that complex systems are more stable than simple systems. In the early 1970s, May (1972; 1973) used mathematical models to argue that diverse communities were less stable than simple systems. Today, some conclude that relatively simple, short food webs that exhibit little omnivory or looping are more stable than longer food webs with much omnivory or looping. A short food web is one with few trophic levels (Polis, 1991). Others hypothesize that high linkage is responsible for making food webs unstable, i.e., stability can develop if numbers of species increase but not if omnivory increases (Pimm et al., 1991; Lawton and Brown, 1993). It is clear from this ongoing debate that it is impossible to generalize the relationship between biodiversity and ecosystem stability. Besides, none of the theories has been tested adequately for application to soil communities.

Factors affecting diversity within trophic groups of the detritus food web include altitude, latitude (Procter, 1984; Rohde, 1992), predation in the presence of strong competitive interactions (Petraitis et al., 1989), and disturbance (Petraitis et al., 1989; Hobbs and Huenneke, 1992). For example, the pervading theory is that the greatest species diversity is found in the tropics and that diversity decreases with increasing latitude (Rohde, 1992). However, the opposite is true for free-living nematodes. Free-living nematodes are more diverse and abundant in temperate than in tropical regions (Procter, 1984; 1990). Nematodes are tolerant of harsh conditions at high latitudes but are not competitive against more-specialized soil fauna in the tropics (Petraitis et al., 1989).

At smaller scales, predators may promote species diversity among competing prey species when they feed preferentially on exceptionally competitive prey (Petraitis et al., 1989). Disturbance also plays a role with the "intermediate disturbance hypothesis" suggesting that taxonomic diversity should be highest at moderate levels of disturbance (Petraitis et al., 1989; Hobbs and Huenneke, 1992). Disturbance is defined as a cause (a physical force, agent, or process, either abiotic or biotic) that results in a perturbation (an effect or change in system state relative to a reference state and system) (Rykiel, 1985). If disturbance is too mild or too rare, then soil communities will approach equilibrium and will be dominated by fewer taxa that can outcompete all other taxa. However, attainment of steady-state equilibria in agricultural or natural ecosystems is uncommon (Richards, 1987). If disturbance is common or harsh, only a few taxa that are insensitive to disruption will persist, therefore decreasing biodiversity (Petraitis et al., 1989). For example, Prostigmatid mites in the Eupodidae, Tarsonemidae, and Tydeidae are among the most abundant in cultivated agroecosystems and their numbers increase rapidly in response to disturbances such as cultivation (Crossley et al., 1992).

AGRICULTURAL DISTURBANCES

Disturbance can alter the diversity of an ecosystem (Atlas, 1984) directly by affecting survivorship of individuals or indirectly by changing resource levels (Hobbs and Huenneke, 1992). Sometimes, diversity measurements reflect the result of disturbance caused by pollution and/or stress. For example, taxonomic diversity of microinvertebrate communities was less in polluted or disturbed than in unpolluted or undisturbed agricultural sites (Atlas et al., 1991). Pollution eliminates sensitive species, reducing competition so that tolerant species proliferate (Atlas, 1984).

The successional status of a soil community may also reflect the history of disturbance. Succession in cropped agricultural fields begins with depauperate soil which acts like an island to which a series of organisms immigrate. First, opportunistic species, such as bacteria and their predators, are colonists of soil. Subsequently, fungi and their predators migrate into the area (Böstrom and Sohlenius, 1986). Microarthropods, such as collembolans, mites, and fly maggots, can colonize nearly bare ground and rise quickly in population density. Top predator microarthropods, such as predaceous mites and nematodes, become established later and may have a function similar to keystone predators in other community food webs (Elliott et al., 1988). Finally, macro- and megafauna, such as earthworms, millipedes, slugs, centipedes, wood lice, sow bugs, and pill bugs, join the soil community (Strueve-Kusenberg, 1982).

Succession can be interrupted at various stages by agricultural practices, such as cultivation and applications of fertilizer and pesticide (Ferris and Ferris, 1974; Wasilewska, 1979). Such interruptions reduce diversity and successional "maturity." Maturity indices are based on the principles of succession and relative sensitivity of various nematode taxa to stress or disruption of the successional sequence (Bongers, 1990). Indices that describe associations within biological communities, such as a maturity index, are less variable than measures of abundance of a single taxonomic or functional group and are, thus, more reliable as measures of ecosystem condition (Neher et al., 1995).

Soil Texture and Compaction

Soil texture may impose physical restrictions on the ability of fauna to graze on microbes; therefore, texture may play a role in faunal-induced mineralization of microbial carbon and nitrogen (van Veen and Kuikman, 1990). Carbon and nitrogen mineralization is generally faster in coarse than in fine-textured soils. In clay soils, organic material is protected physically from decomposers by its location in small pores. In sandy soils, organic matter is protected by its association with clay particles (Hassink et al., 1993a). Nematodes and microarthropods are often less abundant in heavy clay soil than in sandy or peat soil (van de Bund, 1970; Zirakparvar et al., 1980; Verma and Singh, 1989). Euedaphic species such as collembolans in the Onychiuridae and mesostigmatid mite *Rhodacarus roseus* are especially rare in clay soil (Didden, 1987).

Mesofauna are affected adversely by soil compaction (Aritajat et al., 1977a,b). Wheel-induced compaction reduces soil porosity, which is accompanied by a decrease in microbial biomass carbon and the density of Collembola (Heisler and Kaiser, 1995). Collembolans avoid narrow pores to protect their waxy surface from damage (Choudhuri, 1961). Wheel traffic decreased the density of collembolans and predatory mites by 30 and 60%, respectively, compared with noncompacted soil. The number of species was also reduced by compaction (Heisler, 1994).

Cultivation

Cultivation affects biogeochemical cycling by physically rearranging soil particles and changing pore size distribution, patterns of water and gas infiltration, and gas emission (Klute, 1982). Tillage disrupts soil aggregates, closes soil cracks and pores, and promotes drying of the surface soil. Soil fauna become sparse in top layers of cultivated soil because moisture content fluctuates widely and the original pore space network in this layer is destroyed. These physical alterations of the surface layers of soil may persist for many years after cultivation has ceased.

Soils managed by conventional — or reduced — tillage practices have distinct biological and functional properties (Doran, 1980; Hendrix et al., 1986). Plant residue is distributed throughout the plow layer in fields managed with conventional tillage. Under these conditions enhanced by cultivation, organisms with short generation times, small body size, rapid dispersal, and generalist feeding habits thrive (Steen, 1983). These soils are dominated by bacteria and their predators such as nematodes and astigmatid mites (Andrén and Lagerlöf, 1983; Yeates, 1984; Hendrix et al., 1986; Beare et al., 1992) and are considered in an early stage of succession. Oribatid and mesostigmatid mites decrease while other groups such as prostigmatid mites and Collembola tolerate, but do not benefit, from cultivation (Crossley et al., 1992). However, prostigmatid mite communities can be more diverse, containing both fungal- and nematode-feeding taxa in cultivated soils (van de Bund, 1970). Many microarthropods have omnivorous feeding habits in systems cultivated frequently (Beare et al., 1992).

Conservation — or no-till — practices generate more biologically complex soils than conventional tillage; however, in general, no-tillage cultivation does not appear to result in greater concentrations of microarthropods than conventional tillage except under drought stress (Perdue and Crossley, 1989). However, many studies comparing tillage effects are short term. Our knowledge about tillage effects may change as more long-term studies are implemented. Reduced tillage leaves most of the residue of the previous crops on the soil surface, and results in changes in physical and chemical properties of the soil (Blevins et al., 1983). Surface residues retain moisture, dampen temperature fluctuations, and provide a continuous substrate which promotes fungal growth. The increased fungal abundance can be attributed to the ability of fungi to translocate nutrients from soil into surface residues, their tolerance of lower pH and water potentials that often occur in surface residues, and their ability to penetrate and use large detritus particles (Hendrix et al., 1986; Holland and Coleman, 1987). Relative abundances of fungi and their predators, such as nematodes and many microarthropods groups (e.g., uropodid mesostigmatid mites; tarsonemid, eupodid, tydeids, and pygmephorid prostigmatid mites; oribatid mites) (Walter, 1987), in no-till soils represent a more mature successional state than one dominated by bacteria (Yeates, 1984; Böstrom and Sohlenius, 1986; Hendrix et al., 1986; Holland and Coleman, 1987; Neher and Campbell, 1994). Fungal feeding by microarthropods may stimulate microbial growth and enhance decomposition and nutrient immobilization (Seastedt, 1984). However, nutrient mineralization rates are relatively slow with stratification of debris and soil; nutrients are immobilized inside plant debris on the soil surface (Hendrix et al., 1986; Holland and Coleman, 1987).

Fertilization

Fertilization may influence the population abundance or composition of mesofaunal communities in soil. The outcome is a result of factors such as fertilizer quality and/or quantity (Verhoef and Brussard, 1990). Fertilization may, thus, affect the abundance and diversity of soil mesofauna directly or indirectly. These changes in community composition may, in turn, influence ecosystem function.

Nutrients applied to agricultural soils may be derived either from fossil fuels or plant and animal waste products. Nutrients are available in both forms, but organic amendments also contain microbes and their respective food resources. Additions of mineral fertilizers decrease populations of oribatid (cryptostigmatid) and prostigmatid mites, and root- and fungal-feeding, omnivorous and predaceous nematodes (Sohlenius and Wasilewska, 1984). Numbers of root-feeding nematodes may increase with increased nitrogen fertilizer (Wasilewska, 1989). Populations of astigmatid mites (Andrén and Lagerlöf, 1983) and bacterial-feeding nematodes (Sohlenius and Wasilewska, 1984) increase with additions of mineral fertilizer, but even more so when soils are fertilized with manure which simultaneously adds organic matter and microbes (Andrén and Lagerlöf, 1983; Weiss and Larink, 1991). However, in Dutch polder soil, abundances and biomass of nematodes, mites, and Collembola were similar between fields fertilized with manure, crop residues, and green manure and those field soils amended with crop residues and synthetic fertilizer (van de Bund, 1970).

Mesofauna aggregate around manure and plant litter (van de Bund, 1970). Populations of fungal-feeding nematodes (Weiss and Larink, 1991), potworms

(Enchytraeids), collembolans, and sometimes mesostigmatid mites increase with applications of manure (Andrén and Lagerlöf, 1983). Fratello et al. (1989) examined the effects of seven types of organic fertilizer on the microfauna in alfalfa fields. The reactions of populations of microarthropods to the different treatments varied with sample date, illustrating the highly complex interactions that occur in the soil. Poultry manure, sheep manure, worm compost, autoclaved urban sludge, urban sludge, vetch green manure, or straw were added to soil to provide a common level of 4% organic matter. Straw was the only additive that did not depress mite populations. Fewer mites and Collembola were found in plots treated with autoclaved urban sludge than those treated with non-autoclaved urban sludge.

The quality of plant and animal wastes as nutrient sources may be altered by composting. For example, applications of aged compost can increase suppression of plant pathogens by increasing the effectiveness of biocontrol agents. The plant-pathogenic fungus *Rhizoctonia solani* may cause damping-off disease in soil when fresh or immature compost material high in cellulose content is added. However, in aged compost, cellulose is degraded and the biocontrol fungus *Trichoderma* spp. can parasitize the pathogen effectively, thus suppressing disease (Chung et al., 1988).

Large doses of mineral or manure fertilizers can harm mesofauna because of toxicity (e.g., anhydrous ammonia) or high osmotic pressure due to salt (Andrén and Lagerlöf, 1983). The repellent nature of ammonium can affect soil invertebrates adversely (Potter, 1993). The potential for toxic effects can be decreased by applying composted manure and sludge (Ott et al., 1983). However, accumulation of heavy metals from repetitive sludge applications may kill omnivorous and predaceous nematodes (Weiss and Larink, 1991).

Fertilization affects soil microflora and, thus, indirectly impacts soil mesofauna by changing their food resources (Weil and Kroontje, 1979). Additions of nitrogen may acidify soil and inhibit microbial growth and activity. Nitrogen may also affect the quality of microbes as a food source for mesofuana. Booth and Anderson (1979) grew two species of fungi in liquid media with 2, 20, 200, or 2000 ppm nitrogen and determined the fecundity of the collembolan *Folsomia candida* while feeding on the fungi. Fecundity increased with increasing nitrogen content up to 200 ppm. *F. candida* did not show a preference for feeding on fungi with a greater or lesser nitrogen content.

The effect of fertilization on microarthropod species diversity and abundance within taxa and the subsequent impact on decomposition and nutrient mineralization processes are not well understood. For example, synthetic fertilizers increase nematode diversity, but applications of manure decrease nematode diversity (Wasilewska, 1989). The mechanism(s) explaining the differences are not understood. Applications of synthetic nitrogen fertilizer on Swedish arable soils growing spring barley (*Hordeum distichum* L.) changed community composition, but not numbers and biomass of nematodes, Collembola and mites (Andrén et al., 1988). Within a given environment, increased densities of microarthropods have been correlated with increased foliage, root and microbial productivity (Lussenhop, 1981), or increased food resource via fertilization. It is not known at what scale of resolution soil faunal communities respond to changes in ecosystem function.

Pesticides

Pesticides are an integral part of modern farming practice. Pesticides can enter the soil by a variety of routes, e.g., intentional application, spillage, overspraying, runoff, aerial transport with soil, or leaching. Organic matter plays a major role in the binding of pesticides in soil. Fulvic and humic acids are most commonly involved in binding interactions. Pesticides or their degradation intermediates can also be polymerized or incorporated into humus by the action of soil microbial enzymes (Bollag et al., 1992).

Soil fumigation with general biocides such as methyl bromide decreases microbial populations and nearly eliminates nematodes (Yeates et al., 1991). Although recovery occurs, population densities may not return to prefumigation levels even after 5 months (Yeates et al., 1991). Fumigation with general biocides return the successional status of soil to that of a depauperate soil matrix that can only be inhabited by primary colonists. However, within 60 weeks after soil fumigation and manuring, a progression of colonization by early successional species followed by more-specialized, later successional taxa can be observed (Ettema and Bongers, 1993).

Broad-spectrum insecticides that are applied for the control of insect pests can be toxic to predaceous and parasitic arthropods. A single surface application of chlorpyrifos reduced populations of predatory mites in plots of Kentucky bluegrass for 6 weeks and similar applications of isofenphos reduced populations of non-oribatid mites, Collembola, millipedes, and Diplura for as long as 43 weeks (Potter, 1993). Densities of Collembola were lower in aldicarb-treated soil than in untreated soil, but only the collembolans in the suborder Arthropleona were influenced negatively, whereas Symphypleona were not affected or occurred in higher numbers in soil treated with aldicarb (Koehler, 1992). Mesostigmatid mites did not occur at the site for the first 2 months after treatment, and their abundance was reduced for 6 months. After 3 and 4 years, abundance was similar in treated and untreated soil. Koehler (1992) noted a change in species composition associated with aldicarb treatment and categorized three groups of reaction. The most sensitive organisms were absent from 9 months to 1 year after application; other groups showed no reaction to treatment, or a positive reaction. Surface-dwelling microarthropods appeared to be affected less negatively than were soil-dwelling microarthropods.

Badejo and Van Straalen (1992) tested the effects of atrazine on the growth and reproduction of the collembolan *Orchesella cincta*. The lethal concentration (LC_{50}) for atrazine was estimated at 224 μg/g atrazine in food. Mortality and molting frequency increased with increasing concentrations of atrazine. The no observed effect concentration (NOEC) on egg production of *O. cincta* was 40 μg/g. Based on data for five collembolan species, 2.7 μg/g was estimated to be the hazardous concentration for 5% of soil invertebrates, which corresponds to the recommended field rate of 2.5 μg/g. House et al. (1987) investigated the impact of seven herbicides on miroarthropods and decomposition. No effect of any herbicide was observed on numbers of microarthropods, but decomposition of wheat straw was more rapid in soils without than with herbicide.

Generally, phenoxy acetic acid herbicides (e.g., 2,4-D, 2,4,5-T, 2-methyl-4-chlorophenoxyacetic acid) do not depress soil fauna directly with toxic effects, but indirectly through reduced vegetation and smaller additions of organic matter to soil (Andrén and Lagerlöf, 1983). Simazine, a triazine herbicide, is deleterious to most soil fauna (Edwards and Stafford, 1979).

Certain compounds such as the fungicide benomyl and its conversion product carbendazim have negative effects on soil biota even in low concentrations (Andrén and Lagerlöf, 1983). Applications of the fungicide captan to field soil reduce the abundance of saprophytic fungi and fungal-feeding mites compared with untreated field soil (Mueller et al., 1990).

CONCLUDING REMARKS

There are many other factors that influence diversity and function in agricultural soils. Greater diversity and later successional communities of soil fauna such as nematodes are found in soils with perennial crops compared with soils with annual crops (Ferris and Ferris, 1974; Wasilewska, 1979; Freckman and Ettema, 1993; Neher and Campbell, 1994). Root growth is more extensive and less ephemeral with perennial than with annual crops. Differences between soils with perennial (e.g., meadow fescue *Festuca pratensis* L.) and annual (e.g., barley) crops may be less pronounced for perennial crops younger than 3 years old than more mature crops (Böstrom and Sohlenius, 1986).

In fields where annual crops are grown, the diversity of soil fauna is increased with management practices such as crop rotation, polycultures, crop mixtures, trap crops, and intercropping. For example, populations of oribatid (cryptostigmatid) and prostigmatid mites and springtails were greater in soils with crop rotation than without (Andrén and Lagerlöf, 1983). However, diversity of nematode communities in soils in intercropping systems of yellow squash (*Cucurbita pepo* L.) and cucumber (*Cucumis sativa* L.) with alfalfa (*Medicago sativa* L.) or hairy indigo (*Indigofera hirsuta* L.) were not greater consistently than monocultures (Powers et al., 1993). The lack of consistent difference in diversity was attributed to fluctuations in diversity occurring within the growing season. Further studies are needed to elucidate the role of faunal diversity in soils with heterogeneous cropping systems.

Agricultural systems are complex, and most research studies have focused on single factors in an effort to reveal underlying mechanisms. This results in a lack of understanding of how multiple environmental and biotic factors interact to affect soil biodiversity and function. As interest in reducing fossil fuel–based inputs increases, reliance on natural cycles and processes will increase. We should allow the soil to work for us and not work against it (Elliott and Coleman, 1988). More research is needed to determine the impact of multiple and interacting management practices on biodiversity, nutrient cycling, pest populations, and plant productivity. With this information, we can maximize our ability to tailor agricultural practices to optimize crop productivity while positively affecting beneficial soil organisms and the functions they perform.

REFERENCES

Anas, O. and Reeleder, R. D., 1988. Feeding habits of larvae of *Bradysia coprophila* on fungi and plant tissue, *Phytoprotection,* 69:73–78.

Anderson, R. V., Gould, W. D., Woods, L. E., Cambardella, C., Ingham, R. E., and Coleman, D. C., 1983. Organic and inorganic nitrogenous losses by microbivorous nematodes in soil, *Oikos,* 40:75–80.

Andrén, O. and Lagerlöf, J., 1983. Soil fauna (microarthropods, enchytraeids, nematodes) in Swedish agricultural cropping systems, *Acta Agric. Scand.,* 33:33–52.

Andrén, O., Paustian, K., and Rosswall, T., 1988. Soil biotic interactions in the functioning of agroecosystems, *Agric. Ecosyst. Environ.,* 24:57–67.

Aritajat, U., Madge, D. S., and Gooderham, P. T., 1977a. The effect of compaction of agricultural soils on soil fauna. I. Field investigations, *Pedobiologia,* 17:262–282.

Aritajat, U., Madge, D. S., and Gooderham, P. T., 1977b. The effect of compaction of agricultural soils on soil fauna. II. Laboratory investigations, *Pedobiologia,* 17:283–291.

Atlas, R. C., 1984. Use of microbial diversity measurements to assess environmental stress, in *Current Perspectives in Microbial Ecology,* C. J. Klug and C. A. Reddy, Eds., American Society of Microbiology, Washington, D.C., 540–545.

Atlas, R. C., Horowitz, A., Krichevsky, C., and Bej, A. K., 1991. Response of microbial populations to environmental disturbance, *Microb. Ecol.,* 22:249–256.

Badejo, M. A. and Van Straalen, N. M., 1992. Effects of atrazine on growth and reproduction of *Orchesella cincta* (Collembola), *Pedobiologia,* 36:221–230.

Barker, K. R. and Campbell, C. L., 1981. Sampling nematode populations, in *Plant Parasitic Nematodes,* Vol. III, B. M. Zuckerman and R. A. Rohde, Eds., Academic Press, New York, 451–474.

Barker, K. R., Hussey, R. S., Krusberg, L. R., Bird, G. W., Dunn, R. A., Ferris, H., Ferris, V. R., Freckman, D. W., Gabriel, C. J., Grewal, P. S., MacGuidwin, A. E., Riddle, D. L., Roberts, P. A., and Schmitt, D. P., 1994. Plant and soil nematodes: societal impact and focus for the future, *J. Nematol.,* 26:127–137.

Beare, M. H., Parmelee, R. W., Hendrix, P. F., Cheng, W., Coleman, D. C., and Crossley, D. A., Jr., 1992. Microbial and faunal interactions and effects on litter nitrogen and decomposition in agroecosystems, *Ecol. Monogr.,* 62:569–591.

Beare, M. H., Coleman, D. C., Crossley, D. A., Jr., Hendrix, P. F., and Odum, E. P., 1995. A hierarchical approach to avaluating the significance of soil biodiversity to biogeochemical cycling, in *The Significance and Regulation of Soil Biodiversity,* H. P. Collins, G. P. Robertson, and M. J. Klug, Eds., Kluwer Academic, Dordrecht, The Netherlands, 5–22.

Berg, N. W. and Pawluk, S., 1984. Soil mesofaunal studies under different vegetative regimes in North Central Alberta, *Can. J. Soil Sci.,* 64:209–223.

Blevins, R. L., Smith, M. S., Thomas, G. W., and Fry, W. W., 1983. Influence of conservation tillage on soil properties, *J. Soil Water Conserv.,* 38:301–304.

Bödvarsson, H., 1970. Alimentary studies of seven common soil-inhabiting Collembola of southern Sweden, *Entomol. Scand.,* 1:74–80.

Bollag, J.-M., Myers, C. J., and Minard, R. D., 1992. Biological and chemical interactions of pesticides with soil organic matter, *Sci. Total Environ.,* 123/124:205–217.

Bongers, T., 1990. The maturity index: an ecological measure of environmental disturbance based on nematode species composition, *Oecologia,* 83:14–19.

Booth, R. G. and Anderson, J. M., 1979. The influence of fungal food quality on the growth and fecundity of *Folsomia candida* (Collembola: Isotomidae), *Oecologia,* 38:317–323.

Böstrom, S. and Sohlenius, B., 1986. Short-term dynamics of nematode communities in arable soil: influence of a perennial and an annual cropping system, *Pedobiologia,* 29:345–357.

Brian, M. V., 1978. *Production Ecology of Ants and Termites*, Cambridge University, Cambridge, U.K.

Brust, G. E. and House, G. J., 1988. A study of *Tyrophagus putrescentiae* (Acari: Acaridae) as a facultative predator of southern corn rootworm eggs, *Exp. Appl. Acarol.*, 4:335–344.

Chaing, H. C., 1970. Effects of manure applications and mite predation on corn rootworm populations in Minnesota, *J. Econ. Entomol.*, 64:934–936.

Choudhuri, D. K., 1961. Effect of soil structure on Collembola, *Sci. Cul.*, 27:494–495.

Christiansen, K., 1964. Bionomics of Collembola, *Annu. Rev. Entomol.*, 9:147–178.

Chung, Y. R., Hoitink, H. A. H., and Lipps, P. E., 1988. Interactions between organic-matter decomposition level and soilborne disease severity, *Agric. Ecosyst. Environ.*, 24:183–193.

Coleman, D. C., Reid, C. P. P., and Cole, C. V., 1983. Biological strategies of nutrient cycling in soil systems, *Adv. Ecol. Res.*, 13:1–55.

Crossley, D. A., Jr., Mueller, B. R., and Perdue, J. C., 1992. Biodiversity of microarthropods in agricultural soils: relations to processes, *Agric. Ecosyst. Environ.*, 40:37–46.

Curl, E. A. and Truelove, B., 1986. *The Rhizosphere*, Springer-Verlag, New York.

Curl, E. A., Lartey, R., and Peterson, C. C., 1988. Interactions between root pathogens and soil miroarthropods, *Agric. Ecosyst. Environ.*, 24:249–261.

Didden, W., 1987. Reactions of *Onychiurus fimatus* (Collembola) to loose and compact soil — methods and first results, *Pedobiologia,* 30:93–100.

Dindal, D. L., 1990. *Soil Biology Guide*, John Wiley, New York.

Doran, J. W., 1980. Soil microbial and biochemical changes associated with reduced tillage, *Soil Sci. Soc. Am. J.,* 44:765–771.

Edwards, C. A. and Stafford, C. J., 1979. Interactions between herbicides and the soil fauna, *Annu. Appl. Biol.*, 91:132–137.

Elliott, E. T. and Coleman, D. C., 1988. Let the soil work for us, *Ecol. Bull.*, 39:23–32.

Elliott, E. T., Hunt, H. W., and Walter, D. E., 1988. Detrital food web interactions in North American grassland ecosystems, *Agric. Ecosyst. Environ.*, 24:41–56.

Epsky, N. D., Walter, D. E., and Capinera, J. L., 1988. Potential role of nematophagous microarthropods as biotic mortality factors of entomophagous nematodes (Rhabditida: Steinernematidae, Heterorhabditidae), *J. Econ. Entomol.*, 81:821–825.

Ettema, C. H. and Bongers, T., 1993. Characterization of nematode colonization and succession in disturbed soil using the Maturity Index, *Biol. Fertil. Soils,* 16:79–85.

Ferris, V. R. and Ferris, J. C., 1974. Inter-relationships between nematode and plant communities in agricultural ecosystems, *Agro-Ecosystems,* 1:275–299.

Fratello, B., Sabatini, C. A., Mola, L., Uscidda, C., and Gessa, C., 1989. Effects of agricultural practices on soil arthropoda: organic and mineral fertilizers in alfalfa fields, *Agric. Ecosyst. Environ.,* 27:227–239.

Freckman, D. W., 1982. *Nematodes in Soil Ecosystems*, University of Texas, Austin, TX.

Freckman, D. W., 1988. Bacterivorous nematodes and organic-matter decomposition, *Agric. Ecosyst. Environ.,* 24:195–217.

Freckman, D. W. and Ettema, C. H., 1993. Assessing nematode communities in agroecosystems of varying human intervention, *Agric. Ecosyst. Environ.,* 45:239–261.

Gilmore, S. K., 1970. Collembola predation on nematodes, *Search: Agric.,* 1:1–12.

Gilmore, S. K. and Potter, D. A., 1993. Potential role of Collembola as biotic mortality agents for entomopathogenic nematodes, *Pedobiologia*, 37:30–38.

Goodell, P. and Ferris, H., 1980. Plant-parasitic nematode distributions in an alfalfa field, *J. Nematol.*, 12:136–141.

Harding, D. J. L. and Studdart, R. A., 1974. Microarthropods, in *Biology of Plant Litter Decomposition,* D. H. Dickinson and G. J. F. Pugh, Eds., Academic Press, New York, 489–532.

Hassink, J., Bouwman, L. A., Zwart, K. B., Bloem, J., and Brussard, L., 1993a. Relationship between soil texture, physical protection of organic matter, soil biota, and C and N mineralization in grassland soils, *Geoderma,* 57:105–128.

Hassink, J., Bouwman, L. A., Zwart, K. B., and Brussard, L., 1993b. Relationships between habitable pore space soil biota and mineralization rates in grassland soils, *Soil Biol. Biochem.*, 25:47–55.

Heisler, C., 1994. Auswirkungen von Bodenverdichtungen auf die Bodenmesofauna: Collembola and Gamasina — ein dreijähriger Feldversuch, *Pedobiologia,* 38:566–576.

Heisler, C. and Kaiser, E.-A., 1995. Influence of agricultural traffic and crop management on Collembola and microbial biomass in arable soil, *Biol. Fertil. Soils,* 19:159–165.

Hendrix, P. F., Parmelee, R. W., Crossley, D. A., Jr., Coleman, D. C., Odum, E. P., and Groffman, P. M., 1986. Detritus food webs in conventional and no-tillage agroecosystems, *BioScience,* 36:374–380.

Hobbs, R. J. and Huenneke, L. F., 1992. Disturbance, diversity, and invasion: implications for conservation, *Conserv. Biol.*, 6:324–337.

Holland, E. A. and Coleman, D. C., 1987. Litter placement effects on microbial and organic matter dynamics in an agroecosystem, *Ecology,* 68:425–433.

House, G. J., Stinner, B. R., Crossley, D. A., Jr., Odum, E. P., and Longdale, G. W., 1984. Nitrogen cycling in conventional and no-tillage agroecosystems in the southern Piedmont, *J. Soil Water Conserv.*, 39:194–200.

House, G. J., Worsham, A. D., Sheets, T. J., and Stinner, R. E., 1987. Herbicide effects on soil arthropod dynamics and wheat straw decomposition in a North Carolina no-tillage agroecosystem, *Biol. Fertil. Soils,* 4:109–114.

Hunt, H. W., Coleman, D. C., Ingham, E. R., Ingham, R. E., Elliott, E. T., Moore, J. C., Rose, S. L., Reid, C. P. P., and Morley, C. R., 1987. The detrital food web in a shortgrass prairie, *Biol. Fertil. Soils,* 3:57–68.

Imbriani, J. L. and Mankau, R., 1983. Studies on *Lasioseius scapulatus*, a meostigmatid mite predaceous on nematodes, *J. Nematol.*, 15:523–528.

Ingham, R. E., Trofymow, J. A., Ingham, E. R., and Coleman, D. C., 1985. Interactions of bacteria, fungi, and their nematode grazers: effects on nutrient cycling and plant growth, *Ecol. Monogr.*, 55:119–140.

Inserra, R. N. and Davis, D. W., 1983. *Hypoaspis* nr. *aculifer*: a mite predaceous on root-knot and cyst nematodes, *J. Nematol.*, 15:324–325.

Kethley, J., 1990. Acarina: Prostigmata (Actinedida), in *Soil Biology Guide*, D. L. Dindal, Ed., John Wiley, New York, 667–778.

King, K. L. and Hutchinson, K. J., 1976. The effects of sheep stocking intensity on the abundance and distribution of mesofauna in pastures, *J. Appl. Ecol.*, 13:41–55.

Kitazawa, Y., 1971. Biological regionality of the soil fauna and its function in forest ecosystem types, in *Ecology and Conservation No. 4. Productivity of Forest Ecosystems*, Proceedings of the Brussels Symposium 1969, UNESCO, United Nations, New York, 485–498.

Klute, A., 1982. Tillage effects on the hydraulic properties of soil: a review, in *Predicting Tillage Effects on Soil Physical Properties and Processes*, P. W. Unger and D. C. van Doren, Eds., American Society of Agronomy, Madison, WI, 29–43.

Koehler, H. H., 1992. The use of soil mesofauna for the judgment of chemical impact on ecosystems, *Agric. Ecosyst. Environ.*, 40:193–205.

Lal, R., 1991. Soil conservation and biodiversity, in *The Biodiversity of Microorganisms and Invertebrates: Its Role in Sustainable Agriculture,* D. L. Hawksworth, Ed., CAB International, London, U.K., 89–104.

Lawton, J. H. and Brown, V. K., 1993. Redundancy in ecosystems, in *Biodiversity and Ecosystem Function*, E.-D. Schulze and H. A. Mooney, Eds., Springer-Verlag, Berlin, 255–270.

Lee, K. E., 1991. The diversity of soil organisms, in *The Biodiversity of Microorganisms and Invertebrates: Its Role in Sustainable Agriculture,* D. L. Hawksworth, Ed., CAB International, London, U.K., 73–87.

Linn, D. C. and Doran, J. W., 1984. Effect of water-filled pore space on carbon dioxide and nitrous oxide production in tilled and nontilled soils, *Soil Sci. Soc. Am. J.,* 48:1267–1272.

Lussenhop, J., 1981. Microbial and microarthropod detrital processing in a prairie soil, *Ecology,* 62:964–972.

MacArthur, R. H., 1955. Fluctuations of animal populations, and a measure of community stability, *Ecology,* 36:533–536.

Mai, W. F., 1985. Plant–parasitic nematodes: their threat to agriculture, in *An Advanced Treatise on Meloidogyne,* Vol 1*: Biology and Control,* J. N. Sasser and C. C. Carter, Eds., North Carolina State University, Raleigh, NC, 11–17.

May, R. C., 1972. Will a large complex system be stable?, *Nature,* 238:413–414.

May, R. C., 1973. *Stability and Complexity in Model Ecosystems,* Princeton University Press, Princeton, NJ.

Moore, J. C. and de Ruiter, P. C., 1991. Temporal and spatial heterogeneity of trophic interactions within below-ground food webs, *Agric. Ecosys. Environ.,* 34:371–397.

Moore, J. C., Walter, D. E., and Hunt, H. W., 1988. Arthropod regulation of micro- and mesobiota in below-ground detrital food webs, *Annu. Rev. Entomol.,* 33:419–439.

Mueller, B. R., Beare, M. H., and Crossley, D. A., Jr., 1990. Soil mites in detrital food webs of conventional and no-tillage agroecosystems, *Pedobiologia,* 34:389–401.

Neher, D. A. and Campbell, C. L., 1994. Nematode communities and microbial biomass in soils with annual and perennial crops, *Appl. Soil Ecol.,* 1:17–28.

Neher, D. A. and Campbell, C. L., 1996. Sampling for regional monitoring of nematode communities in agricultural soils, *J. Nematol.,* 28:196–208.

Neher, D. A., Peck, S. L., Rawlings, J. O., and Campbell, C. L., 1995. Measures of nematode community structure for an agroecosystem monitoring program and sources of variability among and within agricultural fields, *Plant Soil,* 170:167–181.

Noe, J. P. and Campbell, C. L., 1985. Spatial pattern analysis of plant-parasitic nematodes, *J. Nematol.,* 17:86–93.

Ott, P., Hansen, S., and Vogtmann, H., 1983. Nitrates in relation to composting and use of farmyard manures, in *Environmentally Sound Agriculture,* W. Lockeretz, Ed., Praeger, New York, 145–154.

Paoletti, M. G., Foissner, W., and Coleman, D. C., 1993, *Soil Biota, Nutrient Cycling and Farming Systems,* Lewis, Boca Raton, FL.

Perdue, J. C. and Crossley, D. A., Jr., 1989. Seasonal abundance of soil mites (Acari) in experimental agroecosystems. Effects of drought in no-tillage and conventional tillage, *Soil Tillage. Res.,* 15:117–124.

Peterson, H. and Luxton, C., 1982. A comparative analysis of soil fauna populations and their role in decomposition processes, *Oikos,* 39:287–388.

Petraitis, P. S., Latham, R. L., and Niesenbaum, R. A., 1989. The maintenance of species diversity by disturbance, *Q. Rev. Biol.,* 64:393–418.

Pimm, S. L., 1982. *Food Webs,* Chapman and Hall, London.

Pimm, S. L., Lawton, J. H., and Cohen, J. E., 1991. Food web patterns and their consequences, *Nature,* 350:669–674.

Poinar, G. O., 1979. *Nematodes for Biological Control of Insects,* CRC Press, Boca Raton, FL.

Polis, G. A., 1991. Complex trophic interactions in deserts: an empirical critique of food-web theory, *Am. Nat.*, 138:123–155.

Popovici, I., 1984. Nematode abundance, biomass and production in a beech forest ecosystem, *Pedobiologia,* 26:205–219.

Potter, D. A., 1993. Pesticide and fertilizer effects on beneficial invertebrates and consequences for thatch degradation and pest outbreaks in turfgrass, in *Pesticides in Urban Environments: Fate and Significance*, K. D. Racke and A. R. Leslie, Eds., ACS Symposium Series No. 522, American Chemical Society, Washington, D.C., 331–343.

Powers, L. E., McSorley, R., and Dunn, R. A., 1993. Effects of mixed cropping on a soil nematode community in Honduras, *J. Nematol.*, 25:666–673.

Procter, D. L. C., 1984. Towards a biogeography of free-living soil nematodes. I. Changing species richness, diversity and densities with changing latitude, *J. Biogeogr.*, 11:103–117.

Procter, D. L. C., 1990. Global overview of the functional roles of soil-living nematodes in terrestrial communities and ecosystems, *J. Nematol.*, 22:1–7.

Richards, B. N., 1987. *The Microbiology of Terrestrial Ecosystems*, Longman Scientific and Technical, New York.

Rohde, K., 1992. Latitudinal gradients in species diversity: the search for the primary cause, *Oikos,* 65:514–527.

Rykiel, E. J., Jr., 1985. Towards a definition of ecological disturbance, *Aust. J. Ecol.*, 10:361–365.

Samways, M. H., 1992. Some comparative insect conservation issues of north temperate, tropical and south temperate landscapes, *Agric. Ecosyst. Environ.,* 40:137–154.

Seastedt, T. R., 1984. The role of microarthropods in decomposition and mineralization processes, *Annu. Rev. Entomol.*, 29:25–46.

Seastedt, T. R., James, S. W., and Todd, T. C., 1988. Interactions among soil invertebrates, microbes and plant growth in the tallgrass prairie, *Agric. Ecosyst. Environ.,* 24:219–228.

Small, R. W., 1987. A review of the prey of predatory soil nematodes, *Pedobiologia,* 30:179–206.

Snider, R. J., Snider, R., and Smucker, A. J. M., 1990. Collembolan populations and root dynamics in Michigan agroecosystems, in *Rhizosphere Dynamics,* J. E. Box, Jr. and L. C. Hammond, Eds., Westview, Boulder, CO, 169–191.

Sohlenius, B. and Wasilewska, L., 1984. Influence of irrigation and fertilization on nematode community in a Swedish pine forest soil, *J. Appl. Ecol.*, 21:327–342.

Sohlenius, B., Böstrom, S., and Sandor, A., 1988. Carbon and nitrogen budgets of nematodes in arable soil, *Biol. Fertil. Soils,* 6:1–8.

Steen, E., 1983. Soil animals in relation to agricultural practices and soil productivity, *Swed. J. Agric. Res.*, 13:157–165.

Stirling, G. R., 1991. *Biological Control of Plant Parasitic Nematodes*, CAB International, Wallingford, U.K.

Strueve–Kusenberg, R., 1982. Succession and trophic structure of soil animal communities in different suburban fallow areas, in *Urban Ecology,* R. Bornkamm, J. A. Lee, and C. R. D. Seaward, Eds., Blackwell Scientific, Oxford, U.K., 89–98.

Swift, C. J., Heal, O. W., and Anderson, J. C., 1979. *Decomposition in Terrestrial Ecosystems*, University of California, Berkeley.

Teuben, A. and Verhoef, H. A., 1992. Direct contribution by soil arthropods to nutrient availability through body and fecal nutrient content, *Biol. Fertil. Soils,* 14:71–75.

Thimm, T. and Larink, O., 1995. Grazing preferences of some Collembola for endomycorrhizal fungi, *Biol. Fertil. Soils,* 19:266–268.

Trofymow, J. A. and Coleman, D. C., 1982. The role of bacterivorous and fungivorous nematodes in cellulose and chitin decomposition in the context of a root/rhizosphere/soil conceptual model, in *Nematodes in Soil Ecosystems*, D. W. Freckman, Ed., University of Texas, Austin, 117–138.

van Amelsvoort, P. A. M., van Dongen, C., and van der Werff, P. A., 1988. The impact of Collembola on humification and mineralization of soil organic matter, *Pedobiologia*, 31:103–111.

van de Bund, C. F., 1970. Influence of crop and tillage on mites and springtails in arable soil, *Neth. J. Agric. Sci.*, 18:308–314.

van Veen, J. A. and Kuikman, P. J., 1990. Soil structural aspects of decomposition of organic matter by microorganisms, *Biogeochemistry*, 11:213–233.

Verhoef, H. A. and Brussard, L., 1990. Decomposition and nitrogen mineralization in natural and agro-ecosystems: the contribution of soil animals, *Biogeochemistry*, 11:175–211.

Verma, R. R. and Singh, H. R., 1989. Correlation studies between nematode population and ecological factors in Garhwal Hills, *Agric. Sci. Dig.*, 9:21–22.

Vinciguerra, M. T., 1979. Role of nematodes in the biological processes of the soil, *Boll. Zool.*, 46:363–374.

Walker, B. H., 1992. Biodiversity and ecological redundancy, *Conserv. Biol.*, 6:18–23.

Wallwork, J. A., 1976. *The Distribution and Diversity of Soil Fauna*, Academic Press, New York.

Walter, D. E., 1987. Trophic behavior of (mycophagous) microarthropods, *Ecology*, 68:226–229.

Walter, D. E., 1988. Predation and mycophagy by endeostigmatid mites (Acariformes: Prostigmata), *Exp. Appl. Acarol.*, 4:159–166.

Walter, D. E. and Ikonen, E. K., 1989. Species, guilds, and functional groups: taxonomy and behavior in nematophagous arthropods, *J. Nematol.*, 21:315–327.

Walter, D. E., Hudgens, R. A., and Freckman, D. W., 1986. Consumption of nematodes by fungivorous mites, *Tyrophagus* spp. (Acarina: Astigmata: Acaridae), *Oecologia*, 70:357–361.

Walter, D. E., Hunt, H. W., and Elliott, E. T., 1988. Guilds or functional groups? An analysis of predatory arthropods from a shortgrass steppe soil, *Pedobiologia*, 31:247–260.

Wasilewska, L., 1979. The structure and function of soil nematode communities in natural ecosystems and agrocenoses, *Pol. Ecol. Stud.*, 5:97–145.

Wasilewska, L., 1989. Impact of human activities on nematodes, in *Ecology of Arable Land*, C. Charholm and L. Bergstrom, Eds., Kluwer, Dordrecht, The Netherlands, 123–132.

Wasilewska, L., Jakubczyk, H., and Paplinska, E., 1975. Production of *Aphelenchus avenae* Bastian (Nematoda) and reduction of mycelium of saprophytic fungi by them, *Pol. Ecol. Stud.*, 1:61–73.

Weil, R. R. and Kroontje, W., 1979. Effects of manuring on the arthropod community in an arable soil, *Soil Biol. Biochem.*, 11:669–679.

Weiss, B. and Larink, O., 1991. Influence of sewage sludge and heavy metals on nematodes in an arable soil, *Biol. Fertil. Soils*, 12:5–9.

Whipps, J. M. and Budge, S. P., 1993. Transmission of the mycoparasite *Coniothyrium minitans* by collembolan *Folsomia candida* (Collembola: Entomobryidae) and glasshouse sciarid *Bradysia* sp. (Diptera: Sciaridae), *Annu. Appl. Biol.*, 123:165–171.

Whitford, W. G., Freckman, D. W., Santos, P. F., Elkins, N. Z., and Parker, L. W., 1982. The role of nematodes in decomposition in desert ecosystems, in *Nematodes in Soil Ecosystems*, D. W. Freckman, Ed., University of Texas, Austin, 98–115.

Yeates, G. W., 1984. Variation in soil nematode diversity under pasture with soil and year, *Soil Biol. Biochem.*, 16:95–102.

Yeates, G. W. and Coleman, D. C., 1982. Nematodes in decomposition, in *Nematodes in Soil Ecosystems*, D. W. Freckman, Ed., University of Texas, Austin, 55–80.

Yeates, G. W., Bamforth, S. S., Ross, D. J., Tate, K. R., and Sparling, G. P., 1991. Recolonization of methyl bromide sterilized soils under four different field conditions, *Biol. Fertil. Soils,* 11:181–189.

Zimmerman, G. and Bode, E., 1983. Untersuchungen zur Verbreitung des insektenpathogenen Pilzes *Metarhizium anisopliae* (Fungi Imperfecti, Moniliales) durch Bodenarthropoden, *Pedobiologia,* 25:65–71.

Zirakparvar, M. E., Norton, D. C., and Cox, C. P., 1980. Population increase of *Pratylenchus hexincisus* on corn as related to soil temperature and type, *J. Nematol.*, 12:313–318.

CHAPTER 4

Uses of Beneficial Insect Diversity in Agroecosystem Management

Petr Starý and Keith S. Pike

CONTENTS

1-56670-290-9/99/$0.00+$.50

INTRODUCTION

Beneficial Insects and Their Value

Parasitic and predatory insects occur within a wide range of insect groups, and at times, can be relatively abundant. Some common representatives include predatory carabid, coccinellid, and staphylinid beetles; predatory bugs; lacewings; syrphid, chamaemyiid, and other predatory flies; ants; and parasitic wasps. Related to the beneficial insects are predatory mites and spiders. These different beneficials prey on and reduce phytophagous pest populations and, thus, promote higher standards of crop health and economic returns. They can be highly effective at little or no cost, serving as biotic insecticides in place of chemicals and providing long-term control without the target pests developing significant resistance to them, and with minimal or no harm to humans or the environment (Wilson and Huffaker, 1976).

The value or full value of the insect natural enemies is not always realized because the preferred agent(s) for the target pest(s) are not present, or their abundance or activity is limited by environmental factors, in particular, by human-implemented practices such as clean cultivation, pesticide application, etc. (Johnson and Wilson, 1995). Natural enemies do not act in isolation but within the framework of natural enemy communities comprising individual guilds (Ehler, 1994). These communities manifest definable structure in which species richness and host range are fundamental properties (Hawkins and Sheehan, 1994). Individual members often show marked differences in their utilization of successive life stages of their hosts (Mills, 1994) and manifest certain positions (trophic levels) in the feeding hierarchy (Powell et al., 1996). Also, more or less competitive interactions may occur among species participants (Rosenheim et al., 1995).

The significance of beneficials in agroecosystems is often taken for granted or overlooked, sometimes when they are most effective. At times their significance becomes apparent in their absence or when they have been reduced to ineffective levels allowing the pest to reach crop-injuring levels (Ridgway and Vinson, 1977). The value is also apparent when exotic pest species in a new area rapidly reach pest status, but later are suppressed by adapting indigenous natural enemies or newly released beneficials or both (Clausen, 1978; Nechols et al., 1995).

Biodiversity Crisis

Biodiversity or natural habitat resources are dwindling, in large measure because of urban and agricultural spread and commercial development (LaSalle and Gauld, 1992). Natural enemies that demonstrate an ability to become community members in an agroecosystem have, in general, a much better chance to survive compared with those associated only with natural ecosystems. Through adaptation, at least some beneficials have overcome or are overcoming the biodiversity crisis by moving into or between cultivated landscapes. Plant diversity in cultivated landscapes contributes to overall biodiversity, whereas monocultures, especially large-scale monocultures, usually result in fewer species. In all cases, the diversity of beneficials and their management in agroecosystems should be considered from a dual viewpoint, inclusive of both the agroecologist and nature conservationist (Samways, 1993).

Biodiversity in Farmland

The elements of biodiversity, both floristic and faunistic, that sustain efficacious levels of beneficial insects in farmland settings are challenging to obtain, since farming constitutes a disturbance of the land, and therefore a disturbance of natural systems and a diminishing of the biotic elements. The greater the disturbance, the fewer the opportunities for the natural biota to exist. Present trends in biodiversity development center on equalizing ecological losses through crop diversification, adjacent landscape preservation, and intentional introduction of biotic agents (Michal, 1994). Landscape ecology and biodiversity are ecologically connected and mutually dependent (Carroll, 1990; van Hook, 1994). Overall activities that support and increase diversity and ecological stability in agroecosystems include, but are not limited to, development of biocorridors and biocenters, heterogeneous crops and crop structuring, polycultural crop rotation, biocontrol introductions, and pesticide use modifications (Paoletti et al., 1992; Petr and Dlouhy, 1992).

Biodiversity Monitoring

Biodiversity is described at three fundamental levels — ecosystem diversity, species diversity, and genetic diversity (Office of Technology Assessment, 1987). Changes in the diversity can be monitored by indicator species (Noss, 1990). Monitoring biodiversity in farmlands should account for biota both in the crops and in the soils. Apart from the common crops and livestock in agriculture, some 200,000 other species of plants and animals are involved in agricultural production and perform many essential functions, such as nutrient recycling, waste decay, plant protection, pollination (Pimentel et al., 1989).

Key Beneficials in Agroecosystems

Ideal integrated pest control should reflect ecological approaches that not only target the pest, but also account for the key natural enemies and associated interactions (LaSalle and Gauld, 1992; LaSalle, 1993). The success or lack of success of parasitic and predatory species is commonly linked to not only the target host, but other hosts, bioagents, habitats, and abiotic factors. Understanding the host range, host preferences, seasonal occurrence, interspecies competition and displacement, and habitat and food resource requirements of the beneficials is important to safeguarding them, increasing their numbers, and enhancing their performance. The diversity of beneficials in agroecosystems is often linked to natural or undisturbed environments. Where strong ties exist between biocontrol agents of agriculture and plant communities of natural diversity, it is important that these are identified and that the biodiversity linkages are preserved to undergird and support the existence of the beneficials year-round, and in some cases, for use in redistribution and introduction elsewhere. Some indigenous beneficials, though less important against present pests, may be key to preventing future introduced species from becoming problematic. Environmental diversity should be conserved regardless of what is known about the taxonomy or biology of the flora and fauna. Beneficials or groups of beneficials that are active in individual cropping systems commonly move in and between crops and noncultivated ecosystems. Therefore, any search for beneficials for introduction purposes should include not only the agroecosystems of interest, but also the surrounding habitat (Waage, 1991).

Intraspecific Diversity of Beneficials

A common feature of many agroecosystems is a reduction in species richness coupled with high populations of selected other species. With parasitic Hymenoptera, for example, the phenomenon may affect not only the interspecific relationships within certain parasitic spectra, but also the intraspecific diversity of individual parasitoids (Unruh and Messing, 1993). There may be changes in the gene flow between populations from different host species as well as certain dominance of features in populations from highly populated dominant hosts. These factors are important in the drift of parasitoids between different hosts. One of the main problems in agroecosystems is the lack of hosts for host alternation and relationships (Nemec and Starý, 1984; 1985; Powell, 1986).

Biodiversity of Beneficials in Insect Pest Control Systems

Biodiversity as a factor in pest control varies widely between countries and areas of the world. Introduction strategies in classical biological control typically center on the full range of beneficials attacking the target pest throughout the world, with the aim to find, select, and use the most promissive agents from the world complex.

The introduction of broad-spectrum pesticides starting with DDT contributed to a strong demand for pest-free crops. Resistance to pesticides and pest resurgence connected with natural enemy losses, however, led to the development of integrated pest management (IPM) (Stern et al., 1959; van den Bosch and Stern, 1962; Smith and Reynolds, 1966) and, more recently, alternative pest management emphasizing ecologically adapted and biorationally based approaches to the exclusion of synthetic pesticides uses (U.S. National Research Council, 1989; Vereijken, 1989).

Additionally, sustainable agriculture initiatives stimulated efforts to increase and maintain greater biodiversity through landscape protection of fauna and flora, e.g., introduction of grassland meadows in place of arable land (Petr and Dlouhy, 1992). Diversity, its support and enhancement through species richness, rotations, intercropping, cover crops, etc., is one of the basic principles of agroecology in sustainable agriculture systems (Thrupp, 1996).

IMPORTATION OF BENEFICIALS

Natural Enemy Spectrum, Selection, and Adaptation

The introduction of exotic natural enemies to control exotic pests is the primary approach in classical biological control (DeBach, 1964). Exotic pests, usually inadvertently introduced and without their natural enemies, commonly result in rapid increases and population outbreaks. The lack of natural enemies provides the impetus for purposeful introductions of potentially promising natural enemies. Such enemies are usually sought in the homeland of the target pest, but sometimes are obtained from areas outside the original home of the pest, where secondary adaptation by indigenous species has occurred. Consideration of world populations broadens the scope for selection of useful species and biotypes, and increases the prospect for success (Starý, 1970a; Huffaker et al., 1971; Greathead, 1986; Ehler, 1990; 1992).

New beneficial introductions add to the biodiversity of pest-infested areas, but also they can impact the indigenous biota and, in some instances, stand as biotic contaminants (Samways, 1993) in terms of host displacement and nontarget prey interspecies competition. Ideally, environmental impact determinations or projections on prey specificity should be studied prior to a release (Starý, 1993). With the exception of strict monophagous agents, some adaptation to indigenous biota is expected. Endemic ecosystems may be disrupted or partially disrupted due to various reasons and thus are more easily attacked by exotic invaders (Starý, 1994). An introduced biocontrol agent may not be necessarily harmful to indigenous communities where the spectrum of natural enemies includes numerous broadly oligophagous species (Starý et al., 1988) or where invasive species are suppressed (Samways, 1993).

New pests in an area can be attacked and suppressed to varying degrees by indigenous natural enemies. Here, there is no increase in species richness of natural enemies, simply expanded adaptation and enlargement of the prey/host spectrum. However, where new strains or biotypes occur, differences in intraspecific competition may result.

Once introduced species become established, they become part of the natural enemy spectrum for the target pest and associated area, and thus subject to all of the management approaches applied, such as conservation, augmentation, etc. New agents, however, need to be classified as generally more vulnerable to environmental modifications depending upon the degree of adaptation in their new environment.

Beneficial Introductions and Specificity

Classical biological control efforts are seldom instituted until a pest outbreak occurs, and, even then, there can be a delay because of the time involved to search for, import, quarantine, mass-rear, and release new agents. Although not generally followed, another approach is preventative biocontrol, the introduction of promising exotic biocontrol agents prior to the appearance of a forecasted or expected target pest. Except for strict monophages, oligophagous agents may be introduced for establishment on alternate prey/host species in association with the target agroecosystem (Starý et al., 1993). These then are present to attack the arriving pest, possibly before it is detected by humans. The approach provides a temporal advantage, and more or less limits population outbreak of the invading pest. Switching from a native host to a related introduced species can occur with striking results, and may include the indigenous beneficials (LaSalle, 1993). Understanding the host range of introduced exotics is key to achieving success in preventative biocontrol (Starý et al., 1993).

The host range of introduced beneficials should always be considered, i.e., control of more than one pest using the same regulatory agent. This requires an oligophagous agent. In principle, oligophagous agents can attack several pests in the same or different crops (Starý et al., 1993). Such situations contribute to the stability of the natural enemy interactions across cropping systems. However, such introductions are not without risks, e.g., host preference, poor alternation, or species-specific strains may eventually develop. The strategies, necessary attributes, and opinions for introducing biocontrol agents are widely discussed in the literature (DeBach, 1964; Starý, 1970a; Huffaker et al., 1971; van den Bosch and Messenger, 1973; Clausen, 1978; Croft, 1990; Ehler, 1990; Miller, 1993; Nechols et al., 1995).

CONSERVATION OF BENEFICIALS

Conservation in this discussion means modifying any environmental factors that are adverse to beneficials (DeBach, 1964) and adding requisites (McMurtry et al., 1995). This is a type of environmental insect control (Stern, 1981), a manipulation of the ecosystem to make it less favorable to the pest and more favorable to the natural enemies resulting in reduced pest levels (Mayse, 1983). Such approaches need to be considered in the context of the whole environment, the agroecosystem (target and adjoining crops) and surroundings, as these often have mutual connections. Many insects, whether classified as pests, beneficials, or indifferents, exhibit population drift (Starý, 1978; Bosch, 1987; Vorley and Wratten, 1987; van Emden, 1988). The boundary zone or ecotone where individual crops and noncrops overlap is frequently essential to the conservation and management of beneficials, both indigenous and introduced (DeBach, 1964; van Emden, 1965; Ridgway and Vinson, 1977; Stern, 1981; Powell, 1986; Gross, 1987; Martis, 1988; Altieri et al., 1993; Samways, 1993; Johnson and Wilson, 1995; McMurtry et al., 1995). Strict separation of ecosystems does not occur in nature.

Habitat Management, Crop Structure, and Diversity

Habitat management is viewed as a strategy aimed at designing and constructing "phytocenotic architecture" dominated by plants that support populations of natural enemies (Altieri and Whitcomb, 1979; Altieri, 1983). Diversification of habitat is achieved through crop structure, protective refugia, occurrence of alternative prey/host, and supplementary food resources (nectar, pollen). Crop structure is the agroecosystem and its specific characteristics, its biotic composition, seasonality, etc. Protective refugia are defined as habitats in which beneficials can survive critical periods of the year (principally summer and winter periods) to disperse later to crops. Protected refugia can include a wide array of plant types and setting, e.g., rangeland, weedy field margins, autumn-sown crops, etc.

Alternative prey/hosts and their availability or proximity to crops are important at times when the target pest is low in numbers. Alternative hosts can improve the synchrony between natural enemies and the target pests. For a given beneficial, the alternative host might be a nonpest species feeding on wild plants or it might be a pest species different from the target on another crop (van den Bosch and Telford, 1964; Powell, 1986). For some aphid parasitoids, utilization of alternative hosts in proximity to the target pest and crop is known as multilateral control or the multilateral control approach (Starý, 1972; 1978). Such an approach takes advantage of the oligophagous host range of the bioagent. Switching from one host to another can be effective or ineffective depending upon the biotype or species specificity of the bioagent (Gordh, 1977). Population genetics in relation to host alternation is currently of high research interest. Until recently, host alternation was based solely on field observation and laboratory transfers; now, studies on population molecular genetics are further clarifying the species or species strains of key importance and their host alternation dynamics (Unruh et al., 1983; Nemec and Starý, 1985).

Sources of food such as nectar and pollen are requisite for hymenopterous parasitoid (van Emden, 1962) and syrphid (Hickman and Wratten, 1996) adults to ensure effective reproduction. More often than not, flowering plants in or even around agriculture crops for such uses are not always readily available (van Emden, 1962; Altieri and Whitcomb, 1979; Altieri and Letourneau, 1982; Powell, 1986).

The composition, seasonality, field size, and location of crops and noncrops all affect biodiversity. Semiperennial and perennial crops are generally classified as more stabilizing for biotic diversity in comparison with annuals; nonetheless, population drift can take place across all settings (Gross, 1987; Andow, 1991). Gliessman (1987) reported that whenever two or more crops are planted together, there is increased potential for species interactions, and this would include beneficials.

Diverse landscape mosaics in and around small-sized fields enhance the chances for greater diversity of beneficials. Nonetheless, relatively high biodiversity is thought possible even in intensely cultivated areas as long as crops are arranged together with patches of natural or seminatural areas (Duelli et al., 1989; 1990). Among the most difficult environments for biocontrol to succeed in are the annual crop monocultures. These usually lack the resources for the natural enemies to be efficacious, are grown using cultural practices (e.g., mowing of alfalfa) that often damage the natural enemy populations, and are present for only part of the year (Rabb et al., 1976; Powell, 1986). In such circumstances, it may be necessary through habitat management to maintain small populations of the target pest to ensure survival of the key beneficials (Powell, 1986). Volunteer crops along field margins may be useful in this regard.

Preservation, establishment, or sustainment of small heterogeneous strip habitats, in or neighboring crop farmlands, adds to the overall biodiversity of farm or area (Nentwig, 1993). Also, strip farming, especially intercropping, increases natural enemy opportunities for predation and parasitism over that of strict monocultures (Powell, 1986).

Cultural practices such as full field mowing of perennial legumes and grassy meadows hinder bioagent survival and success. These negative effects can be partially offset by strip mowing which affords arthropods an opening to retreat to uncut portions (Stern et al., 1964; Müller-Ferch and Mouci, 1995) and, thus, increases or strengthens the population stability of the beneficials (Schlinger and Dietrick, 1960; van den Bosch et al., 1967; Starý, 1970b).

Heterogeneous herbaceous strips constitute more suitable habitat for field species than strips of shrubs or trees that tend to harbor ecologically different forest species (Nentwig, 1988). Selected weed species used in strips within crop fields have been shown to attract and aid in the conservation of beneficials (Starý, 1964; van Emden, 1965; Gliessmann, 1987; Nentwig, 1988; 1992; 1993; 1994; 1995; Frei and Manhart, 1992; Weiss and Nentwig, 1992; Hausmann, 1996). Wyss (1995) showed that in apple orchards certain flowering weed strips resulted in more aphidophagous predators and fewer aphid pests than like areas without weeds. Nectar-bearing plants (Chumakova, 1977) and rich undergrowth of wild flowers (Leius, 1967) showed similar beneficial effects. Many natural enemies occur commonly in association with wild or natural habitat not always classified as weeds (van Emden, 1965). Tillage practices (no-till, minimum, conventional), tillage timing, mowing, or other agronomic practices can influence, sometimes significantly, the performance, success, maturation, and dispersal of beneficials (Bugg and Ellis, 1990; Bugg, 1992). Molthan and Ruppert (1988) demonstrated that flowers in wide boundary strips attracted beneficials, some being especially attractive and nutritionally suitable. They further recommended protection or arrangement of boundary strips in the framework of agricultural extensification.

Cultures of medicinal, culinary, and ornamental herbs, such as sweet fennel (*Foeniculum vulgare*) and spearmint (*Menta spicata*), grown in organic market gardens near various vegetable and tree crops are known to attract several adult entomophagous Hymenoptera and flower-visiting beneficials (Sawoniewicz, 1973; Bugg and Wilson, 1989; Bugg et al., 1989; Maingay et al., 1991; Bugg and Waddington, 1994). Some studies on herb attractiveness have listed not only the sampled species, but also have determined or attempted to determine their significance and effect on pests in nearby crops (Bugg and Waddington, 1994).

Pollen can serve as a supplemental or essential food source for beneficials. For example, Ouyang et al. (1992) report that pollen can positively affect polyphagous predacious mites, especially during periods when their arthropod prey is scarce. Green manure crops such as faba bean (*Vicia faba*) can manifest similar positive effects on beneficials (Bugg and Ellis, 1988; Bugg et al., 1989). And, habitat manipulations, such as the addition of mulch and flowers, may enhance spider densities and lower the number of pest insects in a mixed vegetable system (Riechert and Bishop, 1990).

Cover crops in general are known to affect a number of phenomena in orchards. They may harbor pest species, but they can also lead to increased numbers of insect natural enemies and heightened pest biocontrol (Bugg et al., 1990; Bugg and Waddington, 1994). Some covers, or mixtures of cover crops, constitute field insectaries and may be marketable as "insectary crops" or crops that support high densities of beneficials. Different covers may require different management protocols, depending upon whether they are supplementary (alternative prey or hosts, pollen) or complementary (nectar, honeydew) food sources (Bugg, 1992). For a detailed review of cover crop management in temperate zone orchard crops (almond, pecan, walnut, apple, cherry, peach, and citrus), see Bugg and Waddington (1994).

Field margins and crop edges can be highly supportive of beneficials, but their relative abundance varies depending upon the mix of plants at the margins and the adjacent crop (Dennis and Fry, 1992). Some studies have shown that field margins can increase the diversity of arthropods within the crop and that movement between the margin and the field can be significant. Margin habitats provide the stability needed for species that would otherwise not survive across all seasons. The landscape matrix of field margins can be vital for effective field dispersal and conservation success (Röser, 1988; Dennis and Fry, 1992).

Crop edges adjacent to hedges and broad grassy strips bear a rich fauna of beneficials and should not be treated with pesticides or receive fertilizer (van den Bosch and Messenger, 1973; Morris and Weeb, 1987; Basedow, 1988; Holtz, 1988; Klingauf, 1988; Welling and Kokta, 1988). Dover (1991) and Samways (1993) showed that 6-m-wide edges around cereal fields receiving reduced and selective pesticides helped conserve beneficials. Welling and Kokta (1988) reported that wide headlands with a large source of flowering plants guaranteed nutrition for flower-visiting beneficials, served as a refuge for different species (before and after harvest), and acted as a bridge between isolated biotypes.

The plants that compose hedges, hedgerows, and windbreaks vary widely, as does their significance as reservoirs for beneficials (Solomon, 1981). Wide hedgerows or windbreaks composed of trees and shrubs appear to function as a type of biocorridor across the landscape (Forman and Baudry, 1984); the associated beneficials are in part forest-edge species. The hedgerows and windbreaks, together with boundary strips and unsprayed field margins, represent a functional part of the agroecosystem, influencing positively the beneficials (Knauer, 1988), and thus should be encouraged on farms (Basedow, 1988). A number of papers cover hedgerow/windbreak habitats in detail, including the role of beneficials and IPM (Lewis, 1969; Zwölfer et al., 1984; Stechman and Zwölfer, 1988; Welling et al., 1988; Häni, 1989).

Food Sprays and Semiochemicals

Food requirements of predaceous species sometimes vary between life stages. Larval stages may be carnivores, while adults may feed on nectar, honeydew, and pollen. For hymenopterous parasitoids, nectar and pollen requirements are common for adults, but not for parasitic larvae. With some crops, nectar, pollen, and honeydew sources are insufficient or unavailable. By providing an artificial food supplement, beneficials may be retained, arrested, or stimulated to oviposit. Treatments may consist of yeast, sucrose solution, or artificial honeydew (Hagen and Bishop, 1979; Gross, 1987).

Natural enemies are known to respond to a number of environmental cues in the course of locating desired habitats, plants, prey/hosts, and the opposite sex. Behavior-controlling chemicals (semiochemicals) are rather species specific. In theory, synthetically derived semiochemicals may be used to attract increased numbers of natural enemies into a crop, prolong their searching activity, and improve their performance (Vinson, 1977; 1981; Lewis, 1981; Nordlund et al., 1981; Powell, 1986; Gross, 1987; McMurtry et al., 1995).

Modification of Chemical Pest Control Practices

Biodiversity of natural enemies in agroecosystems may be substantially affected by the use of pesticides. Nonselective treatments are toxic to beneficials. They decrease populations, contributing to pest outbreak. For this reason and others, pesticides have begun to be used with greater care. Emphasis is beginning to center more on the development and use of selective pesticides, on target-directed applications, and on applications timed to avoid the direct treatment of beneficials. Information on pesticidal effects on beneficials is extensive.

In some respects, the value of beneficials has been heightened by the overuse of pesticides leading to chemical resistance, secondary pest outbreaks, and environmental pollution. IPM concepts and strategies, from the earliest discussions, have centered on selective treatments to protect beneficials as key components in integrated control (Stern et al., 1959; Smith and Reynolds, 1966; Croft, 1990; McMurtry et al., 1995).

Nonselective pesticides directly and negatively affect natural enemies, sometimes even the larval stages within a host, e.g., parasitoids. Pesticides, both selective and nonselective, indirectly impact beneficials by diminishing prey/host populations and, in turn, force surviving beneficials to disperse to other communities to find food (Petr and Dlouhy, 1992). In a few cases, pesticide-affected systems have led to changes in gene diversity of beneficials through selection of pesticide-resistant strains.

Biocontrol is not the primary approach for some agroecosystems, and may never be, but it is or could be the key component in many agroecosystems. In most natural environments, biocontrol provides common, if not perennial regulation. With the removal of pesticides, diversity of beneficials and restoration of biocontrols are possible (Hagen et al., 1971). Where market demands require blemish-free products or where farm economics dictate treatment to protect investments, chemicals will likely remain a standard defense.

Conservation activities targeting beneficials through habitat diversification may be adversely disrupted by herbicides. Way and Cammell (1981) suggested that insect communities, including natural enemies, in and around agroecosystems are affected more by herbicides than by pesticides. Similarly, fertilizer, especially in heavy dosage, can adversely affect conservation efforts.

AUGMENTATION OF BENEFICIALS

Augmentation, broadly defined, covers all of the activities that improve the effectiveness of beneficials (DeBach, 1964), such as new species releases (inoculative or inundative), planned genetic change, landscape modification, and so on. The addition of a new species or strain of species into a new area increases species diversity, but it can also affect interspecific relationships and population genetic characteristics.

Releases of Mass-Reared Natural Enemies

There are two main types of releases, inoculative and inundative. Inoculative releases utilize species intended for permanent establishment, where progeny of each successive generation continue as regulatory agents. Inundative releases utilize species intended for short-term action (permanent establishment not expected) and act as biotic insecticides.

Pesticide-Resistant Beneficials

Discovery, mass-rearing, and redistribution of endemic resistant populations of beneficials against target pests, or laboratory-planned genetic alteration, hybridization, or artificial selection for pesticide-resistant beneficials and subsequent field release, are measures intended to amplify existing levels of biocontrol. The use of such agents is expanding rapidly. Common cases include predatory mites, aphid predators, and parasitoids in deciduous tree fruit and nut crops (Hoy, 1985; 1995; Croft, 1990; Johnson and Wilson, 1995; McMurtry et al., 1995).

Landscape Ecology

Land management planning and IPM efforts must consider biocontrol issues, the players, their role, protection, and sustainment. Agroecosystems are linked not only to the surrounding natural lands, but also to biological diversification from distant sources (van Hook, 1994). Beneficial insect diversity needs to be dealt with as part of the overall diversity of the landscape. Biodiversity exists in a matrix of habitat patches including managed and natural environments. The landscape and beneficials are ecologically linked and interdependent (Carroll, 1990; van Hook, 1994).

IPM programs, though strongly based on ecological principles, still rely heavily on reductionistic approaches to control single pests on single crops (van Hook, 1994). Discussions on beneficials often center on the interdependency of specific habitats. Although it may not be possible to view all of the biological peculiarities of an ecosystem or to make overall generalizations, greater accounting of multiple factor linkages and features in common (e.g., epigean fauna, role of flowering plants, stability of plants, crop culture specifics, etc.) are necessary to appropriately understand, conserve, and manage beneficials in relation to their phytophagous hosts. Some beneficials may be linked to their prey/host for only a part of their life cycle.

Altieri (1983) divided the crop landscape into two groups: (1) heterogeneous (crop, meadow, woodland, windbreak, etc.) and (2) homogeneous (continuous crop cultures). The latter could be further subdivided into extensive plantings of annual monocultures, (e.g., small grains), semiperennials (e.g., alfalfa), perennials (e.g., grapes), or mixtures of these in small to moderate size.

The most effective beneficials in agriculture are usually species that are perennially present in farmlands. Long-term agricultural landscape optimization should support equilibrium between pest and beneficial, achievable in part or in total, through extended diversification of crops and wild plants, or through development of biocorridors adjoining particular managed habitats (Petr and Dlouhy, 1992). Farmland optimization will vary broadly between areas, but certainly will require long-term ecological planning and commitment with the flexibility to accommodate new discoveries for improved activity and perennial sustainment of desired beneficials.

Today, high-yielding, high-quality, insect-, disease-, and weed-free monocultures require high-energy inputs to produce and harvest. With production costs escalating, interest in enduring and efficient alternatives with less dependence on fossil fuels is escalating (Gross, 1987). Central to this is the management of agricultural communities (Altieri, 1981; Altieri et al., 1993), where selective diversity in, or in association with, cropping systems is used to enhance the stability, survival, and efficiency of the natural community inhabitants (Gross, 1987).

SUMMARY

Natural community beneficials in cropping systems are interlinked with surrounding vegetation matrices. Their biodiversity is affected by numerous farmland approaches and activities. But much of the biological aspects of the beneficials, their specific interrelationships, differences in relative abundance, performance, and management in and between agroecosystems, geographical areas, climatic zones, etc. are yet to be determined or determined in detail. The ecological planning, selection, adaptation, and management of beneficials in a given crop may initially appear costly if just the direct immediate system inputs and profits are considered (Dahlberg, 1992). The measures to enhance diversity and efficiency of beneficials can be difficult, and may not always be feasible. For this reason, each agricultural situation must be assessed separately (Altieri et al., 1993).

Management of beneficials based on current knowledge must be considered as developmental for now. Changes in ecological optimization and stability concepts and implementation will follow further crop structure analysis and landscape diversification research, on-farm trials, and successful IPM examples under variable environments. Legume-based crop rotations, recognized as stabilizing crops, used to improve soils and crop yields, may serve to increase the diversity of beneficials, and may play a significant role in the bilateral drift of beneficials to annuals.

The effects of natural enemies on target pests have economic implications. It is possible to evaluate numerically the ecological value of certain species within a complex net of relationships and connections. We may not appreciate or understand the role or the importance of a given species until after its disappearance, and there may be some insects or insect groups not directly determined to be beneficials in an economic sense, but that may contribute in other ways, such as to the beauty of the environment (Martis, 1988). This could occur in natural systems and in agroecosystems via landscape diversification. Butterflies flying over flowering meadows, hedgerows, weedy ecotones, roadsides, or similar biocorridors improve the often tiring effect of visually plain, monotonous monocultures. Simply stated, biodiversification can be used to improve the aesthetics of the farmland.

This chapter is not an exhaustive review. We have highlighted some key factors, outlined some of the complexities involved in the biology, pathways, and population dynamics of insect natural enemies and their management, cited references for more detail, and mentioned in brief the need for further research.

REFERENCES

Altieri, M. A., 1981. Crop-weed-insect interactions and the development of pest-stable cropping systems, in *Pest Pathogens and Vegetation. The Role of Weeds and Wild Plants in the Ecology of Crop Pests and Diseases, Proc. Univ. of York and Br. Ecol. Soc. and Fed. Br. Plant Path., 1980,* J. M. Thresh, Ed., Pitman, Boston, 459–466.

Altieri, M. A., 1983. Vegetational designs for insect-habitat management, *Environ. Manage.,* 7:3–7.

Altieri, M. A. and Letourneau, D. K., 1982. Vegetation management and biological control in agroecosystems, *Crop Prot.,* 1:405–430.

Altieri, M. A. and Whitcomb, W. H., 1979. The potential use of weeds in the manipulation of beneficial insects, *Hort. Sci.,* 14:12–18.

Altieri, M. A., Cure, J. R., and Garcia, M. A., 1993. The role and enhancement of parasitic Hymenoptera biodiversity in agroecosystems, in *Hymenoptera and Biodiversity,* J. LaSalle and I. D. Gauld, Eds., CAB Int., Wallingford, U.K., 257–275.

Andow, D. A., 1991. Vegetational diversity and arthropod population response, *Annu. Rev. Entomol.,* 36:561–586.

Basedow, T., 1988. Crop edges, boundary strips and hedges — aids for management, *Mitt. Biol. Bundesanst. Land Forstwirtsch. Berlin Dahlem,* 247:129–137.

Bosch, J., 1987. Der Einfluss einiger dominanter Ackerkräuter auf Nutz- and Schadarthropoden in einem Zuckerrübenfeld, *Z. Pflanzenkr. Pflanzenschutz,* 94:398–408.

Bugg, R. L., 1992. Using cover crops to manage arthropods on truck farms, *HortScience,* 27:741–745.

Bugg, R. L. and Ellis, R. T., 1988. Use of green manure crops to subsidize beneficial insects, in *Global Perspectives on Agroecology and Sustainable Agricultural Systems, Proceedings Sixth International Conference of the International Federation of Organic Agricultural Movements,* P. Allen and D. van Dusen, Eds., University of California, Santa Cruz, 553–557.

Bugg, R. L. and Ellis, R. T., 1990. Insects associated with cover crops in Masschusetts, *Biol. Agric. Hortic.,* 7:47–68.

Bugg, R. L. and Waddington, C., 1994. Using cover crops to manage arthropod pests in orchards: a review, *Agric. Ecosyst. Environ.,* 50:11–28.

Bugg, R. L. and Wilson, L. T., 1989. *Ammi visnaga* (L.) Lamarck (Apiaceae): associated beneficial insects and implications for biological control, with emphasis on the bell-pepper agroecosystem, *Biol. Agric. Hortic.,* 6:241–268.

Bugg, R. L., Ellis, R. T., and Carlson, R. W., 1989. Ichneumonidae (Hymenoptera) using extrafloral nectar of faba beans (*Vicia faba* L., Fabaceae) in Massachusetts, *Biol. Agric. Hortic.,* 6:107–114.

Bugg, R. L., Wäckers, F. L., Brunson, K. E., Phatak, S. C., and Dutcher, J. D., 1990. Tarnished plant bug (Hemiptera: Miridae) on seclected cool-season leguminous crop covers, *J. Entomol. Sci.,* 25:463–474.

Carroll, C. R., 1990. The interface between natural areas and agroecosystems, in *Agroecology,* C. R. Carroll, J. H. Vandermeer, and P. M. Rosset, Eds., McGraw Hill, New York, 365–388.

Chumakova, B. M., 1977. Ecological principles associated with augmentation of natural enemies, in *Biological Control by Augmentation of Natural Enemies,* R. L. Ridgway and S. B. Vinson, Eds., Plenum Press, New York, 39–78.

Clausen, C. P., 1978. Introduced parasites and predators of arthropod pests and weeds. A world review, in *Agriculture Handbook,* No. 480, Agriculture Research Service, U.S. Department of Agriculture, Washington, D.C.

Croft, B. A., 1990. Arthropod biological control agents and pesticides, John Wiley, New York.

Dahlberg, K. A., 1992. The conservation of biological diversity and U.S. agriculture: goals, institutions and policies, *Agric. Ecosyst. Environ.,* 42:177–193.

DeBach, P., 1964. The scope of biological control, in *Biological Control of Insect Pests and Weeds,* P. DeBach, Ed., Chapman and Hall, London, 3–20.

Dennis, P. and Fry, G. L. A., 1992. Field margins: can they enhance natural enemy population densities and general arthropod diversity on farmland?, *Agric. Ecosyst. Environ.,* 40:95–115.

Dover, J., 1991. The conservation of insects on arable farmland, in *Conservation of Insects and Their Habitats,* N. M. Collins and J. A. Thomas, Eds., Academic Press, London, 293–318.

Duelli, P., Studer, M., and Marchand, I., 1989. The influence of the surroundings on arthropod diversity in maize fields, *Acta Phytopathol. Entomol. Hung.,* 24:73–76.

Duelli, P., Studer, M., and Marchand, I., 1990. Population movements of arthropods between natural and cultivated areas, *Biol. Conserv.,* 54:193–207.

Ehler, L. E., 1990. Introduction strategies in biological control of insects, in *Critical Issues in Biological Control,* M. Mackauer and L. E. Ehler, Eds., Intercept, Andover, U.K., 111–134.

Ehler, L. E., 1992. Guild analysis in biological control, *Environ. Entomol.,* 21:26–40.

Ehler, L. E., 1994. Parasitoid communities, parasitoid guilds, and biological control, in *Parasitoid Community Ecology,* B. H. Hawkins and W. Sheehan, Eds., Oxford Science Publication, Oxford, 418–436.

Forman, R. T. T. and Baudry, S., 1984. Hedgerows and hedgerow networks in landscape ecology, *Environ. Manage.,* 8:95–510.

Frei, G. and Manhart, C., 1992. Nützlinge and Schädlinge and künstlich angelegten Ackerkrautstreifen in Getreidefeldern, *Agrarökologie*, 4:140.

Gliessman, S. R., 1987. Species interactions and community ecology in low external input agriculture, *Am. J. Alternative Agric.,* 11:160–165.

Gordh, G., 1977. Biosystematics of natural enemies, in *Biological Control by Augmentation of Natural Enemies,* R. L. Ridgway and S. B. Vinson, Eds., Plenum Press, New York, 125–148.

Greathead, D. J., 1986. Parasitoids in classical biological control, in *Insect Parasitoids,* J. Waage and D. J. Greathead, Eds., Academic Press, London, 290–318.

Gross, H. R., Jr., 1987. Conservation and enhancement of entomophagous insects — a perspective, *J. Entomol. Sci.,* 22:97–105.

Hagen, K. S. and Bishop, G. W., 1979. Use of supplemental foods and behavioral chemicals to increase the effectiveness of natural enemies, in *Biological Control and Insect Pest Management,* D. W. Davis, S. C. Hoyt, J. A. McMurtry, and M. T. Aliniazee, Eds., University of California, Davis, 49–60.

Hagen, K. S., van den Bosch, R., and Dahlsten, D., 1971. The importance of naturally occurring biological control in the western United States, in *Biological Control,* C. B. Huffaker, Ed., Plenum Press, New York, 253–293.

Häni, F., 1989. The third way, a research project in ecologically orientated farming systems in Switzerland, *WPRS Bull.,* XII-5:51–66.

Hausmann, A., 1996. The effects of weed strip-management on pests and beneficial arthropods in winter wheat fields, *Z. Pflanzenkr. Pflanzenschuz.,* 103:70–81.

Hawkins, G. H. and Sheehan, W., 1994. *Parasitoid Community Ecology,* Oxford Science Publication Press, Oxford.

Hickman, J. M. and Wratten, S. D., 1996. Use of *Phacelia tanacetifolia* strips to enhance biologiccal control of aphids by hoverfly larvae in cereal fields, *J. Econ. Entomol.,* 89:832–840.

Holtz, F., 1988. Occurrence of aphids on wild plants in crop edges and in boundary strips, *Mitt. Biol. Bundesanst. Land Forstwirtsch. Berlin Dahlem,* 247:7–84.

Hoy, M. A., 1985. Recent advances in genetics and genetic improvement in Phytoseiidae, *Annu. Rev. Entomol.,* 30:345–370.

Hoy, M. A., 1995. Walnut aphid — Part 2, in *Biological Control in the Western United States,* J. R. Nechols, Ed., University of California, Div. Agric. Nat. Res., Davis, CA, 142–144.

Huffaker, K. S., van den Bosch, R., and Dahlsten, D. L., 1971. The importance of naturally occurring biological control in the western United States, in *Biological Control,* C. B. Huffaker, Ed., Plenum Press, New York, 253–287.

Johnson, M. W. and Wilson, L. T., 1995. Integrated pest management: contributions of biological control to its implementation, in *Biological Control in the Western United States,* J. R. Nechols, Ed., University of California, Div. Agric. Nat. Res., Davis, CA, 7–24.

Klingauf, F., 1988. Herbicide- and insecticide-free crop edges: a contribution to an environmentally oriented agriculture, *Mitt. Biol. Bundesanst. Land Forstwirsch. Berlin Dahlem,* 247:14.

Knauer, N., 1988. Herbicide- and insecticide-free crop edges and hedges as compensation areas in agro-ecosystems, *Mitt. Biol. Bundesanst. Land Forstwirsch. Berlin Dahlem,* 247:15–128.

LaSalle, J., 1993. Parasitic Hymenoptera, biological control and biodiversity, in *Hymenoptera and Biodiversity,* J. LaSalle and I. D. Gauld, Eds., CAB Int., Wallingford, U.K., 197–215.

LaSalle, J. and Gauld, I. D., 1992. Parasitic Hymenoptera and the biodiversity crisis, in *Insect Parasitoids: 4th European Workshop,* REDIA, Firenze, 315–334.

Leius, K., 1967. Influence of wild flowers on parasitism of tent caterpillar and codling moth, *Can. Entomol.,* 99:444–446.

Lewis, T., 1969. The diversity of the insect fauna in a hedgerow and neighbouring fields, *J. Appl. Ecol.,* 6:453–458.

Lewis, W. J., 1981. Semiochemicals: their role with changing approaches to pest control, in *Semiochemicals: Their Role in Pest Control,* D. A. Nordlund, R. L. Jones, and W. J. Lewis, Eds., John Wiley, New York, 3–12.

Maingay, H. M., Bugg, R. L., Carlson, R. W., and Davidson, N. A., 1991. Predatory and parasitic wasps (Hymenoptera) feeding at flowers of sweet fennel (*Foeniculum vulgare* Miller var. dulce Battandier and Taabut, Apiaceae) and spearmint (*Mentha spicata* L., Lamiaceae) in Massachusetts, *Biol. Agric. Hortic.,* 7:363–383.

Martis, M., 1988. *Man vs. Landscape,* Horizont, Prague.

Mayse, M. A., 1983. Cultural control in crop fields: a habitat management technique, *Environ. Manage.,* 7:15–22.

McMurtry, J. A., Andres, L. A., Bellows, T. S., Hoyt, S. C., and Hagen, K. S., 1995. A historical overview of regional research projects, in *Biological Control in the Western United States,* J. Nechols, Ed., University of California, Div. Agric. Nat. Res., Davis, CA, 3–5.

Michal, I., 1994. *Ecological Stability,* Veronica Publ. and Ministry of the Environment of the Czech Republic, Prague.

Miller, J. C., 1993. Insect natural history, multi-species interactions and biodiversity in ecosystems, *Biodiversity Conserv.,* 2:233–241.

Mills, N. J., 1994. The structure and complexity of parasitoid communities in relation to biological conrol, in *Parasitoid Community Ecology,* B. H. Hawkins and W. Sheehan, Eds., Oxford Science Publication Press, Oxford, 397–417.

Molthan, J. and Ruppert, V., 1988. Significance of flowering wild herbs in boundary strips and fields for flower-visiting beneficial insects, *Mitt. Biol. Bundesanst. Land Forstwirsch. Berlin Dahlem,* 247:85–99.

Morris, M. G. and Weeb, N. R., 1987. The importance of field margins for the conservation of insects, in *Field Margins,* Thornton Heath, London, 53–65.

Müller-Ferch, G. and Mouci, M., 1995. Influence of mowing on storage compounds of weeds and their insect fauna, *Agrarökologie,* 14:3.

National Research Council, 1989. *The Role of Alternative Farming Methods in Modern Agriculture, 1989,* Alternative Agriculture, Board of Agriculture, Nattional Research Council, National Academy Press, Washington, D.C.

Nechols, J. R., Andres, L. A., Beardsley, J. W., Goeden, R. D., and Jackson, C. G., 1995. *Biological Control in the Western United States,* University of California, Div. Agric. Nat. Res., Davis.

Nemec, V. and Starý, P., 1984. Population diversity centers of aphid parasitoids (Hym., Aphidiidae): a new strategy in IPM, *Ecol. Aphidophaga II,* Zvikov, Podhradie (Abstr.).

Nemec, V. and Starý, P., 1985. Genetic diversity and host alternation in aphid parasitoids (Hymenoptera: Aphidiidae), *Entomol. Gen.,* 10:253–258.

Nentwig, W., 1988. Augmentation of beneficials arthropods by strip-management. Succession of predaceous arthropods and long-term changes in the ratio of phytophagous and predaceous arthropods in a meadow, *Oecologie,* 76:597–606.

Nentwig, W., 1992. Augmentation of beneficials arthropods by sown weed strips in agricultural areas, *Z. Pflanzenkr. Pflanzenschutz. Sonderh.,* XIII:3–40.

Nentwig, W., 1993. Nützlingsförderung in Agrarökosystemen, *Verh. Ges. Ökol.,* Zürich, 22:14.

Nentwig, W., 1994. Wechselwirkungen zwischen Ackerwildpflanzen und der Entomofauna, *Ber. Landwirtsch. N.F.,* 209:23–135.

Nentwig, W., 1995. Ackerkrautstreifen als Systemansatz für eine umweltfreundliche Landwirtschaft, *Mitt. Dtsch. Ges. Allg. Angew. Ent.,* 9:679–683.

Nordlund, D. A., Jones, R. L., and Lewis, W. J., 1981. *Semiochemicals: Their Role in Pest Control,* John Wiley, New York.

Noss, R. F., 1990. Indicators for monitoring biodiversity: a hierarchical approach, *Conserv. Biol.,* 4:355–364.

Office of Technology Assessment, 1987. *Technologies to Maintain Biological Diversity,* U.S. Government Printing Office, Washington, D.C.

Ouyang, Y., Grafton-Cardwell, E. E., and Bugg, R. L., 1992. Effects of various pollens on development, survivorship, and reproduction of Euseius tularensis (Acrari: Phytoseiidae), *Environ. Entomol.,* 21:1371–1376.

Paoletti, M. G., Pimentel, D., Stinner, B. R., and Stinner, D., 1992. Agroecosystem biodiversity: matching production and conservation biology, in *Biotic Diversity in Agroecosystems,* M. G. Paoletti and D. Pimentel, Eds., Elsevier, Amsterdam, 3–23.

Petr, J. and Dlouhy, J., 1992. Ecological agriculture, Zemed. Nakl. Brazda, Prague.

Pimentel, D., Culliney, T. W., Buttler, I. W., Reinemann, D. J., and Beckman, K. B., 1989. Low-input sustainable agriculture using ecological management practices, *Agric. Ecosyst. Environ.*, 27:3–24.

Powell, W., 1986. Enhancing parasitoid activity in crops, in *Insect Parasitoids,* J. Waage and D. J. Greathead, Eds., Academic Press, London, 319–340.

Powell, W., Walton, M. P., and Jervis, M. A., 1996. Populations and communities, in *Insect Natural Enemies: Practical Approaches to Their Study and Evaluation,* M. A. Jervis and N. Kidd, Eds., Chapman and Hall, London, 223–292.

Rabb, R. L., Stinner, R. E., and van den Bosch, R., 1976. Conservation and augmentation of natural enemies, in *Theory and Practice of Biological Control,* C. B. Huffaker and P. S. Messenger, Eds., Academic Press, New York, 233–254.

Ridgway, R. L. and Vinson, S. B., 1977. Biological control by augmentation of natural enemies, Plenum Press, New York.

Riechert, S. E. and Bishop, L., 1990. Prey control by an assemblage of generalist predators: spiders in garden test systems, *Ecology,* 71:1441–1450.

Rosenheim, J. A., Kaya, H. K., Ehler, L. E., Marois, J. J., and Jaffee, B. A., 1995. Intraguild predation among biological-control agents: theory and evidence, *Biol. Control,* 5:303–335.

Röser, R., 1988. *Saum- und Kleinbiotope. Ökologische Funktion, wirtschaftliche Bedeutung und Schutzwürdigkeit in Agrarlandschaften,* Ecomed Verlag, Landsberg.

Samways, M. J., 1993. Insects and biodiversity conservation: some perspectives and directives, *Biodiversity Conserv.,* 2:258–282.

Sawoniewicz, J., 1973. Gasienicznikowate (Ichneumonidae, Hymenoptera) odwiedzajace Kwiaty Goryszu — *Peucedanum oreoselinum* L. (Umbelliferae), *Folia For. Pol. Ser. A,* 21:43–78.

Schlinger, E. I. and Dietrick, E. I., 1960. Biological control of insect pests aided by strip-farming alfalfa in experimental program, *Calif. Agric.,* 14:8–9.

Smith, R. F. and Reynolds, H. T, 1966. Principles, definitions and scope of integrated pest control, in *Proc. FAO Symp. on Integrated Pest Control,* Rome, 1:11–17.

Solomon, M. G., 1981. Windbreaks as a source of orchard pests and predators, in *Pest Pathogens and Vegetation: The Role of Weeds and Wild Plants in the Ecology of Crop Pests and Diseases,* J. M. Thresh, Ed., *Proc. Univ. York and Br. Ecol. Soc. and Fed. Br. Plant Pathol.,* Putmin, Boston, 273–283.

Starý, P., 1964. The foci of aphid parasites (Hymenoptera, Aphidiidae) in nature, *Ekol. Pol. A,* 12:529–554.

Starý, P., 1970a. Biology of aphid parasites (Hymenoptera: Aphidiidae) with respect to integrated control, *Series Entomologia,* 6, Dr. W. Junk, The Hague.

Starý, P., 1970b. Routine alfalfa (*Medicago sativa*) cutting program and its influence on pea aphid (*Acyrthosiphon pisum* Harris) populations, *Oecologia,* (Berlin) 5:347–379.

Starý, P., 1972. Host range of parasites and ecosystem relations, a new viewpoint in multilateral control concept (Hom., Aphididae; Hym., Aphidiidae), *Ann. Soc. Entomol. Fr.,* 8:351–358.

Starý, P., 1978. Seasonal relations between lucerne, red clover, wheat and barley agro-ecosystems through the aphids and parasitoids (Homoptera, Aphididae; Hymenoptera, Aphidiidae), *Acta Entomol. Bohemoslov.,* 75:296–311.

Starý, P., 1993. The fate of released parasitoids (Hymenoptera: Braconidae, Aphidiinae) for biological control of aphids in Chile, *Bull. Entomol. Res.,* 83:633–639.

Starý, P., 1994. Aphid parasitoid fauna (Hymenoptera: Aphidiidae) of the southern beech (Nothofagus) forest, *Stud. Neotrop. Fauna Environ.,* 29:87–98.

Starý, P., Lyon, J. P., and Leclant, F., 1988. Post-colonization host range of *Lysiphlebus testaceipes* (Cresson) in the Mediterranean area (Hymenoptera, Aphidiidae), *Acta Entomol. Bohemoslov.,* 85:1–11.

Starý, P., Gerding, M., Norambuena, H., and Remaudière, G., 1993. Environmental research on aphid parasitoid biocontrol agents in Chile (Hym., Aphidiidae, Aphelinidae), *J. Appl. Entomol.,* 115:292–306.

Stechmann, D. H. and Zwölfer, H., 1988. Die Bedeutung von Hecken für Nutzarthropoden, in *Agrarökosystemen*, R. A. Heft, Ed., Landw. Verlag GmbH, Münster-Hiltrup, Germany, 31–55.

Stern, V. M., 1981. Environmental control of insects using trap crops, sanitation, prevention, and harvesting, in *CRC Handbook of Pest Management in Agriculture,* Vol. 1, D. Pimentel, Ed., CRC Press, Boca Raton, FL, 199–207.

Stern, V. M., Smith, R. F., van den Bosch, R., and Hagen, K. S., 1959. The integration of chemical and biological control of the spotted alfalfa aphid. Part 1. The integrated control concept, *Hilgardia*, 29:81–101.

Stern, V. M., van den Bosch, R., and Leigh, T. F., 1964. Strip cutting alfalfa for Lygus bug control, *Calif. Agric.,* 18:406.

Thrupp, L. A., 1996. *Overview. New Partnerships for Sustainable Agriculture,* World Resource Institute, Washington, D.C.

Unruh, T. R. and Messing, R. H., 1993. Intraspecific biodiversity in Hymenoptera: implications for conservation and biological control, in *Hymenoptera and Biodiversity,* J. LaSalle and I. D. Gauld, Eds., CAB Int., Wallingford, U.K., 27–52.

Unruh, T. R., White, W., Gonzalez, D., Gordh, G., and Luck, R. F., 1983. Heterozygosity and effective size in laboratory populations of *Aphidius ervi* (Hym., Aphidiidae), *Entomophaga*, 28:245–258.

van den Bosch, R. and Messenger, P. S., 1973. *Biological Control,* Educational Publishers, New York.

van den Bosch, R. and Stern, V. M., 1962. The integration of chemical and biological control of arthropod pests, *Annu. Rev. Entomol.,* 7:367–381.

van den Bosch, R. and Telford, A. D., 1964. Environmental modification and biological control, in *Biological Control of Insect Pests and Weeds,* P. DeBach, Ed., Chapman and Hall, London, 459–488.

van den Bosch, R., Lagace, C. F., and Stern, V. M., 1967. The interrelationship of the aphid, *Acyrthosiphon pisum,* and its parasite, *Aphidius smithi,* in a stable environment, *Ecology*, 48:993–1000.

van Emden, H. F., 1962. Observations on the effects of flowers on the activity of parasitic Hymenoptera, *Entomol. Mon. Mag.,* 98:225–236.

van Emden, H. F., 1965. The role of uncultivated land in the biology of crop pests and beneficials insects, *Sci. Hortic.,* 17:121–136.

van Emden, H. F., 1988. The potential for managing indigenous natural enemies of aphids on field crops, *Philos. Trans. R. Soc. London,* 1189:183–201.

van Hook, T., 1994. The conservation challenge in agriculture and the role of entomologists, *Fl. Entomol.,* 77:42–73.

Vereijken, P., 1989. From integrated control to integrated farming, an experimental approach, *Agric. Ecosyst. Environ.,* 26:37–43.

Vinson, S. B., 1977. Behavioural chemicals in the augmentation of natural enemies, in *Biological Control by Augmentation of Natural Enemies,* R. L. Ridgway and S. B. Vinson, Eds., Plenum Press, New York, 237–279.

Vinson, S. B., 1981. Habitat location, in *Semiochemicals: Their Role in Pest Control,* D. A. Nordlund, R. L. Jones, and W. J. Lewis, Eds., John Wiley, New York, 51–77.

Vorley, W. T. and Wratten, S. D., 1987. Migration of parasitoids (Hymenoptera: Braconidae) of cereal aphids (Hemiptera: Aphididae) between grassland, early-sown cereals and late-sown cereals in southern England, *Bull. Entomol. Res.,* 77:555–568.

Waage, J. K., 1991. Biodiversity as a resource for biological control, in *The Biodiversity of Microorganisms and Invertebrates: Its Role in Sustainable Agriculture,* D. I. Hawksworth, Ed., CAB Int., Wallingford, U.K., 149–162.

Way, M. J. and Cammell, M. E., 1981. Effects of weeds and weed control on invertebrate pest ecology, in *Proc. Univ. York and Br. Ecol. Soc. and Feder. Br. Plant Pathol.,* Pitman, London, 443–458.

Weiss, E. and Nentwig, W., 1992. The importance of flowering plants in sown weed strips for beneficial insects in cereal fields, *Mitt. Dtsch. Ges. Allg. Angew. Entomol.,* 8:133–136.

Welling, M. and Kokta, C., 1988. Untersuchungen zur Entomofauna von Feldrainen und Feldrändern in Hindblick auf Nützlingsförderung und Artenschutz, *Mitt. Dtsch. Ges. Allg. Angew. Entomol.,* 6:373–377.

Welling, M., Kokta, C., Molthan, J., Ruppert, V., Bathon, H., Klingauf, F., Langenruch, G. A., and Niemann, P., 1988. Förderung von Nützinsekten durch Wildkräuter im Feld und im Feldrain als vorbeugende Pflanzenschutzmassnahme, R. A. Heft, Ed., Landw. Verlag GmbH, Münstr-Hilltrup, Germany, 56–82.

Wilson, F. and Huffaker, C. B., 1976. The philosophy scope, and importance of biological control, in *Theory and Practice of Biological Control,* C. B. Huffaker and P. S. Messinger, Eds., Academic Press, New York, 3–15.

Wyss, E., 1995. The effects of weed strips on aphids and aphidophagous predators in an apple orchard, *Entomol. Exp. Appl.,* 75:3–49.

Zwölfer, H., Bauer, G., Hensinger, G., and Stechmann, D., 1984. The significance and evaluation of hedgerows for the ecology of animals, *Ber. Akad. Naturschutz Landschaftspflege,* 3(3).

CHAPTER 5

Biodiversity, Ecosystem Function, and Insect Pest Management in Agricultural Systems

Miguel A. Altieri and Clara I. Nicholls

CONTENTS

INTRODUCTION

Today, scientists worldwide are increasingly starting to recognize the role and significance of biodiversity in the functioning of agricultural systems (Swift et al., 1996). Research suggests that, whereas in natural ecosystems the internal regulation of function is substantially a product of plant biodiversity through flows of energy and nutrients and through biological synergisms, this form of control is progressively lost under agricultural intensification and simplification, so that monocultures in order to function must be predominantly subsidized by chemical inputs (Swift and Anderson, 1993). Commercial seedbed preparation and mechanized planting replace natural methods of seed dispersal; chemical pesticides replace natural controls on

1-56670-290-9/99/$0.00+$.50

populations of weeds, insects, and pathogens; and genetic manipulation replaces natural processes of plant evolution and selection. Even decomposition is altered since plant growth is harvested and soil fertility maintained, not through nutrient recycling, but with fertilizers.

One of the most important reasons for maintaining and/or encouraging natural biodiversity is that it performs a variety of ecological services (Altieri, 1991). In natural ecosystems, the vegetative cover of a forest or grassland prevents soil erosion, replenishes groundwater, and controls flooding by enhancing infiltration and reducing water runoff. In agricultural systems, biodiversity performs ecosystem services beyond production of food, fiber, fuel, and income. Examples include recycling of nutrients, control of local microclimate, regulation of local hydrological processes, regulation of the abundance of undesirable organisms, and detoxification of noxious chemicals. These renewal processes and ecosystem services are largely biological; therefore, their persistence depends upon maintenance of biological diversity. When these natural services are lost as a result of biological simplification, the economic and environmental costs can be quite significant. Economically, in agriculture the burdens include the need to supply crops with costly external inputs, since agroecosystems deprived of basic regulating functional components lack the capacity to sponsor their own soil fertility and pest regulation. Often the costs involve a reduction in the quality of the food produced and of rural life in general due to decreased soil, water, and food quality when erosion and pesticide and/or nitrate contamination occurs (Altieri, 1995).

Nowhere are the consequences of biodiversity reduction more evident than in the realm of agricultural pest management. The instability of agroecosystems becomes manifest as the worsening of most insect pest problems is increasingly linked to the expansion of crop monocultures at the expense of the natural vegetation, thereby decreasing local habitat diversity (Altieri and Letourneau, 1982; Flint and Roberts, 1988). Plant communities that are modified to meet the special needs of humans become subject to heavy pest damage, and generally the more intensely such communities are modified, the more abundant and serious the pests. The effects of the reduction of plant diversity on outbreaks of herbivore pests and microbial pathogens are well documented in the agricultural literature (Andow, 1991; Altieri, 1994). Such drastic reductions in plant biodiversity and the resulting epidemic effects can adversely affect ecosystem function with further consequences on agricultural productivity and sustainability (Figure 1).

In modern agroecosystems, the experimental evidence suggests that biodiversity can be used for improved pest management (Altieri and Letourneau, 1984; Andow, 1991). Several studies have shown that it is possible to stabilize the insect communities of agroecosystems by designing and constructing vegetational architectures that support populations of natural enemies or that have direct deterrent effects on pest herbivores (Perrin, 1980; Risch et al., 1983). This chapter analyzes the various options of agroecosystem design which, based on current agroecological theory, should provide for the optimal use and enhancement of functional biodiversity in crop fields.

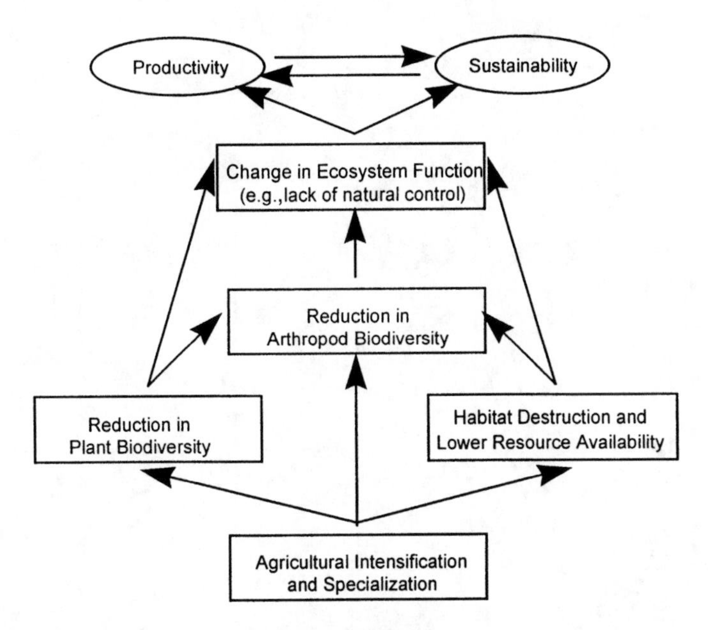

Figure 1 The influence of intensification on biodiversity and function in agricultural ecosystems as it relates to the role of arthropod biodiversity. (Modified from Swift and Anderson, 1993.)

THE NATURE AND FUNCTION OF BIODIVERSITY IN AGROECOSYSTEMS

Biodiversity refers to all species of plants, animals, and microorganisms existing and interacting within an ecosystem. In agroecosystems, pollinators, natural enemies, earthworms, and soil microorganisms are all key biodiversity components that play important ecological roles, thus mediating such processes as genetic introgression, natural control, nutrient cycling, decomposition, etc. (Figure 2). The type and abundance of biodiversity in agriculture will differ across agroecosystems which differ in age, diversity, structure, and management. In fact, there is great variability in basic ecological and agronomic patterns among the various dominant agroecosystems. In general, the degree of biodiversity in agroecosystems depends on four main characteristics of the agroecosystems (Southwood and Way, 1970):

1. The diversity of vegetation within and around the agroecosystem;
2. The permanence of the various crops within the agroecosystem;
3. The intensity of management;
4. The extent of the isolation of the agroecosystem from natural vegetation.

In general, agroecosystems that are more diverse, more permanent, isolated, and managed with low input technology (i.e., agroforestry systems, traditional polycultures) take fuller advantage of work done by ecological processes associated with

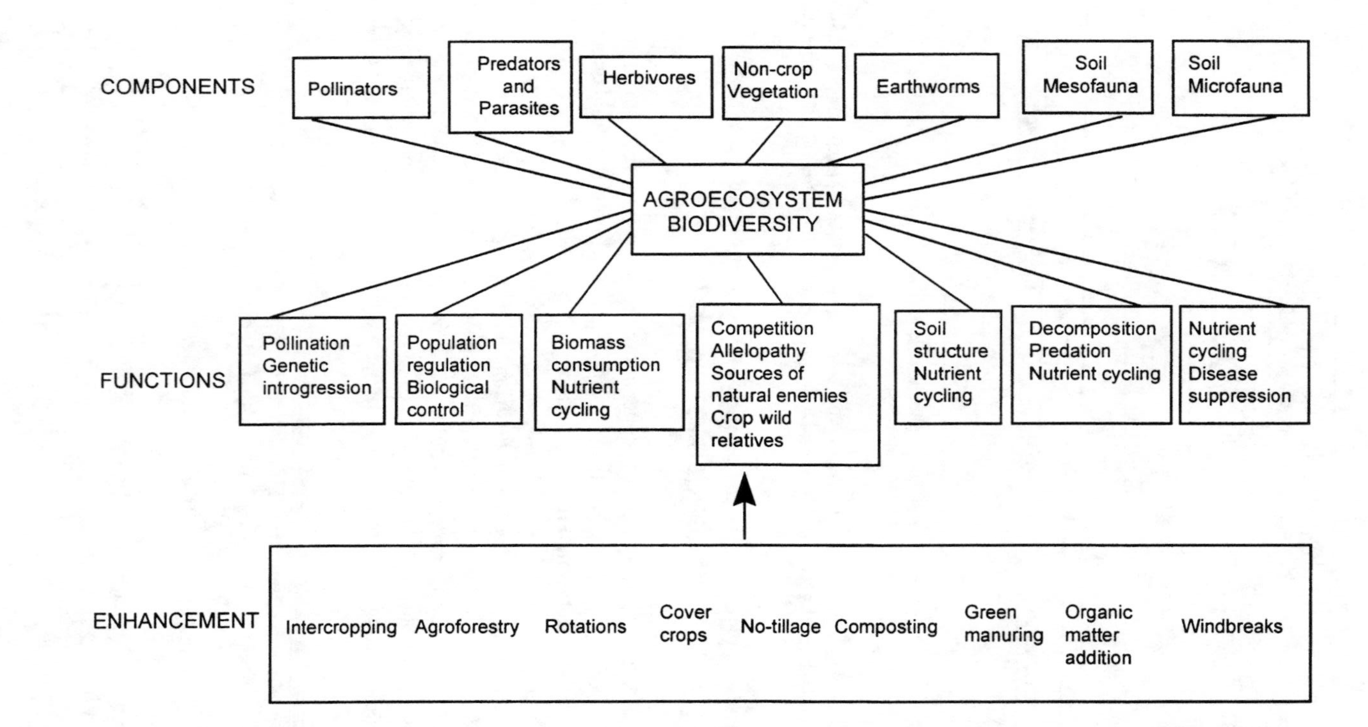

Figure 2 The components, functions, and enhancement strategies of biodiversity in agroecosystems. (From Altieri, M. A., *Biodiversity and Pest Management in Agroecosystems,* Haworth Press, New York, 1994. With permission.)

higher biodiversity than do highly simplified, input-driven, and disturbed systems (i.e., modern row crops and vegetable monocultures and fruit orchards) (Altieri, 1995).

All agroecosystems are dynamic and subject to different levels of management so that the crop arrangements in time and space are continually changing in the face of biological, cultural, socioeconomic, and environmental factors. Such landscape variations determine the degree of spatial and temporal heterogeneity characteristic of agricultural regions, which in turn conditions the type of biodiversity present, in ways that may or may not benefit the pest protection of particular agroecosystems. Thus, one of the main challenges facing agroecologists today is identifying the types of biodiversity assemblages (either at the field or landscape level) that will yield desirable agricultural results (i.e., pest regulation). This challenge can only be met by further analyzing the relationship between vegetation diversification and the population dynamics of herbivore and natural enemy species, in light of the unique environment and entomofauna of each and the diversity and complexity of local agricultural systems.

According to Vandermeer and Perfecto (1995), two distinct components of biodiversity can be recognized in agroecosystems. The first component, planned biodiversity, is the biodiversity associated with the crops and livestock purposely included in the agroecosystem by the farmer, and which will vary depending on management inputs and crop spatial/temporal arrangements. The second component, associated biodiversity, includes all soil flora and fauna, herbivores, carnivores, decomposers, etc. that colonize the agroecosystem from surrounding environments and that will thrive in the agroecosystem depending on its management and structure. The relationship of both biodiversity components is illustrated in Figure 3. Planned biodiversity has a direct function, as illustrated by the bold arrow connecting the planned biodiversity box with the ecosystem function box. Associated biodiversity also has a function, but it is mediated through planned biodiversity. Thus, planned biodiversity also has an indirect function, illustrated by the dotted arrow in the figure, which is realized through its influence on the associated biodiversity. For example, the trees in an agroforestry system create shade, which makes it possible to grow only sun-tolerant crops. So the direct function of this second species (the trees) is to create shade. Yet along with the trees might come small wasps that seek out the nectar in the tree flowers. These wasps may in turn be the natural parasitoids of pests that normally attack the crops. The wasps are part of the associated biodiversity. The trees, then, create shade (direct function) and attract wasps (indirect function) (Vandermeer and Perfecto, 1995).

The key is to identify the type of biodiversity that is desirable to maintain and/or enhance in order to carry out ecological services, and then to determine the best practices that will encourage the desired biodiversity components. As shown in Figure 4, there are many agricultural practices that have the potential to enhance functional biodiversity, and others that negatively affect it. The idea is to apply the best management practices in order to enhance and/or regenerate the kind of biodiversity that can subsidize the sustainability of agroecosystems by providing ecological services such as biological pest control, nutrient cycling, water and soil conservation, etc.

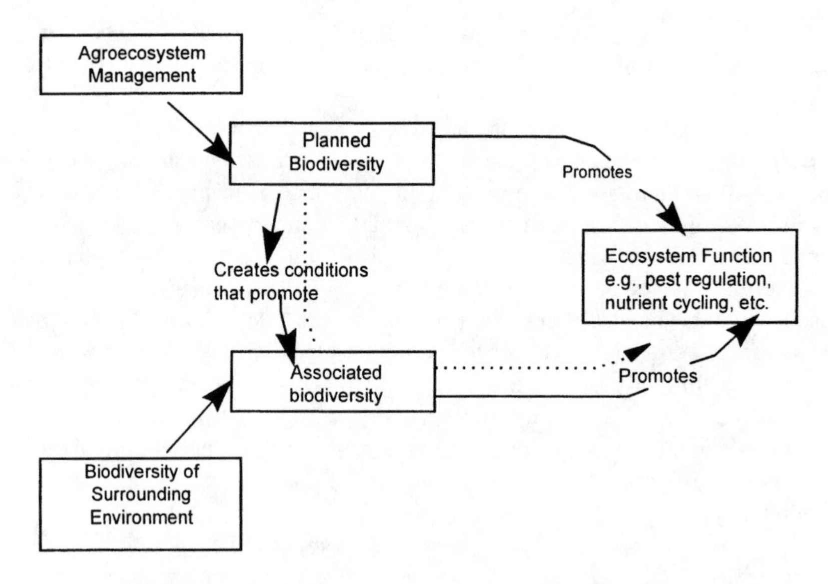

Figure 3 The relationship between planned biodiversity (that which the farmer determines, based on management of the agroecosystems) and associated biodiversity and how the two promote ecosystem function. (Modified from Vandermeer and Perfecto, 1995.)

PATTERNS OF INSECT BIODIVERSITY IN AGROECOSYSTEMS

Arthropod diversity has been correlated with aspects of plant diversity in agroecosystems. A greater variety of plants conforming to a particular crop pattern should lead to a greater variety of herbivorous insect species, and this in turn should determine a greater diversity of predators and parasites (Figure 5). A greater total biodiversity can then play a key role in optimizing agroecosystem processes and function (Altieri, 1984).

Several hypotheses can be offered to support the idea that diversified cropping systems encourage higher arthropod biodiversity (Altieri and Letourneau, 1982):

1. *Heterogeneity hypothesis.* Complex crop habitats support more species than simple crop habitats; architecturally more complex species of plants and heterogeneous plant associations, with greater biomass, food resources, variety and temporal persistence, have more associated species of insects than do architecturally simple crop plants or crop monocultures on an area-for-area basis. Apparently both species diversity and plant structural diversity are important in determining insect species diversity.
2. *Predation hypothesis.* The increased abundance of predators and parasites in rich plant associations (Root, 1973) reduce prey densities, at times to such low levels that competition among herbivores should be reduced. This reduced competition should allow the addition of more prey species, which in turn support new natural enemies.
3. *Productivity hypothesis.* Research has shown that in some situations crop polycultures yield more than monocultures (Francis, 1986; Vandermeer, 1989). This greater

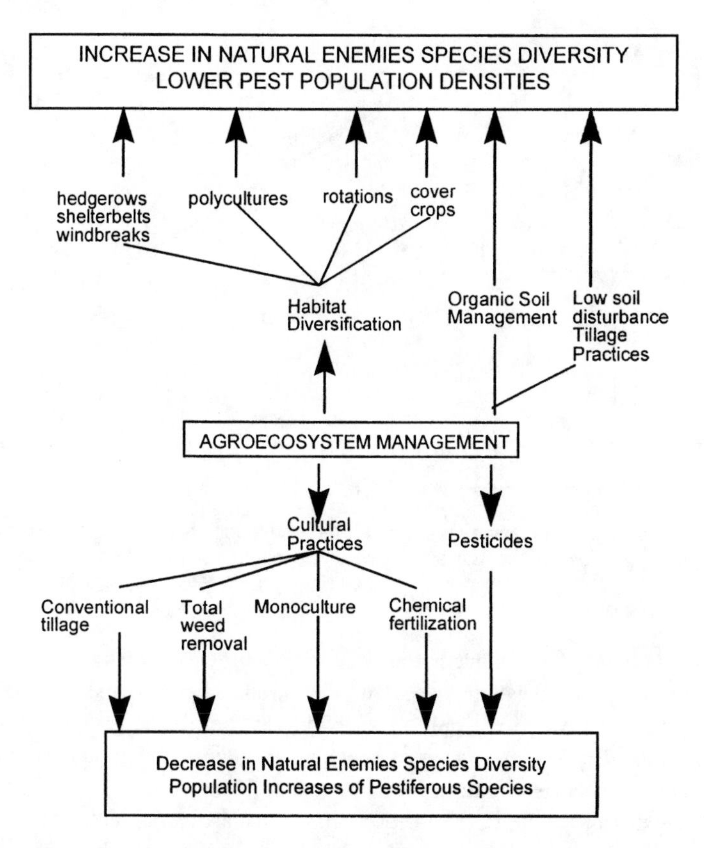

Figure 4 The effects of agroecosystem management and associated cultural practices on the biodiversity of natural enemies and the abundance of insect pests.

productivity can result in greater insect diversity as the number of food resources available for herbivores and natural enemies increases.

4. *Stability and temporal resource-partitioning hypothesis.* This hypothesis assumes that primary production is more stable and predictable in polycultures than in monocultures. This stability of production, coupled with the spatial heterogeneity of complex crop fields, should allow insect species to partition the environment temporally as well as spatially, thereby permitting the coexistence of more insect species.

Further research is needed to clarify whether insect species diversity parallels diversity of vegetation and the productivity of the plant community or simply reflects the spatial heterogeneity arising from the mixing of plants of different structures.

There are several environmental factors that influence the diversity, abundance, and activity of parasitoids and predators in agroecosystems: microclimatic conditions, availability of food (water, hosts, prey, pollen, and nectar), habitat requirements (refuges, nesting and reproduction sites, etc.), intra- and interspecific competition and other organisms (hyperparasites, predators, humans, etc.). The effect of each of these factors will vary according to the spatial and temporary arrangement of crops

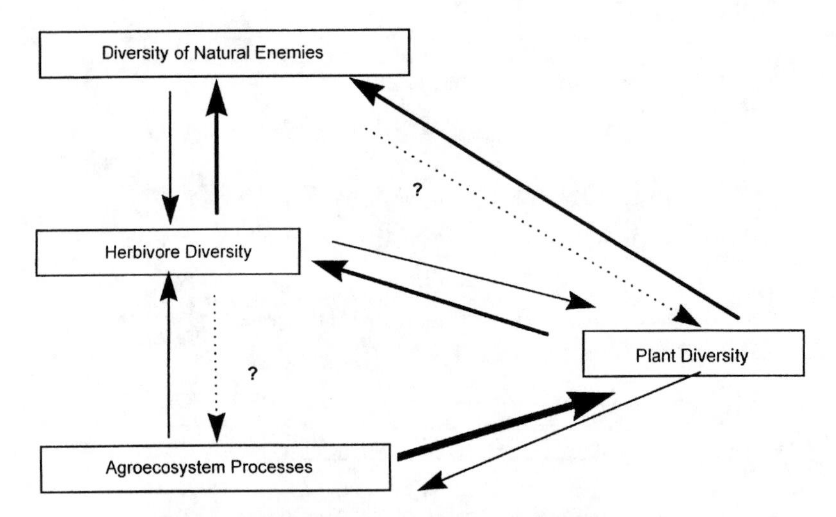

Figure 5 The relationship between plant and arthropod biodiversity and agroecosystem processes. Arrow widths indicate the relative amount of information available on each link; for example, more work has been done on the responses of herbivore populations to plant diversity than on the converse.

and the intensity of crop management, as these features affect the environmental heterogeneity of agroecosystems in several ways (van den Bosch and Telford, 1964).

Although natural enemies seem to vary widely in their response to crop distribution, density, and dispersion, experimental evidence suggests that structural (i.e., crop diversity, input levels, etc.) attributes of agroecosystems influence parasitoid and predator diversity and dynamics. Several of these attributes are related to biodiversity and most are amenable to management (i.e., crop sequences and associations, weed diversity, genetic diversity, etc.). Based on the available information, natural enemy biodiversity can be enhanced and effectiveness improved in the following ways (van den Bosch and Telford, 1964; Rabb et al., 1976; Altieri and Whitcomb, 1979):

- Multiple introductions of parasitoids and predators through augmentative releases for biological control;
- Reducing direct mortality by eliminating pesticide use;
- Provision of supplementary resources other than hosts/prey;
- Increasing adjacent and within-field vegetational diversity;
- Manipulating architectural, genetic, and chemical attributes of host plants;
- Use of semiochemicals (behavioral chemicals such as kairomones) to stimulate host/prey searching behavior and natural enemy retention in the field.

PLANT BIODIVERSITY AND INSECT STABILITY IN AGROECOSYSTEMS

From the early 1970s on, the literature provides hundreds of examples of experiments documenting that diversification of cropping systems often leads to reduced herbivore populations (Andow, 1991; Altieri, 1994). Most experiments that have mixed other plant species with the primary host of a specialized herbivore show that, in comparison with diverse crop communities, simple crop communities have greater population densities of specialist herbivores (Root, 1973; Cromartie, 1981; Risch et al., 1983). In these systems, herbivores exhibit greater colonization rates, greater reproduction, higher tenure time, less disruption of host finding, and lower mortality by natural enemies.

There are various factors in crop mixtures that help constrain pest attack. A host plant may be protected from insect pests by the physical presence of other plants that may provide a camouflage or a physical barrier. Mixtures of cabbage and tomato reduce colonization by the diamondback moth, while mixtures of maize, beans, and squash have the same effect on chrysomelid beetles. The odors of some plants can also disrupt the searching behavior of pests. Grass borders repel leafhoppers from beans, and the chemical stimuli from onions prevent carrot fly from finding carrots (Altieri, 1994).

Alternatively, one crop in the mixture may act as a trap or decoy — the "flypaper effect." Strips of alfalfa interspersed in cotton fields in California attract and trap *Lygus* bugs. There is a loss of alfalfa yield, but this represents less than the cost of alternative control methods for the cotton. Similarly, crucifers interplanted with beans, grass, clover, or spinach are damaged less by cabbage maggot and cabbage aphid. There is less egg laying on the crucifers, and the insect pests are subject to increased predation (Altieri, 1994).

The two hypotheses that have been proposed to explain lower herbivore abundance in polycultures, the resource concentration hypothesis and the enemies hypothesis (Root, 1973), identify key mechanisms of pest regulation in polycultures. They explain why there may be differences in mechanisms between cropping systems, and suggest plant assemblages which enhance regulatory effects and those which do not, and under what management and agroecological circumstances. According to these theories, a reduced insect pest incidence in polycultures may be the result of increased predator and parasitoid abundance and efficiency, decreased colonization and reproduction of pests, chemical repellency, masking and/or feeding inhibition from nonhost plants, prevention of pest movement or immigration, and optimum synchrony between pests and natural enemies (Andow, 1991).

A recently conducted, well-replicated experiment, where species diversity was directly controlled in grassland systems, found that ecosystem productivity was increased and that soil nutrients were utilized more completely when there was a greater diversity of species, leading to lower leaching losses from the ecosystem (Tilman et al., 1996). In agroecosystems, this same pattern applies to insects as herbivore regulation increases with increasing plant species richness. Evidence suggests that as plant diversity increases, pest damage tends to reach acceptable levels, thus resulting in more stable crop yields (Figure 6). Apparently, the more diverse the agroecosystem and the longer this diversity remains undisturbed, the more internal links develop to promote greater insect stability. It is clear, however, that the stability of the insect community depends not only on its trophic diversity but on the actual density-dependence nature of the trophic levels (Southwood and Way,

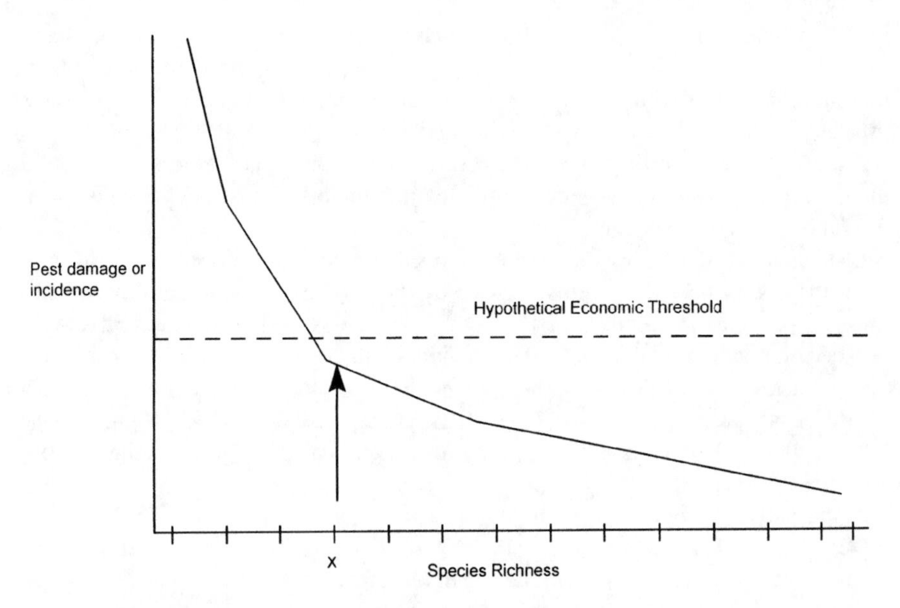

Figure 6 Hypothetical trend of pest regulation or damage reduction as species richness increases in agroecosystems. "X value" represents the level at which a functional assemblage of species with natural control attributes is established.

1970). In other words, stability will depend on the precision of the response of any particular trophic link to an increase in the population at a lower level. Thus, selective diversity, rather than just a random collection of species, is crucial to achieve desired pest regulation (Dempster and Coaker, 1974).

From a practical standpoint, it is easier to design insect manipulation strategies in polycultures using the elements of the natural enemies hypothesis than those of the resource concentration hypothesis. This is mainly because we cannot yet identify the ecological situations or life history traits that make some pests sensitive (i.e., their movement is affected by crop patterning) and others insensitive to cropping patterns (Kareiva, 1986). Crop monocultures are difficult environments in which to induce the efficient operation of beneficial insects because these systems lack adequate resources for the effective performance of natural enemies, and because of the disturbing cultural practices often utilized in such systems. Polycultures already contain specific resources provided by plant diversity and are usually not disturbed with pesticides (especially when managed by resource-poor farmers who cannot afford high-input technology). They are also more amenable to manipulation. In polycultures, the choice of a tall or short, early or late maturing, flowering or nonflowering, legume or nonlegume companion crop can magnify or decrease the effects of particular mixtures on specific pests (Vandermeer, 1989). Thus, by replacing or adding the correct diversity to existing systems, it may be possible to exert changes in habitat diversity that enhance natural enemy abundance and effectiveness.

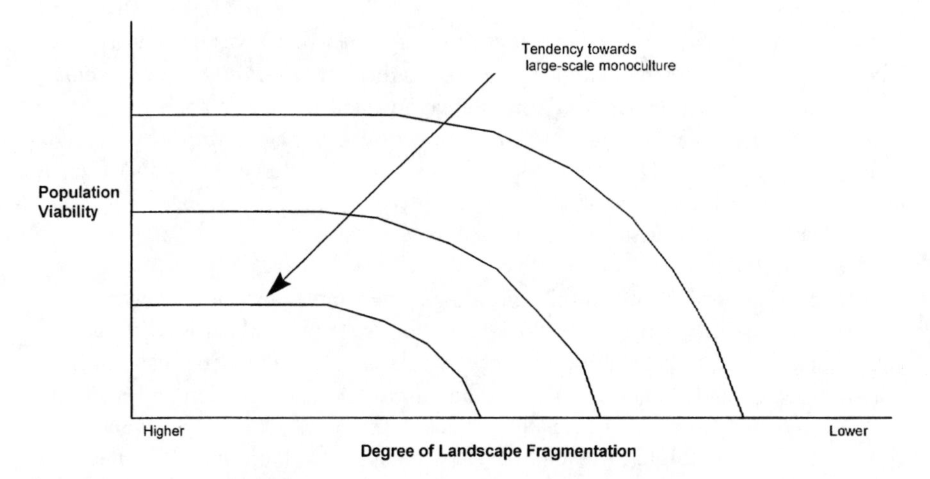

Figure 7 The effects of landscape fragmentation on the expected viability of natural enemy populations in agroecosystems of varying scales and levels of artificialization.

PATTERNS OF LANDSCAPE STRUCTURE AND INSECT BIODIVERSITY

An unfortunate trend accompanying the expansion of agricultural monocultures is that it occurs at the expense of surrounding natural vegetation which serves to add biodiversity to the landscape. One consequence of this trend is that the total amount of habitat available for beneficial arthropods and natural enemies of pests is decreasing at dramatic rates. The hypothetical impact of habitat fragmentation on the survival of natural enemies in agroecosystems is depicted in Figure 7. The implication of this habitat loss for the biological control of pests may be serious in light of scientific data demonstrating increased abundance of insect pests in homogeneous agricultural landscapes (Altieri and Letourneau, 1982). Emerging data demonstrate that there is enhancement of natural enemies and more effective biological control where wild vegetation remains at field edges and in association with crops (Altieri, 1994). These habitats may be more important as overwintering sites for predators, or they may provide increased resources, such as pollen and nectar for parasitoids and predators from flowering plants (Landis, 1994).

Hedgerows and other landscape features have received significant attention in Europe regarding their effects on arthropod distribution and abundance in adjacent crop fields (Fry, 1995). There is wide acceptance of the importance of wild vegetation field margins as reservoirs of natural enemies of crop pests (van Emden, 1965). Many studies have documented the movement of beneficial arthropods from margins into crops, and higher biological control is usually observed in crop rows close to wild vegetation edges than in rows in the center of the fields (Altieri, 1994).

In many cases, weeds and other natural vegetation around crop fields harbor alternative hosts/prey for natural enemies, thus providing seasonal resources to bridge gaps in the life cycles of entomophagous insects and crop pests (Altieri and Whitcomb, 1979). A classic case is that of the egg parasitoid wasp, *Anagrus epos,* whose effectiveness in regulating the grape leafhopper, *Erythroneura elegantula,* was increased greatly in vineyards near areas invaded by wild blackberry (*Rubus* sp.). This plant supports an alternative leafhopper (*Dikrella cruentata*) which breeds in its leaves in winter (Doutt and Nakata, 1973). Recent studies show that French prune orchards adjacent to vineyards provide overwintering refuges for *Anagrus* and early benefits of parasitism are promoted in vineyards with prune trees planted upwind from the vineyard.

Research in northern California showed a considerable amount of movement of entomophagous insects from woodlands into adjacent apple orchards, with organically managed orchards exhibiting a higher rate of natural enemy colonization from bordering woodlands than orchards sprayed with insecticides (Altieri and Schmidt, 1986). Several predators and parasites collected in the woodland edges were intercepted at the orchard interfaces and later collected within the orchards, suggesting that the development of apple orchard beneficial arthropod communities are influenced by the type of surrounding natural vegetation.

Recent developments in temperate agriculture to encourage predators while reducing pesticide applications have included use of beetle banks, flowering strips, and conservation headlands. In Britain, several hundred potentially beneficial species of predators and parasites may live in or by cereal crops. Most of these are killed when the crops are sprayed to control pests. But if the field habitat is manipulated to increase plant diversity, then the need for spraying pesticides can be greatly reduced. When grass strips are constructed across large fields, then predatory beetles proliferate and can get to the field centers, the regions where aphid populations are greatest (Wratten, 1988). The cost of establishing a 400-m bank in a 20-ha field is about $200, including cultivation, grass seed, and loss of crop. One aphid spray costs $750 across the same field, plus the cost of yield reduction due to aphid infestation.

Despite the above findings, no major efforts are under way in the world to diversify agroecosystems at the landscape level with natural edges or windbreaks composed of flowering species that act as insectary plants. Experiments of this sort would fill an information gap on how changes in the physical and biodiversity layout of agroecosystems would affect the distribution and abundance of the whole complex community of pests and beneficial insects.

Determining the dispersal of insects in response to landscape vegetational diversity and whether or not natural plant strips surrounding crop fields serve as a movement corridor for beneficial arthropods in monocultures will have major implications for planning IPM at the landscape level. It is expected that corridors can serve as a conduit for the dispersion of predators and parasites within agroecosystems. Given the high edge-to-area ratio in the corridors, this feature is expected to have a high degree of interaction with adjacent crops, thus providing protection against insect pests within the area of influence of the corridor by allowing distribution of natural enemies within a certain range of the field. By documenting the effects of the corridor on arthropod distribution and abundance, then it may be

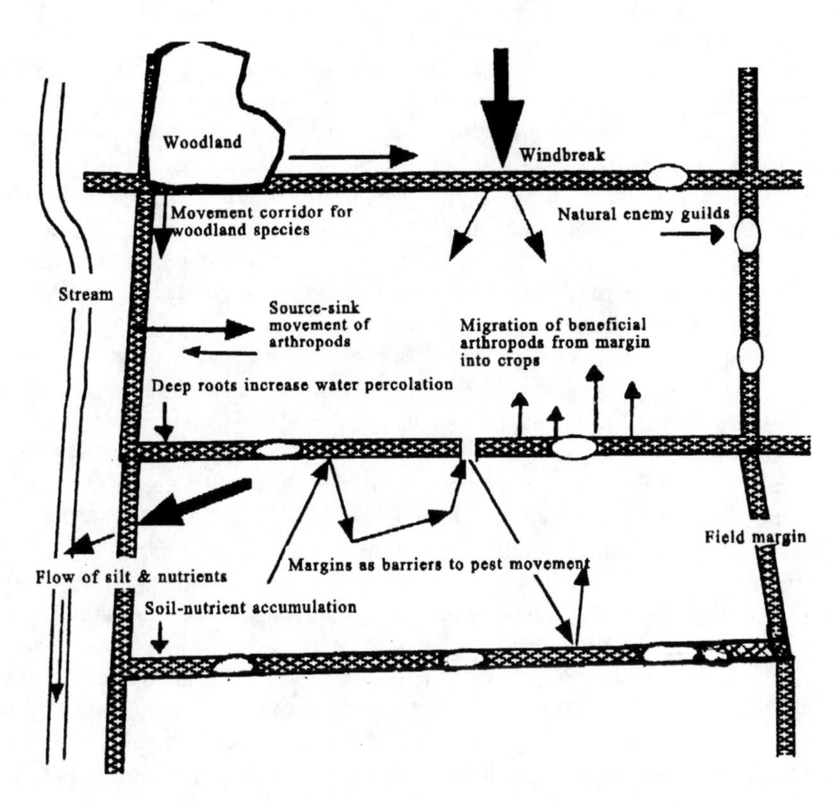

Figure 8 Effects of diversification schemes (field margines, windbreaks, etc.) at the landscape level on agroecosystem functions, with special emphasis on the dynamics of arthropods. (Modified after Fay, 1995.)

possible to determine the length, width, distance, and frequency of corridors needed to maintain a level of functional biodiversity which will provide the necessary crop protection without the need of pesticides. A system of corridors can also have positive effects on the overall system by interrupting inoculum dispersion, serving as barriers to insect pest movement, producing biomass for soil fertility, decreasing outflow of silt and nutrients, and by modifying microclimate through interception of air currents (Figure 8). The most important aspect is that corridor manipulation can be a crucial first step in reintroducing biodiversity into large-scale monocultures, thus facilitating the biological restructuring of agroecosystems for the conversion to agroecological management.

CONCLUSION

Diversified cropping systems, such as those based on intercropping and agroforestry or cover cropping of orchards, have been the target of much research recently. This interest is largely based on the new emerging evidence that these systems are more sustainable and more resource-conserving (Vandermeer, 1995). Much of these attributes are connected to the higher levels of functional biodiversity associated with complex farming systems. In fact, an increasing amount of data reported in the literature documents the effects that plant diversity have on the regulation of insect herbivore populations by favoring the abundance and efficacy of associated natural enemies (Altieri, 1994). Several hypotheses are emerging postulating the mechanisms explaining the relationships between plant species number and the stabilization of agroecosystem processes including the buffering of populations (Tilman et al., 1996). One aspect that is clear is that species composition is more important than species numbers per se. The challenge is to identify the correct assemblages of species that will provide through their biological synergisms key ecological services such as nutrient cycling, biological pest control, and water and soil conservation.

The exploitation of these synergisms in real situations involves agroecosystem design and management and requires an understanding of the numerous relationships among plants, herbivores, and natural enemies. Clearly, the emphasis of this approach is to help to restore natural control mechanisms through the addition of selective biodiversity within and outside the crop field, through a whole array of possible crop arrangements in time and space.

Data and practical experience indicate that it is possible to stabilize the insect communities of agroecosystems by designing and constructing vegetational architectures which support populations of natural enemies or that have direct deterrent effects on pest herbivores (Altieri, 1991). What is difficult is that each agricultural situation must be assessed separately, since herbivore–enemy interactions will vary significantly depending on insect species, location and size of the field, plant composition, the surrounding vegetation, and cultural management. One can only hope to elucidate the ecological principles governing arthropod dynamics in complex systems, but the biodiversity designs necessary to achieve herbivore regulation will depend on the agroecological conditions and socioeconomic restrictions of each area.

REFERENCES

Altieri, M. A., 1984. Patterns of insect diversity in monocultures and polycultures of brussel sprouts, *Prot. Ecol.*, 6:227–232.

Altieri, M. A., 1991. How best can we use biodiversity in agroecosystems, *Outlook Agric.*, 20:15–23.

Altieri, M. A., 1994. *Biodiversity and Pest Management in Agroecosystems,* Haworth Press, New York.

Altieri, M. A., 1995. *Agroecology: The Science of Sustainable Agriculture,* Westview Press, Boulder, CO.

Altieri, M. A. and Letourneau, D. L, 1982. Vegetation management and biological control in agroecosystems, *Crop Prot.,* 1:405–430.

Altieri, M. A. and Letourneau, D. L., 1984. Vegetation diversity and insect pest outbreaks, *CRC Crit. Rev. Plant Sci.,* 2:131–169.

Altieri, M. A. and Schmidt, L. L., 1986. The dynamics of colonizing arthropod communities at the interface of abandoned organic and commercial apple orchards and adjacent woodland habitats, *Agric. Ecosyst. Environ.,* 16:29–43.

Altieri, M. A. and Whitcomb, W. H., 1979. The potential use of weeds in the manipulation of beneficial insects, *HortScience*, 14:12–18.

Andow, D. A., 1991. Vegetational diversity and arthropod population response, *Annu. Rev. Entomol.*, 36:561–586.

Cromartie, W. J., 1981. The environmental control of insects using crop diversity, in *CRC Handbook of Pest Management,* D. Pimentel, Ed., CRC Press, Boca Raton, 223–251.

Dempster, J. P. and Coaker, T. H., 1974. Diversification of crop ecosystems as a means of controlling pests, in *Biology in Pest and Disease Control,* D. P. Jones and M. E. Solomon, Eds., John Wiley & Sons, New York, 106–114.

Doutt, R. L. and Nakata, J., 1973. The *Rubus* leafhopper and its egg parasitoid: an endemic biotic system useful in grape pest management, *Environ. Entomol.,* 2:381–386.

Flint, M. L. and Roberts, P. A., 1988. Using crop diversity to manage pest problems: some California examples, *Am. J. Alternative Agric.,* 3:164–167.

Francis, C. A., 1986. *Multiple Cropping Systems,* MacMillan, New York.

Fry, G., 1995. Landscape ecology of insect movement in arable ecosystems, in *Ecology and Integrated Farming Systems,* D. M. Glen, Ed., John Wiley & Sons, Bristol, U.K., 236–242.

Kareiva, P., 1986. Trivial movement and foraging by crop colonizers, in *Ecological Theory and Integrated Pest Management Practice,* M. Kogan, Ed., John Wiley & Sons, New York, 59–82.

Landis, D. A., 1994. Arthropod sampling in agricultural landscapes: ecological considerations, in *Insect Parasitoids: Handbook of Sampling Methods for Arthropods in Agriculture,* L. P. Pedigo and G. D. Buntin, Eds., Academic Press, London.

Perrin, R. M., 1977. Pest management in multiple cropping systems, *Agroecosystems,* 3:93–118.

Perrin, R. M. 1980. The role of environmental diversity in crop protection. *Prot. Eco.,* 2:77–114.

Rabb, R. L., Stinner, R. E., and van den Bosch, R., 1976. Conservation and augmentation of natural enemies, in *Theory and Practice of Biological Control,* C. B. Huffaker and P. S. Messenger, Eds., Academic Press, New York, 233–253.

Risch, S. J., Andow, D., and Altieri, M. A., 1983. Agroecosystem diversity and pest control: data, tentative conclusions and new research directions, *Environ. Entomol.,* 12:625–629.

Root, R. B., 1973. Organization of a plant-arthropod association in simple and diverse habitats: the fauna of collards (*Brassicae oleraceae*), *Ecol. Monogr.,* 43:95–124.

Southwood, T. R. E. and Way, M. J., 1970. Ecological background to pest management, in *Concepts of Pest Management,* R. L. Rabb and F. E. Guthrie, Eds., North Carolina State University, Raleigh, 6–29.

Swift, M. J. and Anderson, J. M., 1993. Biodiversity and ecosystem function in agroecosystems, in *Biodiversity and Ecosystems Function,* E. Schultz and H. A. Mooney, Eds., Springer-Verlag, New York, 57–83.

Swift, M. S., Vandermeer, J., Ramakrishnan, P. S., Anderson, J. M., Ong, C. K., and Hawkins, B. A., 1996. Biodiversity and agroecosystem function, in *Functional Roles of Biodiversity: A Global Perspective,* H. A. Mooney, Ed., John Wiley & Sons, New York, 261–298.

Tilman, D., Wedin, D., and Knops, J., 1996. Productivity and sustainability influenced by biodiversity in grassland ecosystems, *Nature*, 379:718–720.

van den Bosch, R. and Telford, A. D., 1964. Environmental modification and biological control, in *Biological Control of Insect Pests and Weeds,* P. DeBach, Ed., Chapman and Hall, London, 459–488.

Vandermeer, J., 1989. *The Ecology of Intercropping,* Cambridge University Press, Cambridge, U.K.

Vandermeer, J., 1995. The ecological basis of alternative agriculture, *Annu. Rev. Ecol. Syst.,* 26:201–224.

Vandermeer, J. and Perfecto, I., 1995. *Breakfast of Biodiversity: The Truth about Rainforest Destruction,* Food First Books, Oakland.

van Emden, H. F., 1965. The role of uncultivated land in the biology of crop pests and beneficial insects, *Sci. Hortic.,* 17:121–126.

van Emden, H. F., 1990. Plant diversity and natural enemy efficiency in agroecosystems, in *Critical Issues in Biological Control,* M. MacKauer, L. Ehler, and J. Roland, Eds., Intercept, Andover, 63–80.

Wratten, S. C., 1988. The role of field margins as reservoirs of beneficial insects, in *Environmental Management in Agriculture: The European Experiences,* J. R. Park, Ed., Belhaven Press, New York, 144–150.

CHAPTER 6

Livestock and Biodiversity

Harvey W. Blackburn and Cornelis de Haan

CONTENTS

INTRODUCTION

The manner in which human populations utilize livestock influences biodiversity. If livestock are concentrated too much, the competition between wildlife and livestock increases or livestock alter the physical environment making it unusable by some animal or plant species. But evidence is growing that when livestock are used in balance with environmental resources they can actually enhance habitat for wildlife. Much of the driving force determining how people use livestock, and therefore the impact of livestock on the environment and biodiversity, stems from issues of

1-56670-290-9/99/$0.00+$.50

Table 1 Regional Growth Estimates (%) for Demand of Meat and Cereals from 1990 to 2020

Region	Meat	Cereal
World	60–93	49–65
Developed	17–18	19–33
Developing	123–206	68–91
Sub-Saharan	141–194	136–161
Latin America	76–105	53–77
West Asia and North Africa	104–157	74–100
Rest of Asia	148–255	64–85

human population growth and economic development. There are options that can help mitigate the negative impacts of livestock and biodiversity, and those shall be explored in this chapter. For discussion purposes in this chapter we consider biodiversity to mean not the total number of species present in a specific ecosystem, but rather the presence of critical types of species which permit ecosystems to appropriately function. Given this definition, our contention is that if markets and policies are appropriate, then livestock can help preserve biodiversity.

Globally the demand for livestock products is increasing and will continue to grow (IFPRI, 1995). IFPRI (1995) data in Table 1 present regional growth rates in the demand for meat compared with cereals. This growth in consumption of livestock products is being fueled by economic and population growth throughout the developing world. As livestock numbers grow, there are direct implications for the environment and biodiversity as a subset of any specific environment. There has been concern that livestock have had a detrimental impact on the environment. However, we shall see that this image is often incorrect, as much is dependent upon the human population pressure and how those pressures display themselves. In other words, as human population pressures increase, people can use livestock in a manner which is detrimental to biodiversity.

LIVESTOCK PRODUCTION SYSTEMS

There are three principal types of livestock production systems that interact with biodiversity: grazing systems, mixed farming systems, and industrial systems. All three are found globally. Because these systems are so diverse in structure and environment, it is difficult to make generalized statements about their impact on biodiversity.

Grazing Systems

Grazing systems are defined as animal agricultural systems which are exclusively livestock and have little, if any, crop production grown in conjunction with the grazing of livestock. In these systems, livestock obtain most of the feed from native vegetation. These systems are the most variable and diverse of livestock production systems because of their dependence upon natural vegetation which is controlled by weather changes. Grazing systems tend to be closed systems, where animal waste products (manure) are used by the system.

In arid rangelands, livestock mobility is key to successful maintenance of vibrant livestock and wildlife populations. These regions have tended to be the most controversial areas of livestock use. However, a series of studies does show that the extent of environmental degradation has been exaggerated. Three different works demonstrate this point. R. Mearns (unpublished data) concluded that abiotic factors such as rainfall, rather than livestock density, determine long-term primary plant production and vegetation cover. Tucker et al. (1991) demonstrated the resilience of arid lands using satellite imagery. Their work showed movement of the southern belt of the Sahara depends on rainfall and that the southern boundary of the Sahara is moving north. This movement is occurring after the long droughts which occurred in the 1970s and is contrary to how we normally view the resource base in this region. de Haan et al. (1997) demonstrated that, in the Sahel, livestock productivity in terms of meat production per hectare and per head has increased over the past 30 years. This type of long-term trend would be difficult to obtain if pastoralists constantly degrade the environment they utilize.

Semiarid and subhumid rangelands are more static systems than arid zone rangelands. Due to the higher rainfall levels, these areas contain larger amounts of plant biomass and have considerable biodiversity. These areas also receive heavier livestock grazing and allow opportunities for mixed farming systems to expand. As a result of growing human population pressures, these lands and the biodiversity they contain may be of most concern. For example, data from Mali showed that land degradation in the 600 to 800 mm rainfall area was significantly greater than in the 350 to 450 mm rainfall area. However, such a trend is by no means a global phenomenon. As Milchunas and Lauenroth (1993) demonstrate, in the semiarid and subhumid regions, moderate grazing had no impact on biomass production, species composition, and root development.

Livestock grazing in tropical rain forests (the humid zone) has more than anything else typified the negative effects of livestock development on the environment. Table 2 provides estimates of the main causes of deforestation (Bruenig, 1991). In the humid region, data on biodiversity losses in the rain forest are dramatic. Since 1950, the world has lost about 200 million ha of tropical forest, with the resulting loss of unique plant and animal species. Forested areas of Central America have declined from 29 to 19 million ha since 1950, although after 1990 the rate of deforestation in Central America decreases. In the 1980s rain forest in Central America disappeared at the rate of 30,000 ha/year; this had declined to 320,000 ha

Table 2 Estimated Causes of Deforestation (percent of total deforestation)

Region	Crops	Livestock	Forest Exploitation
South America	25	44	10
Asia	50–60	Negligible	20
Africa	70	Negligible	20

over the period 1990 through 1994. In South America, the deforestation rate in the 1980s was about 750,000 ha/year. It is not known whether or not this rate has declined over the last years. In Brazil, about 70% of the deforested areas is converted into ranch land.

Much of the deforested areas in Latin America went into ranching, after initially being cropped. In Central America, pasture areas have increased from 3.5 million to 9.5 million ha and cattle populations more than doubled. For example, in Central America, livestock have increased from 4.2 million head in 1950 to 9.6 million in 1992 (Kaimonitz, 1995). In Asia and sub-Saharan Africa, the decline in the forest area is mainly the result of crop expansion.

In the temperate zones, extensive grazing livestock are usually produced with some harvested forage during the wintertime. Livestock produced in extensive grazing systems interact most closely with biodiversity in the form of wildlife. In both Australia and the U.S. these systems were heavily used in the late 19th century, and as a result degradation of the resource base occurred. Heavy use during the last century led to legislation controlling the use of the grazing resource and consequently has promoted rangeland health and improved range condition. Key issues driving how people utilize temperate ranges include fuel prices and privatization (Central Asia), heavy levels of fertilization (Europe), and concern over riparian areas (western U.S.).

Mixed Farming Systems

In mixed farming systems, crops and livestock are produced on the same resource base. Globally this system produces the largest share of meat (54%) and milk (90%). Throughout the developing world, mixed farming is the main agricultural production system for smallholders. Developmentally, mixed farming systems provide farmers with an opportunity to reduce financial risk and smooth out production cycles. Farmers are able to take the highs and lows out of their production cycles because they have the capacity to provide livestock with higher-quality forages during winter or dry months of the year. In return, the sale of livestock products helps finance inputs for the farming enterprise; in addition, livestock provide traction for soil preparation. The mixed farming system is also a partially closed system where the manure can be utilized on the farm to build soil fertility while the milk and meat produced in the system flow out to urban markets. In many respects mixed farming systems have the capacity to promote healthy ecosystems and provide for economic development of the farmer, but, due to human population pressure, poverty, and poor infrastructure, these systems can negatively impact biodiversity and the environment at large.

Mixed farming systems tend to be transitional as livestock production shifts from an extensive grazing system to an intensively managed industrial system. McIntire et al. (1992) have documented the role population pressure has in integrating crop and animal agriculture and in promoting the integration of crop–livestock systems. They also discuss how further increases in population pressure drive farmers to become more specialized, therefore, causing the decomposition of the mixed farming system into more-intensive crop or livestock enterprises. Disaggregating livestock and crop agriculture may result in lower levels of biodiversity, for by keeping the two activities biodiversity can be promoted and may prevent the agricultural system from becoming too brittle (Holling, 1995).

Globally mixed farm types and the way livestock are used are extremely diverse. In Southeast Asia, for example, livestock and crop production is very intensive. Cattle are used for draft purposes and in turn consume high proportions of crop aftermath. In contrast, many mixed farming systems in temperate Organization for Economic Cooperation and Development (OECD) countries had and have the potential for being balanced systems. These types of farms have the capacity to produce various crops (e.g., maize) in a rotation with alfalfa, which in turn provides a forage resource for ruminant livestock and helps to replenish soil nutrients extracted in the production of cereal grains. Mixed farming and biodiversity interact at several crucial levels. First is the interaction with wildlife which can be positive or negative. Second, by replenishing soil fertility through manure application, the mixed farm can help provide a viable environment for soil microflora and microfauna. In developed countries there has been a tendency for farmers to focus on monoculture crop production. From a plant and animal perspective this makes these systems more brittle and exposed to major stresses. By maintaining livestock in these farming systems there is an opportunity to keep these systems more robust and encourage more fully the presence of various plants and animals.

Mixed farming systems can be broadly classified into those found in developed and developing countries. Developing country mixed farming systems contain several environmental issues which impact biodiversity. Soil erosion impacts both the people and plant/animal biodiversity in the various mixed systems. Pimental et al. (1995) estimated erosion rates of 30 to 40 ton/ha/year in some Asian, African, and South American systems. Bojos and Casells (1995) determined that in Ethiopia soil loses were 5 ton/ha/year in grazing lands used by crop–livestock farmers while erosion rates on crop lands were 42 ton/ha/year. Here livestock may be critical in maintaining soil fertility and soil organic matter levels. In Southeastern Asia adding pig and ruminant manure together may contribute up to 35% of the soil organic matter requirements, therefore providing an important source of organic matter. This is a crucial contribution because it is the only avenue available for farmers to improve soil organic matter (de Haan et al., 1997).

In developed countries, soil erosion and soil fertility are issues that impact biodiversity. In temperate zones soil losses of up to 15 ton/ha/year have been reported by Pimental et al. (1995). Soil fertility is impacted more by overfertilization than a lack of soil nutrients. Once soils become saturated with excess levels of nitrogen or phosphorus, these nutrients leach into above- and belowground water systems. Driving much of this overfertilization is the ease of importing feed and inorganic fertilizer. By importing these products into the mixed farming system, there becomes less of a need to balance animal feed and cropping activities through rotation and fallow systems.

Industrial System

The industrial system can be the most capital intensive of the livestock production systems. In general, it is a large concentration of livestock (particularly poultry and swine), but can also include small-scale periurban production in developing countries. Industrial systems do not produce their own feed; rather it is imported into the system from other locations within a country or, in some extreme cases, it is imported from other regions of the world. Industrial systems can impact biodiversity locally through the wastes they generate or off-site, where the feed is grown for use in the industrial system.

The industrial system has a threefold effect on biodiversity through:

- Waste production and its effects on terrestrial and aquatic ecosystems. These effects are often geographically confined to areas of high livestock densities. Eutrophication and destruction of habitats is a common phenomenon in parts of northeastern Europe and the U.S. as well as in the densely populated areas of the developing world, in particular Asia and, to a lesser extent, Latin America. Ammonia emission leads to acidification of the environment and negatively affects ecosystem functioning and biodiversity.
- Demand for concentrate feed and resulting changes in land use and cropping intensity. The production of feed grains, in particular, adds additional stress on biodiversity through habitat loss and damages in ecosystem functioning.
- The requirement for extremely uniform animals of similar genetic composition contributes to within-breed erosion of domestic animal diversity.

But the industrial system has many advantages. First, the rapid development of industrial pig and poultry systems helps reduce total feed requirements of the total livestock sector to meet a given demand. Therefore, it may help to alleviate pressures leading to deforestation and degradation of rangelands, such as is happening in parts of Latin America and Asia, thus saving land and preserving biodiversity. Second, the feed-saving technologies developed for this system do not have scale effects and can be successfully transferred to mixed farming systems. The same holds true for animal waste prevention and treatment technologies that have been developed following regulations applied mainly to the industrial system. Therefore, the demand-driven industrial system generates a series of innovations that have spillover effects on the sector as a whole.

LIVESTOCK SYSTEM INTERACTION WITH BIODIVERSITY

The Plant Community

Native plant communities naturally go through a series of successional changes, from low to high to low plant diversities while in the process of obtaining a state of climax (Clements, 1905). Grazing by livestock overlies this natural process. That is, grazing intensity interacts with and can modify the rate at which plant communities move toward climax. In addition, there is some evidence that indicates that for grazing to effect plant communities significantly requires a combination of grazing intensity and rainfall or fire events (Milchunas et al., 1988; Westoby et al., 1989).

An important concept in determining the status of plant community health is that of thresholds (Westoby et al., 1989). The threshold concept proposes that plant communities under grazing pressure do not deteriorate in a linear fashion. Rather, there is a series of levels which a plant community moves to when confronted with a series of pressures. Thresholds separate these levels. Within a level, plant communities can fluctuate in terms of biomass production and species composition, and recovery within a level is more easily accomplished (Archer et al., 1988). If grazing pressure is relaxed prior to a critical level or threshold, plant community recovery becomes less problematic.

Depending upon how livestock graze in a specific environment, biodiversity can increase or decrease. Either heavy or light grazing can lead to a reduction in biodiversity. Moderate grazing tends to promote patchiness of vegetation (CAST, 1996). Increased patchiness allows for diverse plant species to compete in a given environment. Therefore, by moderately grazing native rangelands, plant communities can be manipulated to maintain a desired level of plant diversity.

The semiarid and subhumid areas are some of the world's most important repositories of plant and animal biodiversity. For example, in Africa, Le Houerou (1991) estimates that rangelands contain about 3500 plant species, having a significant role in ruminant nutrition.

For the subhumid savannas, weed invasion is a major problem threatening biodiversity, and the role of livestock is only secondary. For example, the grass *Imperata cylindrica* in the Philippines and Indonesia now has infested more than 5 million ha. Invasion with broad-leafed plants and shrubs is more common in the savannas of Africa and the Americas.

There are a large number of cases that show that in well-balanced grazing systems, especially those using multispecies, plant biodiversity increases. An extensive review of grazing and production data of 236 sites worldwide, including many sites in the semiarid zone, showed no difference in biomass production, species composition, and root development in response to long-term grazing in the field (Milchunas and Lauenroth, 1993).

Wildlife Interactions

There are a variety of ways in which livestock can interact with wildlife communities. These include (Burkholder, 1952; Odum 1971; Mosley, 1994):

1. Neutralism, where neither species affects the other;
2. Direct interference or resource use competition, where both species inhibit each other;
3. Amensalism, where one species is inhibited and the other not affected;
4. Predation, where one species inhibits another by direct attack;
5. Commensalism, where one species is benefited by the presence of another but there is no impact on the second species;
6. Protocooperation, in which the interaction between species is favorable to both species but the association is not obligatory; and

7. Mutualism, the interaction is favorable to both species and the association is obligatory.

The extent to which the interaction between livestock and wildlife is neutral, negative, or positive is dependent upon how domestic livestock are managed in specific situations. Severson and Urness (1994) identify four ways livestock can be used to modify species that can, in turn, develop habitats that are favored by certain wildlife species. Such a modification is achieved by altering the composition of vegetation, increasing the productivity of selected species, increasing nutritive quality of forage, and increasing diversity of habitats by altering plant structure.

Riparian health is an important issue driving the monitoring and use of public grazing lands. However, it is often overlooked that any species of wildlife or livestock can overgraze these critical areas. A key example of such a situation exists today in Yellowstone National Park (YNP), the crown jewel of the U.S. national park system. It has recently been demonstrated that elk are severely overgrazing riparian areas in YNP. In a study comparing riparian areas in YNP and on the summer range of the U.S. Sheep Experiment Station (approximately 30 miles from YNP), it was shown that grazing of sheep had a more beneficial impact on riparian health, as measured by willow populations, a key indicator species (Figure 1). Furthermore, as a result of healthier willow communities on the Sheep Station, beaver populations are also in better condition. This work demonstrates that any grazing animal can cause environmental instability and/or degradation and that by using an appropriate livestock species environmental health can be maintained or increased (Kay and Walker, 1997).

Another key aspect which determines the type of wildlife–livestock interaction is diet preferences. Different types of wildlife and livestock prefer different plant types. For example, cattle select more grass in their diet than sheep which choose a combination of grass, forbs, and browse. The same type of diet selection patterns are evident in wildlife. Murray and Illius (1996) cite examples in the Serengeti of how small-bodied species, such as the Thompson's gazelle, are more selective

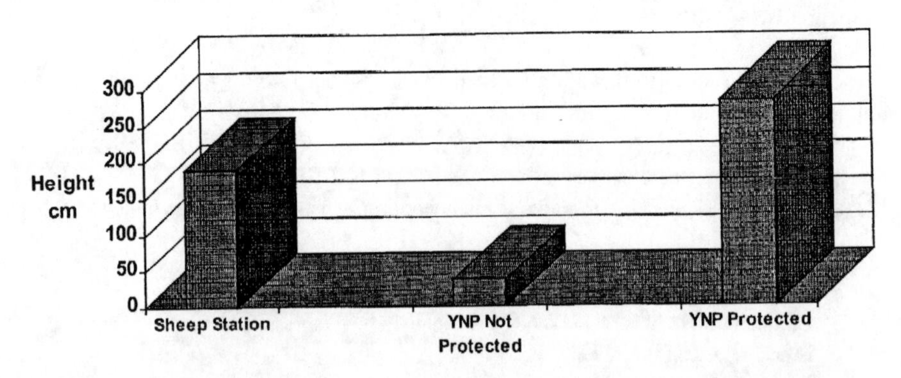

Figure 1 Impact of elk and sheep grazing on willow communities: a measure of ecosystem health and herbivore grazing.

grazers than larger animals, such as topi and buffalo. By having diets which do not overlap helps maintain a broad diversity of plants. They further discuss the fact that grazing pressure in the Serengeti increases the overall spectrum of resource availability to animal communities. By cropping and trampling the tall grasses, larger ungulates increase the range and sward structures providing room for a greater variety of ungulate species.

The interaction between wildlife and livestock in ecosystems can be complex. First, there is increasing evidence of a grazing complementarity between wildlife and livestock. As shown by Schwartz and Ellis (1981), the dietary "overlap" between most wildlife species and livestock is rather limited. Mwangi and Zulberti (1985) and Western and Pearl (1989) showed that the combination of livestock raising and wildlife management resulted in an equal or better species wealth than any of these activities done individually. Furthermore, in national parks in Kenya such as Amboseli, where livestock is not permitted, biodiversity is decreasing, with an increase in unpalatable plant species and bush encroachment (W. K. Ottichilo, unpublilshed data). On the other hand, the same author points out that there are many degraded areas in Kenya due to combined wildlife–livestock pressure.

The driving forces leading to losses in animal biodiversity are habitat destruction, species introduction, and hunting (World Resources Institute, 1994). Habitat destruction is playing an important role in the developing world, especially in the subhumid savannas. In Africa, road construction and human immigration from the drier areas leads to habitat destruction of the vectors of African sleeping sickness. In turn, this lifts the protection of wildlife, which is tolerant to the disease. International agencies, including the World Bank, have also financed extensive vector clearance campaigns in West Africa. Traditionally, these campaigns used a combination of hand and aerial insecticide spraying, initially with organochlorines, to eradicate the tsetse fly. Nagel (1993) argues that pesticides from this period are still notable in some African birds.

Since the mid 1980s, compounds with shorter residual effect, such as synthetic pyrethroids, have been used. These second generation compounds caused substantial initial damage to the flora and fauna, but permanent effects were not observed with single spraying, for example, in the World Bank–funded tsetse clearance project on the Adamaoua Plateau of Cameroon (P. Muller, unpublished data). The permanent damage and high residue levels reported from this project came from repeated spraying in the border areas. Bush encroachment resulting from inappropriate grazing management was the most serious environmental damage. Land-use plans have often been advocated as the essential elements of tsetse clearance, and international financiers have made the preparation of such plans conditional to the financing of the eradication campaigns. However, the experience with the enforcement of such land-use plans has been dismal, as local authorities lacked the authority and means for their enforcement. A critical issue concerning these zones is that traditional land tenure practices have been even less robust in these more humid areas than in the drier areas. This lack of strong traditional tenure practices has been felt in sub-Saharan Africa after tsetse clearance operations, and as a result a rather anarchic settlement pattern developed.

In addition, hunting and culling of wildlife was encouraged in the past, because wildlife was thought to be a reservoir of diseases, such as rinderpest and malignant catarrhal fever, and carriers of disease, such as East Coast fever and trypanosomiasis (Grootenhuis et al., 1991), and competition for scarce grazing resources. However, the control of the above-mentioned diseases has improved considerably and there is a much better understanding of which particular species harbor specific diseases.

The costs and returns from wildlife, compared with livestock and agricultural production, are highly variable. At the level of the national economy, the opportunity cost of wildlife biodiversity conservation in protected areas, in terms of forgone livestock and agricultural production, seems to outweigh the income from tourism and forestry generated by these protected areas. For example, in Kenya, Norton-Griffiths and Southey (1995) estimated the forgone livestock and agriculture from the parks at U.S.$ 203 million, while the revenue from these parks amounts to only U.S.$ 42 million. On the other hand, Engelbrecht and van der Walt (1993) estimated that the Kruger Park in South Africa contributed more than U.S.$ 110 million/year in tourism vs. a forgone production of only U.S.$ 6 million. At household levels, the comparative profitability of wildlife and livestock raising varies greatly, according to the ecological conditions and wildlife use (meat, trophy hunting, tourism). Overall, under present conditions of niche markets for game meat or tourism, wildlife ranching seems financially more attractive, though. For the communal areas, wildlife cannot provide the multiple functions of producing milk for subsistence and providing traction, fertilizer, and investment that livestock can. Without any doubt, the combination of wildlife and livestock is the most appropriate under those conditions.

A recent World Bank study gives comparative income levels from wildlife and livestock production on ranches for four African countries. In Ghana, investments in cattle had an economic rate of return (ERR) of close to zero, whereas private wildlife ranching had an ERR of about 8. In Kenya, financial return on investment (FRR) from game ranching was estimated at 7 to 12% vs. about 6 to 8% from livestock ranching. In Namibia, the study gave a wide range of net returns varying from 0 to 0.28 rands/ha under livestock, to 0.28 to 1.50 rands/ha under wildlife. In Zimbabwe, the FRR for livestock was about 2%, compared with 10% for wildlife. Economic returns per hectare were higher for ranching than for wildlife. All wildlife enterprises benefited from the special niche market, either through higher meat prices, or through revenue from tourism or trophy hunting (Bojos, 1996).

In fostering sustainable wildlife–livestock integration on communal areas, institutional constraints play a big role. Traditionally, there has been a rigid centralized and regulatory attitude of public institutions in the protection of wild animals. This was especially the case in East Africa, where wildlife management was typically organized by central administrations in a rather military fashion. There was no benefit sharing with the local population, whereas wildlife causes high financial cost to the local population because of crop damage and livestock loss due to predators and diseases. This has led to antagonistic reactions from many herding and farming communities. In addition, the interdiction, still in effect in many countries, of sport hunting and consumption precludes benefits from wildlife to be realized as an important part of the potential benefits (W. K. Ottichilo, unpublished data).

PROMOTION OF LIVESTOCK AND BIODIVERSITY

Literature cited in this chapter demonstrates that livestock can have either a positive or negative impact on biodiversity. The driving forces are human population growth, economic development, and how societies within a country value their natural resource base and, in particular, biodiversity. Given that livestock can have either a positive or negative impact on biodiversity and the use of livestock is based upon larger policy and economic issues, what are the avenues that promote a positive symbiotic relationship between livestock and plant and other animal species? Is there a group of technologies and policies that can be implemented or practiced which can promote an agenda involving economic growth and maintenance of biodiversity? The use and development of such technologies and/or policies can be critical to sustaining a rich array of biodiversity.

Technologies

Increasingly, the scientific community is determining that there are a variety of technologies that can promote economic growth of the livestock sector and encourage or enhance biodiversity. For example, in Southeast Asia as well as in the North American Pacific northwest, there is a growing appreciation that sheep can be used in forestry systems (rubber or timber, respectively) to control undesirable plant species. The significance of this approach is that it reduces the amount of herbicide that has to be applied. The development of the above practices demonstrates that there is capacity to develop technologies and practices that successfully blend livestock production with biodiversity issues.

The following list presents ways which could facilitate the development of technologies that both increase livestock productivity and promote biodiversity. Generalized technologies needed for grazing systems include

- Better quantification of the global economic costs and benefits of livestock use and development with conserving biodiversity;
- Methodologies for identification of appropriate indicators that provide reliable information on plant and animal trends;
- Utilization of mixed species grazing (cattle, sheep, goats with wildlife) at appropriate levels — mixed species grazing has the advantage of increasing plant community diversity while concurrently improving the habitat of wild ungulates, but there is a need to develop methodologies which can more accurately determine optimal combinations of domestic and wild ungulates;
- Design of sustainable drought preparedness plans in arid and semiarid areas which not only take into account the needs of livestock and their owners but the impact their activities will have on plant and wildlife communities;
- Development and use of breeds of livestock which are well suited to the environment in which they are expected to perform, such as indigenous breeds.

Generalized technologies needed for mixed farming and industrial systems include

- Improved soil cover through use of alternative crops for mulching;
- Improved feed production and quality to reduce the pressure on grazing areas and improved internal nutrient transfers;
- Reduced nutrient losses from manure and improved efficiency of application (both actions promote biodiversity either through the contribution of manure to soil organic matter or by preventing the over application);
- Improved feed formulation which better balances animal dietary needs and reduces nitrogen excretion;
- Improved animal management through better matching the nutrient needs and use of genotypes which best match the environment;
- Development and utilization of precision agriculture to determine the amount, type, and benefit of inputs needed.

Policies

Both livestock sector and biodiversity issues are affected by broader policy issues that a country may deem necessary to implement. For example, a country may decide to import cereal grains for livestock during times of drought rather than support a rational destocking program which would lead to more productive livestock sectors and support a quicker range recovery after the drought has ended. Biodiversity is compromised by the value the society of a country places upon its natural resource base. In other words, demand for food and economic growth can overwhelm any concern for the environment. Cheap food policies and a desire to achieve self-sufficiency, particularly with cereals, have been important factors in determining livestock sector growth and biodiversity conservation. Furthermore, overvalued exchange rates in sub-Saharan Africa and Latin America have favored importation of cheap food from the industrialized world, thus competing against local production and providing few incentives for local producers to intensify into mixed crop–livestock systems and to practice soil conservation.

The following list of policy options could contribute to a better blending of livestock development and conservation of biodiversity:

- Eliminate overvalued exchange rates which favor importation of cheap food from the industrialized world, thus competing against local production and providing no incentives for local producers to intensify into mixed crop–livestock systems and practice soil conservation.
- Strengthen land tenure security, especially in the rain-fed mixed farming systems of the developing world that will provide an incentive for investment in long-term soil fertility improvements, such as the use of inorganic fertilizers and the use of green manure and leguminous fodder crops in the crop rotation.
- In dry areas, improve infrastructure, roads, and markets to facilitate movement of goods and services especially in drought. Carefully introduce water development so that vegetation is not negatively impacted and aquifers are not depleted.
- Develop institutional capacity that blends biodiversity and livestock needs and provides a basis for analyzing and evaluating economic and environmental needs.
- Develop more effective benefit-sharing mechanisms for communities practicing livestock production in a manner which helps promote biodiversity.

- Phase out subsidies on feed, fertilizers, and mechanization to promote a tighter integration of crop–livestock systems and remove opportunities to overfertilize.
- Tax inorganic fertilizer, set maximum application limits, and regulate time of application to reduce leaching and volatilization.

CONCLUSION

As research continues on the relationship between livestock and the environment, and specifically on biodiversity, there is a growing awareness that livestock can have positive impacts on biodiversity as well as a negative relationship. Driving the livestock–biodiversity relationship are the pressures human populations place on the natural resource base. Negative impacts on plant and wildlife communities occur when livestock either outcompete wildlife (usually through high stock rates) or overgraze an area for long periods of time, thus causing permanent damage to a plant community. However, appropriate stocking rates have been shown to improve biodiversity in both the plant and animal communities involved. CAST (1996) states that wildlife habitat has generally improved in conjunction with improving range conditions in the western U.S.

Due to the increased needs of human populations on a global scale, agricultural pressures, including livestock, will continue to mount and potentially deplete some portion of biodiversity. But as de Haan et al. (1997) discuss, it is imperative to have appropriately formulated policies which lessen the negative relationships between livestock and biodiversity and strengthen the positive aspects of the relationship.

REFERENCES

Archer, S., Scifres, C. J., and Bassham, C. R., 1988. Autogenic succession in a subtropical savanna; conversion of grassland to thorn woodland, *Ecol. Monogr.,* 80:272–276.

Bojos, J., 1996. The Economics of Wildlife: Case Studies from Ghana, Kenya, Namibia and Zimbabwe, AFTES Working Paper No. 19. World Bank, Washington, D.C.

Bojos, J. and Casells, D., 1995. Land Degradation and Rehabilitation in Ethiopia; A Reassessment, AFTES Working Paper No. 17. World Bank, Washington, D.C.

Bruenig, J. 1991. Tropical Forest Report, Government of Federal Republic of Germany, Bonn, Germany, 118.

Burkholder, P. R., 1952. Cooperation and conflict among primitive organisms, *Am. Sci.,* 40:601–631.

CAST, 1996. Grazing on Public Lands, Task Force Report No. 129, Council for Agricultural Science and Technology, Ames, IA.

Clements, F. E. 1905. *Research Methods in Ecology,* University of Nebraska Publishing Company, Lincoln.

de Haan, C., Steinfield, H., and Blackburn, H., 1997. *Livestock and the Environment: Finding a Balance,* European Commission, Brussels.

Engelbrecht, W. and van der Walt, P., 1993. Notes on the Economic Use of the Kruger National Park, *Koedoe,* 36:113–119.

Grootenhuis, J. G., Njuguan, S. G., and Kat, P. W., 1991. Wildlife Research for Sustainable Development: Proceedings of an International Conference held by the Kenya Agricultural Institute, KARI, Nairobi.

Holling, C. S., 1995. Sustainability: The cross-scale dimension, in *Defining and Measuring Sustainability, The Biogeophysical Foundations,* M. Munasinghe and W. Shearer, Eds., The International Bank for Reconstruction and Development. Washington, D.C., 65–76.

IFRPI, 1995. Global Food Projections to 2020. Implications for Investment, Food, Agriculture and Environment Discussion Paper 5, International Food Policy Research Institute, Washington, D.C.

Kaimonitz, D. 1995. Livestock and Deforestation in Central America, EPTD Discussion Paper No. 9, IFPRI, Washington, D.C., and IICA Coronado, Costa Rica.

Kay, C. E. and Walker, J. W., 1997. A comparison of sheep- and wildlife-grazed willow communities in the greater Yellowstone ecosystem, *SID Sheep Goat Res. J.,* 13:6–14.

Le Houerou, H. N., 1989. The shrublands of Africa, in *The Biology of Utilization of Shrubs,* Academic Press, New York, 119–143.

McIntire, J., Burst, D., and Pingali, P., 1992. *Crop-Livestock Interaction in Sub-Saharan Africa,* World Bank, Washington, D.C.

Milchunas, D. G. and Lauenroth, W. K., 1993. Quantitative effects of grazing on vegetation and soils over a global range on environments, *Ecol. Monogr.,* 63:327–366.

Milchunas, D. G., Sala, O. E., and Laurenroth, W. K., 1988. A generalized model of the effects of grazing by large herbivores on grassland community structure, *Am. Nat.,* 132:87.

Mosley, J. C., 1994. Prescribed sheep grazing to enhance wildlife habitat on North American range lands, *Sheep Res. J.,* Special Issue, 79–91.

Murray, M. G. and Illius, A. W., 1996. Multispecies in the Serengeti, in *The Ecology and Management Grazing Systems,* J. Hodgson and A. W. Illius, Eds., CAB International, Wallingford, Oxon, U.K., 247–272.

Mwangi, Z. J. and Zulberti, C. A., 1985. *Optimization Wildlife and Livestock Production. Wildlife/Livestock Interfaces on Rangelands,* Inter-African Bureau of Animal Resources, Nairobi.

Nagel, 1993. L'Incidence de la lutte contre la tse-tse sur les resources naturelles, in *Réunion sur les Aspects Techniques et Développement du Programmes de la Lutte contre la Trypanosomiase Tropicale,* FAO, Rome.

Norton-Griffiths, M. and Southey, C., 1995. The opportunity costs of biodiversity conservation in Kenya, *Ecol. Econ.,* 12:125–139.

Odum, E. P., 1971. *Fundamentals of Ecology,* 3rd ed., W.B. Saunders Co., Philadelphia, PA.

Pimental, D., Harvey, C., Resosudarmo, P., Sinclair, K., Kurz, D., McNair, M., Crist, S., Shpritz, L., Fitton, L., Saffouri, R., and Blair, R., 1995. Environmental and economic costs of soil erosion and conservation benefits, *Science,* 267:1117–1122.

Schwartz, C. C. and Ellis, J. E., 1981. Feeding ecology and niche separation in some native and domestic ungulates on the shortgrass prairie, *J. Appl. Ecol.,* 18:343–353.

Severson, K. E. and Urness, P. J., 1994. Livestock grazing: a tool to improve wildlife habitat, in *Ecological Implications of Livestock Herbivory in the West,* M. Vavra, W. A. Laycock, and R. D. Pieper, Eds., Society of Range Management, Denver, CO, 232–249.

Tucker, C. J., Dregne, H. E., and Newcomb, W. W., 1991. Expansion and contraction of the Sahara Desert from 1980–1990, *Science,* 253:299.

Westoby, M., Walker, B., and Noy-Merr, I., 1989. Opportunistic management for rangelands not at equilibrium, *J. Range Manage.,* 42:266.

Western, D., 1989. Conservation without parks: wildlife in the rural landscape, in *Conservation for the Twenty-First Century,* D. Western and M. Pearl, Eds., Oxford University Press, New York.

Western, D. and Pearl, M. C. 1989. *Conservation for the Twenty-First Century,* Oxford University Press, New York.

World Resources Institute, 1994. *World Resources 1994–95. A Guide to Global Environment,* Oxford University Press, New York.

CHAPTER 7

Managing for Biodiversity of Rangelands

Neil E. West

CONTENTS

INTRODUCTION

Under accelerating extinctions within world biota and increasing invasion of exotics, consequent to the expansion of human populations and their increased

1-56670-290-9/99/$0.00+$.50

demands for space devoted to producing their immediate needs and aspirations (Vitousek et al., 1997), numerous environmental interest groups have clamored for more consideration of natural biotic wealth of all kinds. In the past these efforts focused on creation of reserves. Activists, however, now realize that increasing the size of existing reserves and demarcation of new ones will not conserve all the biodiversity many would like. Furthermore, changing climates mean that fixed boundary reserves will not guarantee that suitable habitat will be available for organisms to migrate to (Harte et al., 1992). Conservation biologists (e.g., Noss and Cooperrider, 1994) are thus shifting some of their attention to nonreserve lands of all kinds and attempting to alter land-use policies such that biodiversity is provided for over a larger fraction of the Earth.

Rangelands, where native biota intermingle with humans and their domestic livestock, involve a huge fraction of the Earth's surface (about 70% by the estimate of Holechek et al., 1989). Increasing conflict between graziers and conservation biologists seems inevitable, especially in the developed world where people have at least the short-term luxury of considering wildlife and other amenities over production of food and fiber. The fact that the wildlife are owned by most states, whereas most habitats are owned by individuals or local communities (Cumming, 1993), is the major reason for biodiversity issues providing clashes between private rights and public values, particularly on publicly owned lands.

DEFINITIONS

Before we go further, we need to define some critical terms. First, one needs to realize that biodiversity entails many different things to different interest groups (West, 1995). To some, it is mainly genetic material. To others, it is taxonomic richness, usually species, of biota within plots or more abstract communities and landscapes. To still others, it is properly functioning ecosystems, including indigenous human cultures living in sustainable ways. All these views are legitimate and have to be respected in democratic societies.

Even though scientists of various kinds are pushing broadened views of biodiversity, the public activists are, as reflected in legislation, budgets, and activity, favoring the charismatic megafauna, the warm, fuzzy, and appealing organisms, particularly the vertebrates, not the little things that run the world (Wilson, 1987). Administration of the Endangered Species Act (ESA), the strongest environmental law in the U.S., currently only impacts what can be done to listed species and their habitats, including activities on privately held lands and waters.

It is becoming obvious that far more than scientific information is involved in what is being done about biodiversity. Stances about biodiversity inevitably involve one's personal and professional ethics (Coufal, 1997). Thus, this is a topic that will inevitably cause philosophical reflection, as well as scientific and managerial action.

The second term deserving further definition is rangelands. Some prefer a strictly use-oriented definition. In that sense, rangelands are agroecosystems since they are all lands with self-sown vegetation used for livestock grazing. That is the oldest definition that still prevails in developing countries. This traditional definition also applies to a wide array of ecosystem types where livestock grazing has and could occur, including recently cut forests, tundras, and marshes. The majority of rangelands, however, occur where grasslands, shrub steppes, deserts, woodlands, or savannas prevail, in other words, most of the untilled or undeveloped western U.S. (about 70% of the area). Rangeland managers and scientists are thus more familiar with drier and less fertile systems than most foresters, wildlife biologists, and agronomists. Whereas most of such lands were recently seen primarily as sources of food and fiber, in developed countries many of them are being increasingly dedicated to sustaining other values that are now prized more highly in industrialized societies. We thus have to contrast how rangeland biodiversity is being considered in the developed compared with the developing world.

My focus here will be on the drier parts of the world where self-sown vegetation is managed extensively based on ecological principles. Agronomic principles rarely apply to these lands: the costs of attempting to till, seed, fertilize, treat with pesticides, and use other means of strong manipulative control to enhance production of food and fiber rarely justify their expenditure because plant responses are fundamentally low due to meager precipitation, salty, steep, and rocky soils, etc. The previous lack of such treatments is the major reason that rangelands are now seen as valuable repositories of biodiversity. That is, most of these rangelands have not yet been simplified and homogenized by intensive agricultural activities (Matson et al., 1997). There are some important exceptions, however, such as the Conservation Reserve Lands (Allen, 1995), which are former croplands that could become rangelands and/or wildlife reserves, depending on Congress' budget setting.

A CASE STUDY IN BIODIVERSITY

Sagebrush Steppe

A thorough review of all aspects of biodiversity in all kinds of rangelands around the world would be impossible for several reasons. First of all, not all aspects of biodiversity have been thoroughly studied in all kinds of rangelands. The genetics of even dominant plants and vertebrates, and anything about invertebrates, microbes, ecosystem functions, and feedbacks, have rarely been studied. Second, even the information that does exist cannot all be summarized in the space available here. Therefore, what I have chosen to do is exemplify how biodiversity issues interact with science and policy in one ecosystem type (sagebrush steppe) well known to the author. I will bring in ideas and experimental results from other contexts as well and discuss how they might apply to sagebrush steppe. In that way I can give a more-focused introduction to the topic at hand.

Shrub steppes are ecosystems with organisms and life-forms of both deserts and grasslands. Although, on average, they are drier than most grasslands and wetter than deserts, the variation in climate is high (coefficients of variation in total annual precipitation usually exceed 30%). Thus, some years have grasslandlike climate whereas other years are desertlike. This climatic variation is probably the main reason for the mix of grassland and desert life-forms in making up shrub steppes. Another result of the high climatic variation is the inherently low stability of these systems under disturbance (Archer and Smeins, 1991).

Because the environmental conditions of the sagebrush steppe are harsh and highly variable over time and space, the dominant organisms are few and widely distributed. This belies the probable high degree of intraspecific ecotypic and genetic variation, which has barely been studied. Once these patterns are understood, variations in autecological and ecophysical responses and synecological interactions will be more comprehendible.

Location, Ownership, and Land Uses

Sagebrush steppe occurs wherever there is or once was vegetation with shared dominance by sagebrushes (woody *Artemisia* spp.) and bunchgrasses (West and Young, 1998). This system occurs mostly in the lowlands of the northern part of the Intermountain West. Sagebrush steppe once occupied about 45×10^6 ha there (West and Young, 1998). About 20% of this ecosystem type passed into private ownership with the Euroamerican settling of the West (Yorks and McMullen, 1980). The remaining 80% is managed by various agencies of the U.S. and state governments. This circumstance makes the management of these lands much more difficult than those under private ownership. Many interest groups, including those championing biodiversity, can and do politically influence management policies on these public lands.

About half of the original sagebrush steppe area now in private ownership has been converted to either dryland or irrigated agriculture over the past 150 years. The approximately 90% remaining untilled lacks irrigation water or is too steep, rocky, or shallow soiled for annual cultivation. The dominant historical uses of these wildlands by human societies have been first hunting and gathering and then livestock grazing.

Climate

The prevailing climate in sagebrush steppe is temperate, semiarid (mean annual precipitation of 20 to 40 cm) and continental (cool, wet winters and springs and warm, drier summers and autumns). Mean annual temperatures range from 4 to 10°C. Winters are cold enough so that snow packs of 50 to 100 cm are common. Snowmelt is usually gradual and thus most of the moisture therein becomes stored at depth in the soil. Native plant growth occurs largely from April to July, the only part of the year when both temperatures and soil moisture are favorable. Summer precipitation is rarely enough to carry herbaceous plant growth throughout the summer. Early fall precipitation is not dependable and by October temperatures are usually too cool to allow much regreening of grasses (West and Young, 1998).

Primary Producers

The major woody dominants here are woody *Artemisia*, collectively known as the sagebrushes. These are shrubs derived from progenitors which came from Eurasia over the Bering Land Bridge and have subsequently radiated into about 13 species (McArthur, 1983). Furthermore, the major species, *Artemisia tridentata* (big sagebrush), has at least five relatively easily recognizable subspecies that should be used in separating out different ecological sites (McArthur, 1983).

The sizes and degrees of dominance of the sagebrush species vary greatly with both site and disturbance history. Sagebrush density is generally greater, but height lower, on more xeric sites. Sagebrush also increases in abundance following excessive livestock grazing in the spring (West and Young, 1998). Livestock grazing also reduces the chance of fires by removal of fine fuels in the interspaces connecting the clumps of shrubs. Fire formerly kept the sagebrush steppe more frequently burned (60 to 110 year return interval) (Whisenant, 1990) and less dominated by sagebrush because most species of sagebrush do not resprout after fire, but have to regenerate from seed (Blaisdell et al., 1982).

Even when sagebrush is dominant, a moderate number of other plant species are found associated with it. On relict (naturally ungrazed by livestock) sites in central Washington, Daubenmire (1970) found an average of 20 vascular plant species in 1000-m plots. Tisdale et al. (1965) found a range of 13 to 24 vascular plant species on three relict stands in southern Idaho. Zamora and Tueller (1974) found a total of 54 vascular plant species in a set of 39 late seral stands in the mountains of northern Nevada. Mueggler (1982) found between 24 and 41 vascular plant species in a set of 68 0.05-ha lightly grazed macroplots in sagebrush steppe of western Montana.

The vertical and horizontal plant community structures are remarkably similar in all relatively undisturbed examples of this ecosystem type. The shrub layer reaches approximately 0.5 to 1.0 m in height. The shrubs have a cover of about 10 to 80%, depending on site and successional status. The grass and forb stratum reaches to about 30 to 40 cm during the growing season. Herbaceous cover also varies widely depending on site and successional status. On relict sites, the sum of cover values usually exceeds 80%, and can approach 200% on the most mesic sites (Daubenmire, 1970).

The herbaceous life-forms most prevalent on relict sites are hemicryptophytes (Daubenmire, 1975). The proportion of therophytes increases markedly with disturbance. The proportion of geophytes is around 20%. A microphytic crust dominated by mosses, lichens, and algae is commonly found where litter from perennials is not excessive (West, 1990). Sagebrushes have both fibrous roots that can draw water and nutrients near the surface and a taproot that can function from deep in the soil profile. Near the end of the growing season for grasses, sagebrushes nocturnally water from more than 90 cm and excrete it in the upper part of the soil profile at night (Caldwell and Richards, 1990). This hydraulic can help the grasses stay active longer than possible on their own.

Perennial grasses associated with *Artemisia* vary greatly throughout the region. The C bunchgrasses (*Agropyron spicatum*, *Festuca idahoensis*, *Stipa* spp., *Sitanion hystrix*, *Poa* spp.) dominate the herbaceous layer in the north and western parts of the type. C sod grasses (e.g., *Agropyron smithii, Hilaria jamesii*) become more common in the south and east where more growing season precipitation occurs (West, 1979).

Total aboveground standing crop phytomass within the sagebrush steppe type varies between about 2000 to 12,000 kg/ha, depending on site differences, successional status, and age of the brush (West, 1983). Litter standing crops are about one half the live nonwoody material (West, 1985). Belowground phytomass is similar in magnitude to that aboveground. Annual net aboveground primary production varies between about 100 and 1500 kg/ha, depending on site, successional status, stand age, and preceding climatic conditions (Passey et al., 1982).

Plant ecologists have long assumed that communities that are floristically richer stabilize primary production in the face of variable climate (Chapin et al., 1997). Indeed, Passey et al. (1982) in their discussion of long-term data gathered from ungrazed sagebrush steppe relicts conclude that each year brings both unique dominance–diversity and production relationships. They attribute this to differing phenologies, rooting patterns, and green leaf persistence. Harper and Climer (1985) reanalyzed the Passey et al. (1982) data set and concluded that variation in plant community production was more positively related with floristic richness than either average precipitation or precipitation of a given year. Tilman et al. (1996) have shown that greater species richness in tall grass prairie leads to greater production during drought than in more depauperate stands created by adding nutrients.

Any landscape within which sagebrush steppe is the matrix is a patchwork of stands of differing species composition and shrub or other growth form dominance. The mix of plant species and growth forms is dependent on ecological site potential and time since particular disturbances. Fires, grazing by both native and introduced vertebrates and invertebrates, as well as unusual climatic events such as deep soil freezing before snowpack accumulation and unusually heavy precipitation and consequent soil anoxia, all contribute to resetting the successional clock (West and Young, 1998). Livestock grazing on these rangelands usually takes place in large paddocks with only one or a few watering points. The parts most distant from water thus are less grazed and of higher seral status (Hosten and West, 1996). This creates a patchwork of differing seral statuses across the landscape (Laycock et al., 1996).

Consumers

The native vertebrates using this ecosystem type are a mixture of grassland and desert species. Maser et al. (1984) grouped the vertebrates of sagebrush steppe in southeastern Oregon into 16 life-forms and related them to vegetation structure and other features of habitat. The vertebrate community is more diverse when the vegetation has the greatest structural diversity (Parmenter and MacMahon, 1983). Neither shrub-dominated nor grass-dominated situations favor as many different kinds of vertebrates as do the mixtures. A few such as voles (*Microtus montanus*) can influence the structure by girdling the shrubs (Mueggler, 1967; Parmenter et al., 1987).

Over 1000 species of insects have been observed on a sagebrush–grass site in southern Idaho (Bohart and Knowlton, 1976). Wiens et al. (1991) recently identified 76 taxa of invertebrates on sagebrush alone in central Oregon. Relatively little is known about the habitat preferences, trophic relationships, and other aspects of the roles of invertebrates in this ecosystem type. Only a few — thrips, webworms, grasshoppers, cicadas, aphids, and coccids (Kamm et al., 1978; West, 1983) — are known to be irruptive and visibly alter vegetation structure.

Decomposers and Nutrient Cycling

Very little is known about microbes and the decomposition process in this ecosystem type. Initial studies of the nitrogen (West and Skujins, 1978) and phosphorus (West et al., 1984a) cycles showed that available forms of these elements may limit plant production in wetter than average years. Allelochemics from sagebrush and the high C:N ratios of its litter may inhibit some decomposition and nitrogen-cycling processes, perhaps indirectly strengthening sagebrush dominance in this ecosystem type (West and Young, 1998). Changes in litter quality can lead to degradation of soil organic matter in such systems (Lesica and DeLuca, 1996). Global environmental changes may produce some unexpected interactions among plants, soil microbes, and soil degradation (West et al., 1994).

Interactions among Plants, Animals, and Humans

The pristine sagebrush steppe evolved with large browsers (megafauna), most of which had disappeared by about 12,000 years ago (Mehringer and Wigand, 1990; Burkhardt, 1996). The loss of the megafauna is inextricably linked to simultaneous increases in human hunting and climatic warming (Grayson, 1991). Remaining graminivores were few in the pre-European system (Mack and Thompson, 1982; Harper, 1986). The small populations of aboriginal hunters and gatherers of the mid-Holocene probably influenced the vegetation largely by burning. It took European colonization to change drastically the native vegetation and the wildlife habitat it provides (Young, 1989).

The pre-European era livestock grazing capacity, when shrubs were fewer and grasses more prevalent, was estimated to be 0.83 animal unit months (AUM)/ha (McArdle and Costello, 1936). Because sagebrushes are usually unpalatable to livestock, whereas herbs are palatable, uncontrolled livestock use led to a decline of herbs and increase in brush. Carrying capacities declined to an average of 0.27 AUM/ha in the 1930s (McArdle and Costello, 1936), but had improved slightly to 0.31 AUM/ha by 1970 (Forest-Range Task Force, 1972).

Livestock populations built up rapidly near the end of the 19th century. Griffiths (1902) judged that the grazing capacity of these rangelands had been exceeded by 1900. Hull (1976) examined historical documents and concluded that major losses of native perennial grasses and expansion of shrubs took only 10 to 15 years after a site was first grazed by livestock.

The native grasses are extremely palatable, especially when green. They die easily when grazed heavily in the spring (Miller et al., 1994). In addition, they rarely produce good seed crops (Young, 1989).

The only time the grasses and forbs have an advantage over brush is when sites are burned. However, on the sites with heavy historical livestock use, both remaining native herbaceous perennials and their seed reserves have been greatly diminished (Hassan and West, 1986). In addition to tall, thicker sagebrush, grazing-induced freeing of space and resources gave opportunities for the invasion of aggressive Eurasian plants. The advent of introduced winter annual grasses, notably *Bromus tectorum* in the 1890s (Mack, 1981), and the continuous, fine, and early-drying fuels they provide has led to seasonally earlier, more frequent (less than 5 years), and larger fires (Whisenant, 1990). After repeated fires, combined with unrestricted grazing, any remaining native vegetation becomes easily replaced by other, even more noxious introduced annuals, such as medusahead (*Taeniatherum caput-medusae*), knapweeds (*Centaurea* spp., *Acroptilion* spp.), and yellow star thistle (*Centaurea solstitialis*). The result has been a considerable decrease in plant species structural and floristic diversity, average forage production, and nutritional value to vertebrates (Billings, 1990; Whisenant, 1990). This simplification of self-sown vegetation results in much more frequent bare ground and accelerated wind and water erosion (Hinds and Sauer, 1974). Variability in plant production goes up several orders of magnitude after replacement with annuals (Rickard and Vaughn, 1988).

Wildlife responds dramatically to these changes in vegetation structure (Maser et al., 1984). For instance, the pigmy rabbit (*Brachylagus idahoensis*) is a threatened species that prefers the tallest, densest stands of Basin big sagebrush. Sites occupied by this plant have been widely converted to intensive agriculture. Thus, the range of this sensitive animal has been reduced and its abundance greatly diminished.

Another native herbivore of special interest in the sagebrush steppe is the sage grouse (*Centrocercus urophosianus*). This is a large galliform with a unique digestive system that has coevolved with *Artemisia*. The mature birds survive the less hospitable times of the year by eating the twigs of sagebrushes, especially the low sagebrushes found on windswept ridges. There are, however, other requirements during other parts of their life cycle. During March and April, the males gather on open areas without brush (called leks) and display themselves to the females. Only about half of the males survive raptor predation and intraspecific fighting during this about 2-week mating period. The females fly to the most productive interfluvial areas to nest and raise the chicks. For the first 6 weeks of life, the young birds require a high protein diet made up of insects and forb buds. These are most abundant in fresh burns and in riparian corridors.

Sage grouse were very abundant in the region when Europeans first arrived and have remained abundant enough to be an important game bird until recent decades. Unfortunately, they are now being considered for placement on the endangered lists in several Intermountain states. Wildlife and conservation biologists find it tempting to single out the range livestock industry for causing this problem. However, sheep, which prefer forbs over other types of forage, were much more abundant on these rangelands up to about 1960, but have since declined to a tiny fraction of their former abundance. Sheep do, however, eat some sagebrush, particularly in the fall and winter. The amount of time cattle are permitted on public lands of the sagebrush steppe has also been declining since about 1964, well before sage grouse populations crashed. The amount of perennial cover on much remaining sagebrush steppe has been increasing of late because of reduced livestock grazing and more effective fire control. There is now probably more sagebrush than necessary for optimum sage grouse use in most portions of the sagebrush steppe. Several other possible influences have also been increasing of late, such as vehicular access and nonhuman predators. Coyotes, foxes, skunk, racoons, corvids (jays, magpies, crows, and ravens), and raptors (eagles, hawks, and owls) have all been increasing because of less shooting and pesticide use and could be taking more eggs and chicks, as well as adults. The thickened brush could be making predator stalking and capture easier.

Because of passage of laws such as the ESA and National Forest Management Act, the interests of wildlife, particularly the rare, endangered, and threatened vertebrates, can take precedence over optimal livestock grazing on publicly owned rangelands in the U.S. This is the reason that the U.S. Forest Service and Bureau of Land Management currently strives to leave about 15 to 20% of the mature sagebrush cover intact across the landscape rather than burning or using herbicides to reach the 100% kill they once strived for in the 1940s and 1950s when the nation demanded more red meat.

There has already been a vast replacement of native plant species by Eurasian plant invaders in sagebrush steppes. More is expected, especially if global warming materializes. Controlling fires entirely is an impossibility. Reductions or even complete removal of livestock will not result in a rapid return to the vegetation that occurred before European colonization (Miller et al., 1994). Sheep, grazed during the fall, because they utilize some sagebrushes and can do little damage to the herbaceous understory during that time of year, can actually enhance floristic richness (Bork et al., 1998).

Our major means of obtaining greater dependability of forage production and soil protection on severely degraded sagebrush steppe sites, while at the same time reducing the chance of fire, has been to plant Eurasian wheatgrasses and ryegrasses (Asay, 1987). However, this can only be done easily on relatively level sites with deep, largely rock-free soils. Environmental and archaeological interest groups have recently stopped these procedures, however. Environmentalists object to using any introduced species, regardless of their ability to grow rapidly and protect the soil. Archaeologists object to the physical disturbances to archaeological objects and strata. Native species have been repeatedly tried in plantings, but rarely grow early and rapidly enough to outcompete the introduced annuals. Because environmentalists have prevailed, public land managers are no longer daily involved in proactive management or ecosystem repair here.

Let us now turn to other possible ways to conserve remaining community diversity, alter existing stands, or rehabilitate degraded sagebrush steppe stands. Figure 1 will be used to guide the following discussion. This figure is a state-and-transition model (Laycock, 1995) thought to accommodate better our current understanding of degradation and successional processes in sagebrush steppe than the simpler, linear models of the past with one end point (the climax).

Preservation of Relatively Unaltered Ecosystems

Pristine, relictual areas (State I in Figure 1) no longer exist nor are probably recoverable. The reasons for this view are

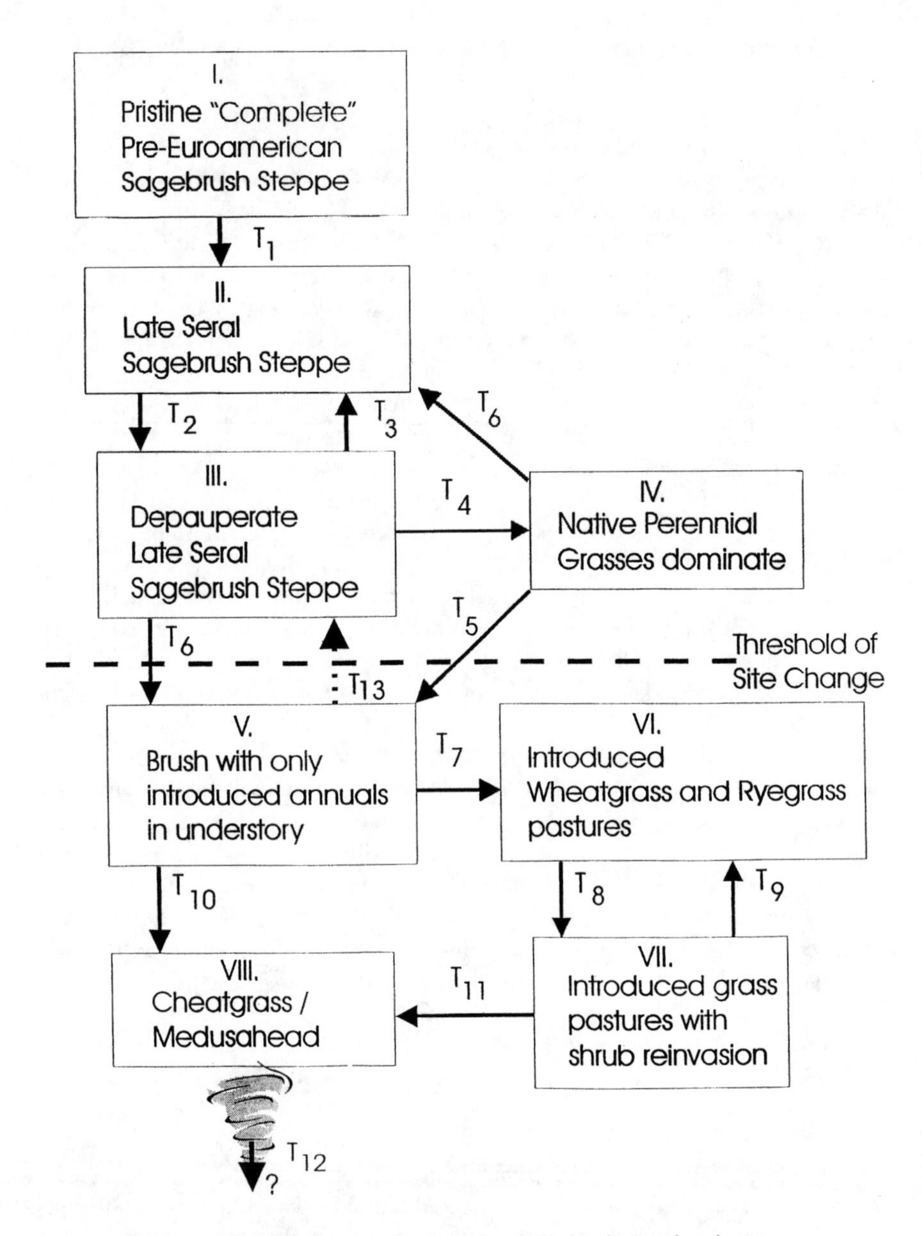

Figure 1 State-and-transition model of successional change in sagebrush steppe.

1. Humans (indigenous peoples) are no longer hunting, gathering, and burning these areas. The previous fire regimes are no longer in place and as the vegetation changes in response to less frequent fires, the hydrologic and nutrient cycles are being altered, as is the habitat for numerous animals and microbes.
2. The present climate is warmer and drier than the cooler, wetter Little Ice Age climate which prevailed up to about 1890. Thus, only heat- and drought-tolerant species may thrive now under global warming.

3. Atmospheric CO has increased about 20% during the past century, altering the competitive balances in this vegetation as well as changing the nutritional qualities of the phytomass and litter (Polley, 1997).
4. About 15% of the flora is now new to the region.

Since we can reverse none of these influences, at least in the short term, we should learn to live with what remains and manage it toward the desired plant communities we choose for each circumstance.

There are, however, some remnants of these landscapes that have escaped direct human influences. These relics exist because they have no surface water, are surrounded by difficult topography, or protected in special-use areas (e.g., military reservations). I place these in State II of Figure 1. Tisdale et al. (1965) describe an example. I estimate that less than 1% of the sagebrush steppe that remains has avoided the direct impact of any livestock. Even these relicts are, however, incomplete because of lack of indigenous humans and lengthened fire frequencies. Relicts are influenced by air pollutants, climatic change, and invasion by exotics (Passey et al., 1982). Most of the existing late seral sagebrush steppe (State II in Figure 1) has had light livestock use. Even light livestock use puts inordinate pressure on a few highly palatable species (ice cream plants), partially explaining the lack of a return arrow from State I to State II.

In some places, feral horses and burros now put considerable pressure on such rangelands, but are protected by federal law on most public lands. I estimate that about 20% of the remaining sagebrush steppe is in State II.

The perceived will of a majority of Americans now is to identify these remaining State II areas, especially those on public lands, and protect them from being developed. Some advocate all such areas be reserved (Kerr, 1994), whereas others (Bock et al., 1993) propose that 25% have livestock excluded. Rose et al. (personal communication) have, however, recently demonstrated that lightly grazed sagebrush steppe has higher species richness than adjacent exclosures dating to 1937. Others propose restoration efforts to bring further-degraded systems back to States I or II (Dobson et al., 1997). State II areas serve as the parts catalogue for restoration efforts. The Gap Analysis Program (GAP) of the U.S. Fish and Wildlife Service (Scott et al., 1993) and the various natural heritage programs initiated by the Nature Conservancy are well under way to put these views in action. These efforts are, however, not without attack from both political and scientific groups (Machlis et al., 1994; Short and Hestbeck, 1995).

I expect to see physical modifications to enhance production of food and fiber (formerly called range improvements) to be more spatially limited than in the past because such actions on public lands or with public monies require environmental assessments or impact statements and thus public scrutiny and debate. The remaining relatively unaltered areas on public lands will probably be consciously protected to provide the later seral condition patches necessary to hold a broader spectrum of all species, and meet the special requirements for some featured species such as sage grouse and pigmy rabbit (Call and Maser, 1985). Of special concern are other sagebrush bird obligates that are also apparently declining: sagebrush sparrow

(*Amphispiza belli*), sage thrasher (*Oreoscoptes montanus*), and Brewer's sparrow (*Spizella breveri*).

Rangeland managers in the past strove to reduce the limitations of the land for producing livestock. These limitations were mainly topography, forage availability, and water. For example, trails were constructed into areas where topographic breaks limited livestock access. Natural water was supplemented by development of springs, building stock tanks and small dams, drilling wells, piping and hauling water. Fences were constructed and salt distributed to control livestock movement and institute grazing management systems (e.g., rest–rotation grazing). All these improvements were designed to distribute livestock utilization more uniformly across the land, gain greater efficiency of food and fiber production, and divert livestock from the especially sensitive riparian areas (Elmore and Kauffman, 1994; Laycock et al., 1996). The net result has been progressively more widespread intensive use of a landscape that has become partially tamed from the wild. These assumptions need to be reexamined in the light of biodiversity concerns. Let us continue our consideration of these relationships in the sagebrush steppe.

Alteration of Existing Heavily Grazed Stands

Because livestock grazing of native sagebrush steppe usually avoids the unpalatable forages, particularly woody species, they are freed from competition and dominance becomes concentrated in the few woody plants on areas with a history of heavy livestock grazing (T_2), but not recent fire (State III, Figure 1). About 30% of this ecosystem type is estimated to exist currently in this state. Most of these stands can stay stagnated for decades (Rice and Westoby, 1978; West et al., 1984b; Sneva et al., 1984; Winward, 1991). The dense, competitive stands of excess sagebrush prevent the herbaceous species from recovering. Such brush-choked stands are usually chosen by both livestock and wildlife managers for manipulation to diversify vegetation structure. This enhances it for livestock or native animals in spots, concentrating livestock use, reducing their pressure elsewhere, while simultaneously advantaging some wildlife species through vegetation modifications via grazing systems, prescribed burning, brush beating, or chaining (T_3). For example, sheep grazing in the fall, because they consume more sagebrush then (Bork et al., 1998), can be used to obtain a reversal from State III to State II. Prescribed burning (Harniss and Murray, 1973) can also be applied to stands with sufficient remnant populations of native herbs to quickly recover following brush kill. Rest from livestock use, such as with a rest–rotation grazing system or winter only use (Mosely, 1996), will often allow a slower return to State II from State III. Reduction of brush also enhances water yields (Sturges, 1977), and some seeps, springs, and streams reappear. When phenoxy herbicides are used alone (Evans et al., 1979) (T_4) or in conjunction with fire, the community becomes dominated with native grass (State IV, Figure 1) because the chemicals impact all broad-leafed species. The conversion only slowly returns (T_6) to State II with judicious grazing and a secondary treatment with prescribed burning. About 5% of the remaining sagebrush steppe is now estimated to be in State IV. This is a short-lived state, especially under heavy grazing (T_5). Mueggler (1982) found enhanced alpha diversity in moderately grazed sagebrush steppe communities in western Montana following prescribed fire, 2,4-D, and brush-beating treatments. Summer fires can damage some of the grasses (Young, 1983), but encourage the resprouting rabbitbrushes (*Chrysothamnus* spp.) and horsebrushes (*Tetradymia* spp.) (Anderson et al., 1996).

If accelerated soil erosion does not ensue and the fundamental potential of the site does not change, then State III can be maintained or managed toward States II or IV. However, as herbaceous plants and litter in the interspaces between perennials are reduced, soil aggregate stability declines, infiltration of precipitation diminishes, overland flow increases, and soil erosion frequently increases (Blackburn et al., 1992). When a probable threshold of use is exceeded, the site can irreversibly change to one of lesser potential. This explains the dashed line and downward arrows below States III and V as the only believable transitions. This is where the syndrome of desertification is most evident. All the former states can be dealt with via soft energy management approaches. Once this threshold is exceeded, however, subsequent management requires expensive, risky, hard energy solutions. Unfortunately, it is often easier to get political attention after major damage has been done rather than getting budgets and personnel to plan, monitor, and tweak the higher-condition, more-natural systems at opportune times.

The desertified sites with thickened brush have largely introduced annuals in their understory. I estimate that State V comprises about 30% of the current sagebrush steppe. Reduction or removal of livestock only hastens further degradation from State V because livestock remove part of the fuel load and thus reduce the chance of fire destroying the sagebrush and the spots of soil it protects.

If insufficient amounts of native herbs remain on sagebrush steppe, the usual land management agency response has previously been to replace them mechanically (T_7) with introduced wheatgrasses and ryegrasses, especially crested wheatgrass (Asay, 1987). This has been done because the introduced perennial grasses are much more easily established than the native grasses and they grow quickly to provide more forage with a higher nutritional plane. The introduced perennial grass stands are also much more tolerant of subsequent heavy livestock use and have lasted for many decades (Johnson, 1986). There are some long-range concerns, however (Lesica and DeLuca, 1996), because the introduced perennial grasses suppress the return of natives and richer plant species assemblages. Some large treatment areas have monocultures of Eurasian perennial grasses prevailing (State VI, Figure 1). I estimate about 5% of the original sagebrush steppe has already been transformed to State VI.

Wildlife biologists have noted declines in the numbers of birds (Olson, 1974; Reynolds and Trost, 1979; 1981), small mammals (Reynolds and Trost, 1980), and large reptiles (Reynolds, 1979) on such seedings of introduced grasses. It should be noted, however, that such studies present a worst-case scenario because samples came from the center of large treatments. Provision for increased diversity near edges (Thomas et al., 1979) is not usually mentioned in such studies. Present-day more-sensitized planners would provide for optimum edge effect and patchiness (McEwen and DeWeese, 1987). When society makes the investment in repairing severely damaged sagebrush steppe, creating perennial grass–dominated pastures of much greater productivity of species palatable to livestock (T_7), this should compensate for livestock reductions and other management restrictions on lands where

States II, III, and IV (Figure 1) predominate. Because introduced grass pastures take heavy degrees of utilization in the spring, the native shrub steppe can support fall and winter grazing with less impact, especially on the native herbaceous perennials.

Introduced perennial grass plantings in the sagebrush steppe region, especially if grazed by livestock, will eventually experience shrub reinvasion (T_8 to State VII) largely in response to intensity of livestock grazing. I estimate that about 5% of the remaining sagebrush steppe region is currently represented by shrub-reinvaded introduced wheatgrass/ryegrass pastures (State VII). Not all brush is now eliminated by re-treatment (T_9). Herbicide use on public lands in the Pacific Northwest has been suspended by judicial decree. Prescribed burning of the coarser, introduced grasses is difficult and leaves patches where the shrubs prevail. There are, therefore, chances to enhance edge effects in large areas that were formerly homogenized. As in the untilled native areas, we could enhance wildlife habitat by providing a mix of successional stages or stand conditions, providing both cover and forage for either featured species or total species richness (Maser et al., 1984). For example, some success has been attained in creating alternate leks for sage grouse breeding following disturbances of development (Eng et al., 1979).

Rehabilitation of Burned Sagebrush Steppe

Despite greatly increased attention to fire prevention and control, much of the most-depauperized sagebrush steppe (State V) has been burned (T_{10}) at least once during the past three decades and is now dominated by introduced annuals, mainly grasses such as cheatgrass and medusahead (State VIII, Figure 1). The Bureau of Land Management (M. Pellant, personal communication) estimates that about 3 million acres of public lands in Idaho, Utah, Oregon, and Nevada are now dominated by cheatgrass and medusahead. I estimate that about 25% of the total original sagebrush steppe has made this transition (T_{10}, T_{11}). Because of their short stature, restricted nutritional characteristics (short period of aboveground greeness), and greater susceptibility to recurring fires than sagebrush steppe, such areas are undesirable from all viewpoints. Without nutritional supplementation, livestock can graze State VIII only during the short, early spring plant-growing season (winter use is possible in the lower elevation areas near the Columbia River; Mosely, 1996). Only the most generalist animals, such as the introduced chukars (*Alectoris chukar*), horned larks (*Eremophila alpestris*), grasshoppers, and deer mice (*Peromyscus maniculatus*) seem to thrive on the annual grasslands (Maser et al., 1984). When such areas burn in early summer, soils are bared to wind and water erosion during the convectional storms of summer. The consequent needs for revegetation after fire are increasing while the budgets of federal land management agencies decline and pressure from environmentalists against active management increases.

Land dominated by annuals may provide fair watershed protection during years without fire and actually appears to be more productive of total plant tissues than the original sagebrush–native perennial grass and forb combination (Rickard and Vaughan, 1988). This is likely, however, to be only a temporary situation based on the priming effect of decomposing litter (Lesica and DeLuca, 1996) and the mineralization of nutrients from the enormous belowground necromass of the original system. When these reserves of nutrients and soil organic matter are finally respired away, the annual grasslands are likely to become much less productive. Similar transitions happened in the Middle East several millennia ago (Zohary, 1973). Many other, more noxious weeds from that region could find their way here, and we could witness a downward spiral of further degradation (T_{12}).

Rather than allowing the annual grasslands derived from former sagebrush steppe (State VIII, Figure 1) to remain and the land to degrade further, some land managers are attempting to intervene. A joint U.S. Forest Service, Bureau of Land Management, Agricultural Research Service, and University of Idaho program is under way to reduce these threats (Pellant, 1990). The most notable component of this effort is the green-striping program most evident in southern Idaho. The basic approach is to begin breaking up the now vast stretches of cheatgrass and other annual dominance that have developed as fires have become earlier, larger, and more frequent (Whisenant, 1990). Land managers are attempting to break the area into smaller, burnable units, especially nearer to cities and towns. The approaches used thus far include planting strips of vegetation that stay green (and thus wetter and less burnable) longer than cheatgrass.

Although the introduced wheatgrasses and ryegrasses do stay green longer and burn less readily, because of coarser aboveground structure, they are not native and thus are rejected by some interest groups. Because the genetic biodiversity of the native plants is so primitively understood, the best that can be done is to gather such seed locally and plant it on comparable sites. Such seed sources are undependable, however; thus a root-sprouting big sagebrush is seen as a potentially better keystone species to put back in this area. A few sagebrushes may actually help sustain perennial grasses by harboring the predators on black grass bugs (*Labops* spp.) (Haws, 1987). Furthermore, total plant community production can be enhanced (Harniss and Murray, 1973) because sagebrushes help trap blowing snow (Sturges, 1977) and scattered sagebrushes moderate temperatures (Pierson and Wight, 1991), benefit the reestablishment of native herbs, and protect them from excessive utilization (Winward, 1991). Sagebrushes also harbor mycorrhizal fungi (Wicklow-Howard, 1989), which helps them extract nutrients from deep in the soil and recycle them to the surface through litter production (Mack, 1977; West, 1991).

Whether or not we can accomplish restoration of sagebrush steppe (T_{13}, between States V and III in Figure 1) is highly questionable. Even where money is less limiting and topsoil is stockpiled on coal strip mines, early results are only partially encouraging (Hatton and West, 1987). We will have to know much more about how sagebrush steppe ecosystems are structured and function and obtain vast budgets and more trained personnel before such efforts are routinely successful. It is cheaper and more feasible to foster good stewardship of land having late seral vegetation (manage while in States I, II, III, or IV of Figure 1) rather than rely on restoration efforts after degradation has taken place (States V, VI, VII, and VIII of Figure 1).

Regional Considerations

Because biodiversity issues in sagebrush steppe are interconnected to multiple impacts and other ecosystem types over the entire region, the federal management agencies are attempting to address them in a holistic fashion. An important example of this is the proposal for ecosystem management in the interior Columbia Basin (Haynes et al., 1996). The documents generated (Quigley et al., 1997) appear to favor restoration practices. Environmentalists (e.g., Belsky, 1997), however, perceive little change in livestock grazing practices and intend to test the process judicially.

GUIDELINES FOR A NEW STYLE OF RANGELAND MANAGEMENT SENSITIVE TO BIODIVERSITY

Recent happenings in the Interior Columbia Basin are symptomatic of the start of a new era in land management. When human populations were lower and demands on resources were less, we could encourage development without much concern for other species or equity to the future. It is becoming obvious now that more consideration for present neighbors and future generations must be consciously given. Environmental impacts no longer have only local consequences. Biodiversity can be viewed as a natural treasure and as having a role in the maintenance, cleansing, and repair of ecosystems at local to global scales (Chapin et al., 1997). Development plans of the U.S. Agency for International Development and the World Bank now require consideration of biodiversity within their environmental impact sections.

We are seeing enhanced efforts to inventory, monitor, and zone with biodiversity in mind. National, regional, and local rankings of organisms and system rarity and endangerment (e.g., the GAP analysis, Scott et al., 1993) are leading to plans to create core reserves, buffer, corridor, multiple-use, and intensive-use zones. Some graziers and other rangeland consumptive users are bound to be either displaced or have their activities altered by these designations. The consequences could be complete removal of livestock in some places of special sensitivity. In most other situations, more thoughtful and careful pastoralism can complement conservation (Friedel, 1994). Areas too small and dispersed to be managed efficiently within a reserve could instead be managed by the permittee on public rangelands. Such areas are called Excised Management Units in Australia (Morton et al., 1995). Where livestock use is critical only at certain times, Restricted Use Units may be designated. Where such designations cause economic hardship, land trades or subsidization may help ease the transition. It is becoming clear that no modern government or nongovernmental entity can afford the expense of buyouts of increasingly greater blocks of reserves. Furthermore, unless reserves are well managed, they can be just as deleterious to the conservation of biodiversity as have been exploitative pastoral systems. Such unmanaged areas can quickly become havens for predators, feral animals, and noxious weeds (Friedel and James, 1995).

In cases where restoration is being attempted, it seems only right that displaced graziers be employed to stay on the land and actively work to heal it. After all, these are people who best understand the local environment. Their children should be assisted in training for other jobs and professions.

Not all healing of degraded rangeland ecosystems requires complete displacement of livestock grazing. Fleischner (1994) and Noss and Cooperrider (1994, Table 7.1) provide a comprehensive discussion of the negative ways management of rangelands and attendant activities (e.g., irrigating winter fodder, predator control, etc.) influence biodiversity. Brussard et al. (1994; 1995) and Brown and McDonald (1995) point out the imbalances of Fleischer's and Noss and Cooperrider's presentations. To help further balance those discussions (see Laycock et al., 1996), I wish to add some positive aspects of the interactions of rangeland livestock husbandry and biodiversity.

First, full-time ranching provides daily contact with the land at all seasons and thus provides experience and a degree of attention to the land that occasional field visits by agency personnel and intermittently interested environmentalists can never replace. Indeed, the Nature Conservancy is calling on such full-time ranchers to manage some of their properties actively with continued livestock grazing, yet with enhanced sensitivity to biodiversity. The Nature Conservancy realizes that simply buying up key properties and eliminating direct human influence is not a viable way to conserve biodiversity on a grander scale. It realizes that humans are part of the ecosystem and that it could not purchase and preserve all the desirable properties anyway. Instead, it understands that encouraging management in economically, as well as ecologically, sustainable ways is the long-term answer to holding on to the maximum biodiversity across the rangelands of the western U.S. The conservancy intends to lead by developing examples that neighboring ranchers will emulate. It has already established worthy examples in places such as Red Canyon Ranch, Wyoming.

Conservation biologists often mistakenly assume that, because mismanaged livestock have done much damage to rangelands in the past (a fact that even livestock-oriented scientists don't deny, e.g., Pieper, 1994), simply removing them permanently will automatically result in the return to similitudes of a romanticized pre-Columbian Eden. In many ways, Euroamericans have enhanced biodiversity by their activities on rangelands (Johnson and Mayeaux, 1992). So many aspects of environment and biota have and are currently undergoing change that it is fruitless for us ever to expect equilibrial scenarios henceforth (Botkin, 1990; Pimm, 1991; Allen and Hoekstra, 1992; Vitousek et al., 1997). The only rational choice is to monitor and adjust through adaptive resource management (Kessler et al., 1992).

First, trade-offs must be made between maximum production of livestock and the best possible wildlife habitat, watershed, and soil protection under the ecosystem management philosophy (Kessler et al., 1992). No matter what the manager does or does not do, habitat of some species will be enhanced and that of others simultaneous diminished (West, 1993). For instance, in a recent rangeland study in Australia (James et al., 1997), about one quarter of the biota was disadvantaged by livestock grazing, about half was neutral to grazing, and about one quarter was increased. Humans have to make the choices of what is favored. Grazing by livestock can be advantageous or disadvantageous to wildlife and other land uses depending on species, uses and values, and their ecosystem context (West and Whitford, 1995). Retention of total species richness, featured species, and stenotypic species cannot usually be simultaneously maximized on the same small piece of land. Such objectives have to be managed for on landscape and regional bases (Friedel and James, 1995), thus inevitably involving many landowners and institutions.

Midseral and even early-seral conditions are not detrimental to all wildlife species. Some species require these conditions to complete their life cycle (West, 1993). Maser et al. (1984) found that the greatest species numbers and highest population levels of most featured vertebrate species of sagebrush steppe are in midseral condition. Thus, maintenance of disturbances to create early to midseral conditions is desirable on at least some parts of a landscape. This is one reason why total removal of livestock from public rangelands would not necessarily lead to optimal habitat for either featured species or total species richness. Furthermore, prohibiting livestock grazing would lead to more fine herbaceous fuels and thus hotter, more frequent, and larger fires and eventual takeover by introduced annual plants in sagebrush steppe (Figure 1) as well as many other kinds of adjacent rangelands.

Domestic livestock management directly affects wildlife in two major ways: (1) consumption of forage that could be used by wildlife and (2) alteration of vegetation as it influences escape and thermal and protective cover (Noss and Cooperrider, 1994). Most animal species are more adapted to gross vegetation structure for thermal and hiding cover than they are to particular plant species for food (Dealy et al., 1981). Structural diversity of rangeland vegetation also relates positively to wildlife species richness, except if the mosaic is on a scale too small to meet the home range needs of species that require large blocks of uniform vegetation. For example, optimum spacing between stands of big sagebrush and crested wheatgrass for black-tailed jackrabbits (*Lepus californica*) requires that the wheatgrass openings be no more than 600 m across because rabbit use of wheatgrass occurs mostly within 300 m of the type edge (Westoby and Wagner, 1973).

Domestic livestock management indirectly affects wildlife by (1) human and livestock presence, (2) fencing, (3) salting, (4) water developments, (5) roads, (6) trails, (7) predator control and other physical and chemical manipulations, such as prescribed burning, chaining, cabling, root plowing, brush beating, reseeding, and herbicidal application. The latter treatments usually simplify and homogenize habitat structure, but mosaics and edge can be increased with planning and plant species richness enhanced by interplanting in areas with large expanses of currently homogeneous vegetation.

Better planning and management could result in retention of livestock and their use as tools for constructive improvement of wildlife habitat and watersheds. Deseret Ranch in northern Utah is an example of a commercially viable operation that has derived income from both consumable and nonconsumable wildlife while increasing both livestock use and stabilizing range condition (Wolfe et al., 1997). Roads can be closed and off-road vehicles prohibited, especially at critical times. Scattered trees could either be retained or planted to provide shade and storm cover for both livestock and wildlife, simultaneously enhancing their overall distribution.

Use of electronic sensors on livestock and thoughtful placement of invisible electronic boundaries offer promise to foregoing building more fences and possibly even to removing the existing ones eventually (Fay et al., 1989). We should recognize, however, that fences serve as perch posts for some birds (Graul, 1980), and, thus, unintended impacts of fence removal and pole line installation could ensue.

Some feel that there is promise in either selecting domestic animals that naturally spend less time in sensitive areas (e.g., riparian zones) or training them to avoid such areas through adversive conditioning. Indeed, some progress has been recently made in doing just that (Howery et al., 1998).

Some water sources can be completely or selectively closed off to favor certain species. Development of naturally occurring springs and seeps through installation of perforated pipe and water troughs at a distance from the water source has made more water available for drinking by both ungulates and birds. Unfortunately, the natural wetlands around the original springs have often been highly altered. We can, however, pipe out the overflow and create new fenced-out wetlands to replace those altered (Kindschy, 1978).

We could even fertilize certain portions of some plant communities to increase and freshen (make more palatable) some areas to draw animals to them to enhance fire control, strutting, feeding, and nesting grounds. Guzzlers (artificial water catchments) can be built in areas with limited free water for drinking by both wildlife and livestock.

CONCLUSIONS

I have demonstrated that application of some knowledge, logic, planning, sensitivity, and compromise could allow us to continue using most rangelands for traditional values as well as provide for preservation and even enhancement of biodiversity. Because of their rarity, desirability to research, and in guiding management, most areas that have escaped livestock use thus far should probably be protected. This will provide maximum landscape diversity, reference areas for monitoring and basic research, and materials for restoration efforts. For the much larger fraction of the rangelands that have had, and continue to have, extensive use, but are still dominated by native plant species, we should continue to apply our increasing knowledge of individual species responses, community dynamics, and ecosystem feedbacks in devising low-input ways to direct succession toward desired sustainable outcomes. Prescribed burning, behavioral modification of animals, and improved ways of distributing animals without fences should be further developed. If some compromise is deemed possible, livestock grazing systems compatible with wildlife, recreational use, watershed and soil protection, as well as biodiversity, can be devised. It is much cheaper and satisfying to prevent such seminatural areas from slipping over the brink of irreversible trends toward desertification than trying to rehabilitate or restore areas that have already been seriously degraded. We should not let the possibility of artificial recovery prevent us from concentrating on trying to develop sustainable use strategies for the rangelands still relatively intact.

REFERENCES

Allen, A. W., 1995. Agricultural ecosystems, in *Our Living Resource,* E. T. Laurie, Ed., U.S. Dept. Interior, Natl. Biol. Survey, Washington, D.C., 423–426.

Allen, T. F. H. and Hoekstra, T. W., 1992. *Toward a Unified Ecology,* Columbia University Press, New York.

Anderson, J. E., Ruppel, K. T., Glennon, J. M., Holte, K. E., and Rope, R. C., 1996. Plant Communities, Ethnoecology, and Flora of the Idaho National Engineering Laboratory, Environ. Sci. and Research Found. Rep. 005, Idaho Falls, ID.

Archer, S. and Smeins, F. E., 1991. Ecosystem-level processes, in *Grazing Management: An Ecological Perspective,* R. K. Heitschmidt and J. W. Stuth, Eds., Timber Press, Portland, OR, 109–139.

Asay, K. H., 1987. Revegetation in the sagebrush ecosystem, in *Integrated Pest Management on Rangeland: State of the Art in the Sagebrush Ecosystem,* J. A. Onsager, Ed., U.S. Dept. Agric., Agric. Research Service, ARS-50, Washington, D.C., 19–27.

Belsky, J., 1997. Quotation in Durbin, K., New plan draws hisses, boos, *High Country News,* 23 June.

Billings, W. D., 1990. *Bromus tectorum,* a biotic cause of ecosystem impoverishment in the Great Basin, in *The Earth in Transition: Patterns and Processes of Biotic Impoverishment,* G. M. Woodwell, Ed., Cambridge University Press, New York, 301–322.

Blackburn, W. H., Pierson, F. B., Hanson, C. L., Thurow, T. L., and Hansen, A. L., 1992. The spatial and temporal influence of vegetation on surface soil factors in semi-arid rangelands, *Trans. Am. Soc. Agric. Eng.,* 35:479–486.

Blaisdell, J. P., Murray, R. B., and McArthur, E. D., 1982. Managing Intermountain Rangelands — Sagebrush Grass Ranges, U.S. Dept. Agric. Forest Service, Intermountain Forest and Range Expt. Sta. Gen. Tech. Rep. INT-134, Ogden, UT.

Bock, C. E., Bock, J. H., and Smith, H. M., 1993. Proposal for a system of federal livestock exclosures on public rangelands in the western U.S., *Conserv. Biol.,* 7:731–733.

Bohart, G. N. and Knowlton, G. E., 1976. Invertebrates, in Final Environmental Impact Statement for the Sodium Cooled Class Three Experimental Reactor, App. D-IV, Idaho Nat. Engineering Laboratory, D-47–D-58.

Bork, E. W., West, N. E., and Walker, J. W., 1998. Three-tip sagebrush steppe responses to long-term seasonal sheep grazing, *J. Range Manage.,* 51:293–300.

Botkin, D. B., 1990. *Discordant Harmonies: A New Ecology for the 21st Century,* Oxford University Press, New York.

Brown, J. H. and McDonald, W., 1995. Livestock grazing and conservation on southwestern rangelands, *Conserv. Biol.,* 9:1644–1647.

Brussard, P. F., Murphy, D. D., and Tracy, C. R., 1994. Cattle and conservation biology — another view, *Conserv. Biol.,* 8:919–921.

Brussard, P. F., Murphy, D. D., and Tracy, C. R., 1995. Letter, *Conserv. Biol.,* 9:239.

Burkhardt, J. W., 1996. Herbivory in the Intermountain West, Bull. 58, Idaho Forest Wildlife and Range Expt. Sta., University of Idaho, Moscow, ID.

Caldwell, M. M. and Richards, J. H., 1990. Hydraulic lift: water efflux from upper roots improves effectiveness of water uptake by deep roots, *Oecologia,* 79:1–5.

Call, M. W. and Maser, C., 1985. Wildlife Habitats in Managed Rangelands — the Great Basin of Southeastern Oregon: Sagegrouse, Gen. Tech. Rep. PNW-187, U.S. Department of Agriculture, Forest Service, Pacific Northwest Forest and Range Expt. Sta., Portland, OR.

Chapin, F. S. III, Walker, B. H., Hobbs, R. J., Hooper, D. U., Lawton, J. H., Sala, O. E., and Tilman, D., 1997. Biotic control over the functioning of ecosystem, *Science,* 277:500–504.

Coufal, J. E., 1997. Biodiversity and environmental ethics: a personal reflection, *J. Am. Water Res. Assoc.,* 33:13–19.

Cumming, D. H. M., 1993. Summary: biodiversity in rangelands and grasslands, in *Grasslands for Our World,* Vol. 17, M. J. Baker, Ed., SIR Publ., Wellington, New Zealand, 791–793.

Daubenmire, R. F., 1970. Steppe Vegetation of Washington, Bull. 62, Washington Agric. Expt. Sta., Pullman.

Daubenmire, R. F., 1975. An analysis of structural and functional characteristics along a steppe forest catena, *N.W. Sci.,* 49:122–140.

Dealy, J. E., Lechenby, D. A., and Concannon, D. M., 1981. Wildlife Habitats in Managed Rangelands of the Great Basin of Southeastern Oregon. Plant Communities and Their Importance to Wildlife, Gen. Tech. Rep. PNW-160, U.S. Department of Agriculture Forest Service, Pacific Northwest Forest and Range Expt. Sta., Portland, OR.

Dobson, A. P., Bradshaw, A. D., and Baker, A. J. M., 1997. Hopes for the future: restoration ecology and conservation biology, *Science,* 277:515–522.

Elmore, W. and Kauffman, B., 1994. Riparian and watershed systems: degradation and restoration, in *Ecological Implications of Livestock Herbivory in the West,* M. Vavra, Ed., Society for Range Management, Denver, CO, 212–231.

Eng, R. L., Pitcher, E. J., Scott, S. J., and Greene, R. J., 1979. Minimizing the effect of surface coal mining on a sagegrouse population by a directed shift of breeding activities, in *The Mitigation Symposium,* Gen. Tech. Rep. RM-65, U.S. Department of Agriculture, Forest Service, Rocky Mtn. Forest and Range Expt. Sta., Ft. Collins, CO, 464–468.

Evans, R. A., Young, J. A., and Eckert, R. E., Jr., 1979. Herbicides as a management tool, in *The Sagebrush Ecosystem: A Symposium,* Utah State University, Logan, 110–116.

Fay, P. K., McElligott, V. T., and Havstad, K. M., 1989. Containment of free-ranging goats using pulsed radio-wave activated shock collars, *Appl. Anim. Behav. Sci.,* 23:165–171.

Fleischner, T. L., 1994. Ecological costs of livestock grazing in western North America, *Conserv. Biol.,* 8:629–644.

Forest-Range Task Force, 1972. The Nation's Range Resources — A Forest-Range Environmental Study, Forest Resource Rep. 19, U.S. Department of Agriculture, Forest Service, Washington, D.C.

Friedel, M., 1994. Can pastoralism and biodiversity coexist in rangelands? in *Proc. 8th Biennial Conference,* Australian Rangeland Society, Katherine, N.T.

Friedel, M. H. and James, C. D., 1995. How does grazing of native pastures affect their biodiversity?, in *Conserving Biodiversity, Threats and Solutions,* R. A. Bradstock, T. D. Auld, D. A. Keith, R. T. Kingsford, D. Lunney, and D. P. Sivertsen, Eds., Surrey, Beatty & Sons, Sydney, Australia, 249–259.

Graul, W. D., 1980. Grassland management practices and bird communities, in *Management of Western Forests and Grasslands for Nongame Birds,* Gen. Tech. Rep. INT-86, R. M. DeGraaf and N. G. Tilghman, Eds., U.S. Department of Agriculture, Forest Service Rocky Mtn. Forest and Range Expt. Sta., Ft. Collins, CO, 38–47.

Grayson, D. K., 1991. Late Pleistocene mammalian extinctions in North America: taxonomy, chronology and explanations, *J. World Hist.,* 5:193–232.

Griffiths, D., 1902. Forage conditions on the northern border of the Great Basin, being a report upon investigations made during July–August, 1902, in the region between Winnemucca, Nevada and Ontario, Oregon, Bull. 15, U.S. Department of Agriculture, Bureau of Plant Industry, Washington, D.C.

Harniss, R. O. and Murray, R. B., 1973. 30 years of vegetal change following burning of sagebrush-grass range, *J. Range Manage.*, 26:322–325.

Harper, K. T., 1986. Historical environments, in *Handbook of North American Indians,* Vol. II, *Great Basin,* W. L. D'Azeredo, Ed., Smithsonian Institution Press, Washington, D.C., 51–63.

Harper, K. T. and Climer, C. S., 1985. Factors affecting productivity and compositional stability of *Artemisia* steppes in Idaho and Utah, U.S.A., in *Proceedings XV International Grassland Congress,* Kyoto, Japan, 592–594.

Harte, J., Torn, M., and Jensen, D., 1992. The nature and consequences of indirect linkages between climate change and biological diversity, in *Global Warming and Biological Diversity,* R. L. Peters and T. E. Lovejoy, Eds., Yale University Press, New Haven, CT.

Hassan, M. A. and West, N. E., 1986. Dynamics of soil seed pools in burned and unburned sagebrush semi-deserts, *Ecology,* 76:269–272.

Hatton, T. J. and West, N. E., 1987. Early seral trends in plant communities on a surface coal mine in southwestern Wyoming, *Vegetatio,* 73:21–29.

Haws, B. A., 1987. The status of IPM strategies for controlling grass bugs infesting introduced grassland monocultures, in *Integrated Pest Management on Rangeland: State of the Art in the Sagebrush Ecosystem,* J. A. Onsager, Ed., U.S. Department of Agriculture, Agriculture Research Service, ARS-50, 67–72.

Haynes, R. W., Graham, R. T., and Quigley, T. M., 1996. A Framework for Ecosystem Management in the Interior Columbia Basin, Gen. Tech. Rep. PNW-GTR-374, U.S. Department of Agriculture, Forest Service, Pacific N.W. Research Sta., Portland, OR.

Hinds, W. T. and Sauer, R. H., 1974. Soil erodibility, soil erosion and revegetation following wildfire in a shrub-steppe community, in *Proc. Symposium on Atmospheric-Surface Exchange of Particulate and Gaseous Pollutants,* U.S. Energy Research and Develop. Admin., Richland, WA.

Holechek, J. L., Pieper, R. D., and Herbel, C. H., 1989. *Range Management: Principles and Practices,* Prentice-Hall, Engelwood Cliffs, NJ.

Hosten, P. E. and West, N. E., 1996. Using a piosphere approach to examine change in sagebrush steppe along gradients of livestock impact in north Laidlaw Park, Idaho, in *Proc. Vth International Rangeland Congress,* Vol. 1, Society for Range Management, Denver, CO, 248–249.

Howery, L. D., Provenza, F. D., Banner, R. E., and Scott, C. B., 1998. Social and environmental manipulations by man and nature affect cattle dispersion on rangeland, *Appl. Anim. Behav. Sci.,* 55:231–244.

Hull, A. C., Jr., 1976. Rangeland use and management in the Mormon West, in *Symposium on Agriculture, Food, and Man – A Century of Progress,* Brigham Young University, Provo, UT, 21–31.

James, C. D., Landsberg, J., Morton, S. R., Stol, J., Drew, A., Tongway, H., and Hobbs, T., 1997. Changes in biodiversity along gradients of grazing intensity in Australian rangelands, *Bull. Ecol. Soc. Am.,* 78:116.

Johnson, H. B. and Mayeaux, H. S., 1992. A view on species additions and deletions and the balance of nature, *J. Range Manage.,* 45:322–333.

Johnson, K. L., 1986. *Crested Wheatgrass: Its Values, Problems, and Myths,* Utah State University, Logan.

Kamm, J. A., Sneva, F. A., and Rittenhouse, L. M., 1978. Insect grazers on the cold desert biome, in *Proc. 1st International Rangelands Congress,* Society for Range Management, Denver, CO, 479–483.

Kerr, A., 1994. Don't try to improve grazing; abolish it! *High Country News,* June 13:15.

Kessler, W. B., Salwasser, H., Cartwright, C. W., Jr., and Caplan, J. A., 1992. New perspectives for sustainable natural resources management, *Ecol. Appl.,* 2:221–225.

Kindschy, R. R., 1978. Rangeland management practices and bird habitat values, in Proceedings of a Workshop on Non-game Bird Habitat in the Coniferous Forests of the Western U.S., Gen. Tech. Rep. PNW-64, U.S. Department of Agriculture, Forest Service, Pacific Northwest Forest and Range Expt. Sta., Portland, OR, 66–69.

Laycock, W. A., 1995. New perspectives on ecological condition of rangelands: can state and transition or other models better define condition and diversity?, in *Proc. Int. Workshop on Plant Genetic Resources, Desertification and Sustainability,* L. Montes and G. E. Oliva, Eds., INTA-EEA, Rio Gallegos, Argentina.

Laycock, W. A., Loper, D., Obermiller, F. W., Smith, L., Swanson, S. R., Urness, P. J., and Vavra, M., 1996. *Grazing on Public Lands,* Council for Agricultural Science and Technology, Ames, IA.

Lesica, P. and DeLuca, T. H., 1996. Long-term harmful effects of crested wheatgrass on Great Plains grassland ecosystems, *J. Soil Water Conserv.,* 51:408–409.

Machlis, G. E., Forester, D. J., and McKendry, J. E., 1994. Biodiversity Gap Analysis: Critical Challenges and Solutions, Contribution No. 736, Idaho Forest Wildlife and Range Experiment Station, University of Idaho, Moscow.

Mack, R. N., 1977. Mineral return via litter of *Artemisia tridentata, Am. Midl. Nat.,* 97:189–197.

Mack, R. N., 1981. Invasion of *Bromus tectorum* L. into western North America: an ecological chronicle, *Agroecosystems,* 7:145–165.

Mack, R. N. and Thompson, J. N., 1982. Evolution in steppe with few large-hooved mammals, *Am. Nat.,* 119:757–773.

Maser, C., Thomas, J. W., and Anderson, R. G., 1984. Wildlife Habitats in Managed Rangelands — The Great Basin of Southeastern Oregon. The Relationship of Terrestrial Vertebrates to Plant Communities, Parts 1 & 2, Gen. Tech. Rep. PNW-172, U.S. Department of Agriculture Forest Service, Pacific N.W. Forest and Range Expt. Sta, Portland, OR.

Matson, P. A., Parton, W. J., Power, A. G., and Swift, M. J., 1997. Agricultural intensification and ecosystem properties, *Science,* 277:504–509.

McArdle, R. F., 1936. The whiteman's toll, in Letter from the Secretary of Agriculture to the U.S. Senate — A Report on the Western Range — A Great but Neglected Resource, *Senate Document,* 199:81–116.

McArdle, R. F. and Costello, D. F., 1936. The virgin range, in Letter from the Secretary of Agriculture to the U.S. Senate — A Report on the Western Range — A Great but Neglected Resource, *Senate Document,* 199:71–80

McArthur, E. D., 1983. Taxonomy, origin, and distribution of big sagebrush and allies (subgenus Tridentatae), in *Proceedings of the First Utah Shrub Ecology Workshop,* K. L. Johnson, Ed., Utah State University, Logan, 3–13.

McEwen, L. C. and DeWeese, L. R., 1987. Wildlife and pest control in the sagebrush ecosystem: basic ecology and management considerations, in *Integrated Pest Management on Rangeland: State of the Art in the Sagebrush Ecosystem,* J. A. Onsager, Ed., U.S. Department of Agriculture, Agric. Research Service, ARS-50, Washington, D.C., 76–85.

Mehringer, P. J. and Wigand, P. E., 1990. Comparison of late Holocene environments from woodrat middens and pollen: Diamond Craters, Oregon, in *Packrat Middens: The Last 40,000 Years of Biotic Change,* J. L. Betancourt, Ed., University of Arizona Press, Tucson, AZ, 294–325.

Miller, R. F., Svejcar, T. J., and West, N. E., 1994. Implications of livestock grazing in the Intermountain sagebrush region: plant composition, in *Ecological Implications of Livestock Herbivory in the West,* M. Vavra, Ed., Society for Range Management, Denver, CO, 101–146.

Morton, S. R., Stafford Smith, D. M., Friedel, M. H., Griffin, G. F., and Pickup, G., 1995. The stewardship of arid Australia: ecology and landscape management, *J. Environ. Manage.,* 43:195–217.

Mosely, J. C., 1996. Prescribed sheep grazing to suppress cheatgrass: a review, *Sheep Goat Res. J.,* 12:74–81.

Mueggler, W. F., 1967. Vole damage big sagebrush in southwestern Montana, *J. Range Manage.,* 20:88–90.

Mueggler, W. F., 1982. Diversity of western rangelands, in *Natural Diversity in Forest Ecosystems,* J. L. Cooley and J. H. Cooley, Eds., University of Georgia, Athens, 211–217.

Noss, R. F. and Cooperrider, A. Y., 1994. *Saving Nature's Legacy: Protecting and Restoring Biodiversity,* Island Press, Washington, D.C.

Olson, R. A., 1974. Bird populations in relation to changes in land use in Curlew Valley, Idaho and Utah. M.S. thesis, Idaho State University, Pocatello.

Parmenter, R. R. and MacMahon, J. A., 1983. Factors determining the abundance and distribution of rodents in a shrub-steppe ecosystem: the role of shrubs, *Oecologia,* 59:145–156.

Parmenter, R. R., Mesch, M. R., and MacMahon, J. E., 1987. Shrub litter production in a sagebrush-steppe ecosystem: rodent population cycles as a regulating factor, *J. Range Manage.,* 40:50–54.

Passey, H. B., Hugie, V. K., Williams, E. W., and Ball, D. E., 1982. Relationships between soil, plant community, and climate on rangelands of the Intermountain West, Tech. Bull. 1669, U.S. Department of Agriculture Soil Conservation Service.

Pellant, M., 1990. The cheatgrass wildfire cycle — are there any solutions? In Proceedings of the Symposium on Cheatgrass Invasion, Shrub Die-Off and Other Aspects of Shrub Biology and Management, E. D. McArthur, Ed., Gen. Tech. Rep. INT-276, U.S. Department of Agriculture Forest Service Intermountain Forest and Range Expt. Sta., Ogden, UT, 11–18.

Pieper, R. D., 1994. Ecological implications of livestock, in *Ecological Implications of Livestock Herbivory in the West,* M. Vavra, W. A. Laycock, and R. D. Pieper., Eds., Society for Range Management, Denver, CO, 177–211.

Pierson, F. B. and Wight, J. R., 1991. Variability of near-surface temperature on sagebrush rangeland, *J. Range Manage.,* 44:491–497.

Pimm, S. L., 1991. *The Balance of Nature? Ecological Issues in the Conservation of Species and Communities,* University of Chicago Press, Chicago.

Polley, H. W., 1997. Implications of rising atmospheric carbon dioxide concentration for rangelands, *J. Range Manage.,* 50:561–577.

Quigley, T. M., Lee, K. M., and Arbelbide, S. J., 1997. Evaluation of EIS Alternatives by the Science Integration Team, Vols. I & II, Report PNW-GTR-406, U.S. Department of Agriculture, Forest Service, Pacific N.W. Research Station, Portland, OR.

Reynolds, T. D., 1979. Response of reptile populations to different land management practices on the Idaho National Engineering Laboratory site, *Great Basin Nat.,* 39:255–262.

Reynolds, T. D. and Trost, C. H., 1979. The effect of crested wheatgrass plantings on wildlife on the Idaho Nat. Engineer. Lab site, in The Mitigation Symposium: A National Workshop on Mitigating Losses of Fish and Wildlife Habitats, Gen. Tech. Rep. RM-65, U.S. Department of Agriculture Forest Service, Rocky Mtn. Forest & Range Expt. Sta., Ft. Collins, CO, 665–666.

Reynolds, T. D. and Trost, C. H., 1980. The response of native vertebrate populations to crested wheatgrass planting and grazing by sheep, *J. Range Manage.,* 33:122–125.

Reynolds, T. D. and Trost, C. H., 1981. Grazing crested wheatgrass and bird populations in southeastern Idaho, *Northwest Sci.,* 55:225–234.

Rice, B. and Westoby, M., 1978. Vegetative responses of some Great Basin shrub communities protected against jackrabbits or domestic stock, *J. Range Manage.,* 31:28–34.

Rickard, W. H. and Vaughn, B. E., 1988. Plant communities: characteristics and responses, in *Shrub-Steppe: Balance and Change in a Semi-Arid Terrestrial Ecosystem,* W. H. Rickard, Ed., Elsevier, Amsterdam, 109–179.

Scott, J. M., Davis, F., Csuti, B., Noss, R., Butterfield, B., Groves, C., Anderson, H., Caicco, S., D'Erchia, F., Edwards, T. C., Jr., Ulliman, J., and Wright, J. G., 1993. Gap analysis: a geographic approach to protection of biological diversity, *Wildlife Monogr.,* 12:1–41.

Short, H. L. and Hestbeck, J. B., 1995. National biotic resource inventories and GAP analysis, *BioScience,* 45:535–539.

Sneva, F. A., Rittenhouse, L. R., Tueller, P. T., and Reece, P., 1984. Changes in Protected and Grazed Sagebrush-Grass in Eastern Oregon 1937–1974, Oregon Agric. Expt. Sta. Bull. 663.

Sturges, D. L., 1977. Snow accumulation and melt in sprayed and undisturbed big sagebrush vegetation, Res. Note RM-348, U.S. Department of Agriculture Forest Service, Rocky Mtn. Forest and Range Expt. Sta., Ft. Collins, CO.

Thomas, J. W., Maser, C., and Rodisk, J. E., 1979. Wildlife Habitats in Managed Rangelands. The Great Basin of Southeastern Oregon: Edges, Gen. Tech. Rep. PNW-85, U.S. Department of Agriculture, Forest Service, Pacific Northwest Forest and Range Expt. Sta., Portland, OR.

Tilman, D., Wedin, D., and Knops, J., 1996. Productivity and sustainability influenced by biodiversity in grassland ecosystems, *Nature,* 379:718–720.

Tisdale, E. W., Hironaka, M., and Fosberg, M., 1965. An area of pristine vegetation in Craters of the Moon National Monument, Idaho, *Ecology,* 46:349–352.

Vitousek, P. M., Mooney, H. A., Lubchenko, J., and Melillo, J. M., 1997. Human domination of earth ecosystems, *Science,* 277:494–499.

West, N. E., 1979. Basic synecological relationships of sagebrush-dominated lands in the Great Basin and Colorado Plateau, in *The Sagebrush Ecosystem,* Utah State University, Logan, 33–41.

West, N. E., 1983. Western intermountain sagebrush steppe, in *Temperate Deserts and Semi-deserts,* N. E. West, Ed., Elsevier Science Publishers, Amsterdam, 351–374.

West, N. E., 1985. Aboveground litter production of three temperate semidesert shrubs, *Am. Midl. Nat.,* 113:158–169.

West, N. E., 1990. Structure and function of soil microphytic crusts in wildland ecosystems of arid to semi-arid regions, *Adv. Ecol. Res.,* 20:179–223.

West, N. E., 1991. Nutrient cycling in soils of semiarid and arid regions, in *Semiarid Lands and Deserts: Soil Resource and Rehabilitation,* J. Skujins, Ed., Marcel Dekker, New York, 295–352.

West, N. E., 1993. Biodiversity of rangelands, *J. Range Manage.,* 46:2–13.

West, N. E., 1995. Biodiversity on rangelands: definitions and values, in *Biodiversity on Rangelands,* N. E. West, Ed., Utah State University, College of Natural Resources, Logan, 1–4.

West, N. E. and Skujins, J., 1978. *Nitrogen in Desert Ecosystems,* Dowden, Hutchinson and Ross, Stroudsburg, PA.

West, N. E. and Whitford, W. G., 1995. The intersection of ecosystem and biodiversity concerns in the management of rangelands, in *Biodiversity of Rangelands,* N. E. West, Ed., Natural Resource Studies IV. College of Natural Resources, Utah State University, Logan, 72–79.

West, N. E. and Young, J. A., 1998. Vegetation of Intermountain valleys and lower mountain slopes, in *North American Terrestrial Vegetation,* 2nd ed., M. A. Barbour and W. D. Billings, Eds., Cambridge University Press, New York, (in press).

West, N. E., Griffin, R. A., and Jurinak, J. J., 1984a. Comparison of phosphorus distribution and cycling between adjacent native and cultivated grass-dominated ecosystems, *Plant Soil,* 81:151–164.

West, N. E., Provenza, F. D., Johnson, P. S., and Owens, N. K., 1984b. Vegetation change after 13 years of livestock grazing exclusion on sagebrush semi-desert in central Utah, *J. Range Manage.,* 37:262–264.

West, N. E., Stark, J., Johnson, D., Abrams, M., Wight, R., Heggem, D., and Peck, S., 1994. Effects of climatic change on the edaphic features of arid and semi-arid lands of western North America, *Arid Soils Res. Rehabil.,* 8:307–351.

Westoby, M. and Wagner, F. H., 1973. Use of crested wheatgrass seedings by black-tailed jackrabbits, *J. Range Manage.,* 26:349–351.

Whisenant, S. G., 1990. Changing fire frequencies on Idaho's Snake River Plains; ecological and management implications, in Proceedings Symposium on Cheatgrass Invasion, Shrub Die-off, and Other Aspects of Shrub Biology and Management, E. D. McArthur, Ed., Gen. Tech. Rep. INT-270, U.S. Department of Agriculture Forest Service, Intermountain Research Sta., Ogden, UT, 4–10.

Wicklow-Howard, M., 1989. The occurrence of vesicular-arbuscular mycorrhizal in burned areas on the Snake River Birds of Prey Area, Idaho, *Mycotaxon,* 34:253–257.

Wiens, J. A., Cates, R. G., Rotenberry, J. T., Cobb, N., Van Horne, B., and Redok, R. A., 1991. Arthropod dynamics on sagebrush (*Artemisia tridentata*): effects of plant chemistry and avian predation, *Ecol. Monogr.,* 61:299–321.

Wilson, E. O., 1987. The little things that run the world, *Conserv. Biol.,* 1:344–346.

Winward, A. H., 1991. A renewed commitment to management of sagebrush-grasslands, in *Management in the Sagebrush Steppe,* Oregon Agric. Expt. Sta. Spec. Rep. 880, 2–7.

Wolfe, M. L., Simonds, G. E., Danver, R., and Hopkin, W. J., 1997. Integrating livestock production and wildlife in a sagebrush-grass ecosystem, in Ecosystem Disturbance and Wildlife Conservation in Western Grasslands: A Symposium Proceedings, D. M. Finch, Ed., Gen. Tech. Rep. 285, U.S. Department of Agriculture, Forest Service, Rocky Mountain Forest and Range Expt. Sta., Ft. Collins, CO, 73–77.

Yorks, T. P. and McMullen, C., 1980. *Western Shrub and Grassland,* Vol. II of the descriptive supplement to *FRODAS: An Integrated Resource Data Analysis System for the Forest-Range Natural Resource Ecology Lab,* Colorado State University, Ft. Collins, CO.

Young, J. A., 1989. Intermountain shrub steppe plant communities — pristine and grazed, in *Proceedings of the Western Raptor Management Symposium and Workshop,* B. G. Pendleton, Ed., Natl. Wildl. Fed. Sci. & Tech. Series No. 12, Washington, D.C., 3–14.

Young, R. P., 1983. Fire as a management tool in rangelands of the Intermountain Region, in *Managing Intermountain Rangelands — Improvement of Range and Wildlife Habitats,* Gen. Tech. Rep. INT-157, U.S. Department of Agriculture Forest Service, Intermountain Forest and Range Expt. Sta., Ogden, UT, 18–31.

Zamora, B. and Tueller, P. T., 1974. *Artemisia arbuscula, A. longiloba,* and *A. nova* habitat types in northern Nevada, *Great Basin Nat.,* 33:225–242.

Zohary, M., 1973. *Geobotanical Foundation of the Middle East,* Gustav Fischer Verlag, Stuttgart, Germany.

CHAPTER 8

Agroforestry for Biodiversity in Farming Systems

Roger R. B. Leakey

CONTENTS

Abstract — Agroforestry can be used to diversify and intensify farming systems through the integration of indigenous trees producing marketable timber and nontimber forest products and is described in terms of an agroecological succession, in which climax agroforests are biodiverse, highly productive, and profitable. The role of biodiversity in agroecosystem function is one of the keystones of sustainability. Complex agroforests that combine profitability with biodiversity are presented as a model worthy of expansion. However, little is known ecologically about how best to integrate agroforestry into the landscape, or to what extent agroforestry can be used to link forest patches and expand biogeographical islands.

1-56670-290-9/99/$0.00+$.50

Tree domestication is one way to diversify and intensify agroforestry systems and to make them profitable. A wise domestication strategy for indigenous trees will involve the capture and maximization of intraspecific genetic diversity and so benefit both production and the environment.

INTRODUCTION

The numbers of plant and animal species on Earth represents only 0.1% of those that have existed since life appeared on this planet, the other 99.9% being already extinct as a result of five episodes of mass extinction over geological time (Leakey and Lewin, 1996) and the current period of extinctions (30,000/year against a background of 0.25/year) arising from our population growth and lifestyle. Land-use changes associated with colonization have been the major cause of these species losses over the last few centuries. Agriculture, which has been described as the "engine of economic growth" because of its powerful role in facilitating and stimulating growth of other sectors of the economy (Mellor et al., 1987), started as a subsistence activity 8000 years ago. The early subsistence systems are generally considered to have been sustainable, while large-scale, capital-intensive, modern agriculture has traded innate sustainability for chemical and other inputs and is characterized by deforestation and a decrease in the overall numbers of associated wild plants and animals to favor the growth of the planted crop. Typically, agriculture systems and forestry plantations are monocultures of staple food or tree crops with the almost total disappearance of the biodiversity and spatial complexity of natural ecosystems. Characteristically, these monocultures are also based on the few plant species that have been domesticated (Leakey and Tomich, 1998) and which also have a narrow genetic base. In recent years, these developments have given rise to concerns about deforestation, the loss of biodiversity, and the sustainability of our lifestyle and, particularly, the crucial food production systems that are essential to prevent famine and malnutrition in the tropics.

SUSTAINABLE PRODUCTION

Many of the means to increased productivity and profitability are now perceived by society as carrying too high a cost in social disruption, human inequity, and environmental degradation. The problem in trying to address this is how to define and quantify sustainability. Izac and Swift (1994) developed an operational framework to assess sustainability based on the premise that "a cropping system is sustainable if it has an acceptable level of production of harvestable yield which shows a non-declining trend from cropping cycle to cropping cycle over the long term." Their framework is based on the assessment of key ecological and economic parameters at the field, farming system, and village/catchment scales and the concept that a sustainable system never reaches threshold levels of irreversibility and that it achieves a sufficient level of economic efficiency and social welfare. One of the requirements identified by Izac and Swift (1994) is that the by-products (soil and water quality, biological diversity, etc.) of agricultural activities must not disrupt the biological functions of the system to the extent that the capacity of the system to absorb these disruptions is surpassed.

Sustainability thus involves a symbiosis between the properties of the ecosystem and the management activities that results in nondeclining and relatively stable outcomes. Izac and Swift (1994) consider that the key to this symbiosis lies in the assumed positive relationship of agroecosystem function to biodiversity and complexity. Biodiversity therefore is a keystone in sustainability, and its loss has been one of the common outcomes of agricultural intensification (Figure 1).

Agroforestry, through the replenishment of soil fertility and the domestication of indigenous trees producing marketable forest products, has been proposed as one way of diversifying and intensifying agroecosystems in a way that is beneficial to the environment and can maintain and perhaps enhance biodiversity (Sanchez and Leakey, 1998; Sanchez et al., 1998). In its *Medium Term Plan 1998–2000,* the International Centre for Research in Agroforestry (ICRAF, 1997) foresees that agroforestry can contribute to human welfare and environmental resilience, with improved systems providing:

1. Tree products that both increase food and nutritional security and generate cash income for poverty alleviation and
2. Services that support and enhance ecosystem function (Figure 2).

The relevant services of trees are those that increase the crop yields (nitrogen fixation, increased soil organic matter, nutrient cycling, soil conservation, etc.), create environmental resilience (niche diversification, food web complexity, carbon sequestration, reduced greenhouse gas emissions, etc.), and provide social benefits (boundary delineation, shade, etc.). Of these, the least is known about the ways in which trees enhance the environment, although the body of information is increasing (see Ingram, 1990; Swift et al., 1996).

AGROFORESTRY

Agroforestry, where it has been practiced traditionally, such as in the damar agroforests of Sumatra and Jungle Rubber on Kalimantan (Michon and de Foresta, 1996) and in the home gardens of Sri Lanka (Jacob and Alles, 1987), Nigeria (Okafor and Fernandes, 1987), and Tanzania (Fernandes et al. 1984), is a mixed and often apparently haphazard polyculture of indigenous trees and crops that form a complex, multistrata system somewhat like a natural forest. Interestingly, recent findings in the damar agroforests of Sumatra show that these complex multistrata agroforests contain over 50% of all the regional pool of resident tropical forest birds, most of the mammals, and about 70% of the plants (Table 1). They are also a major source of resins, fruits, and timber for domestic use and for export. Thus, these agroforests are potentially a sustainable resource, a valuable compromise between conservation

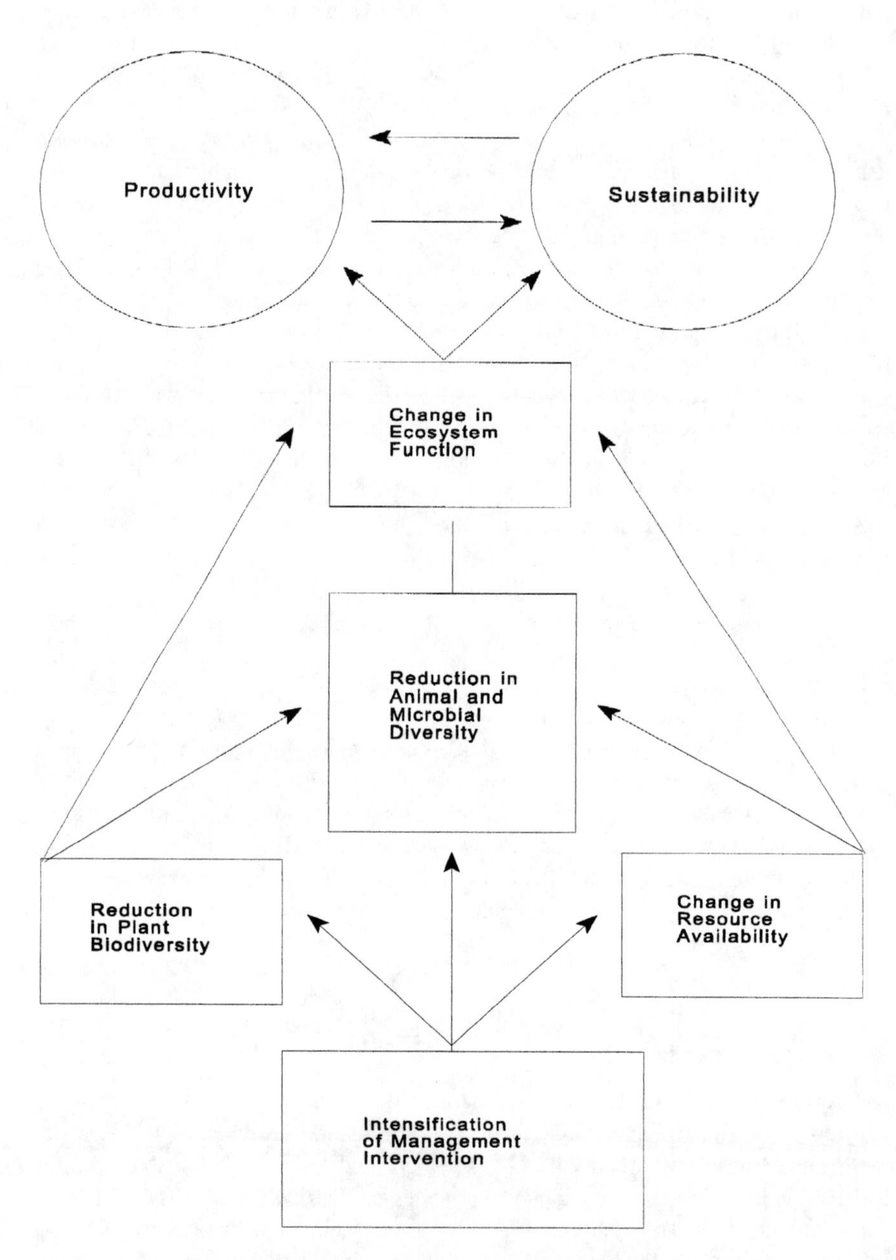

Figure 1 The impact of agricultural intensification on an agroecosystem. (From Swift, M. J. and Anderson, J. M., 1993. *Biodiversity and Ecosystem Function,* Schulze, E.-D. and Mooney, H. A., Eds., Springer-Verlag, Berlin. With permission.)

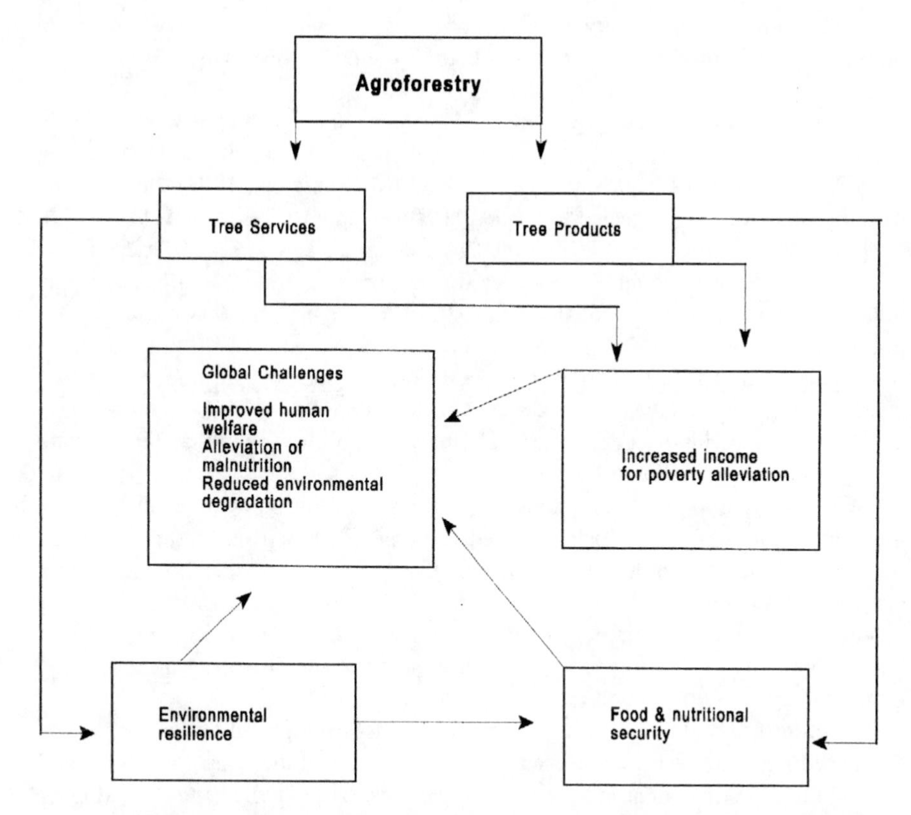

Figure 2 The relationship between the two functions of trees and the three goals of agroforestry to meet three global challenges. (Modified from ICRAF 1997, Int'l Centre for Research in Agroforestry Medium Term Plan 1998–2000. With permission.)

Table 1 Biodiversity in Indonesia Agroforests: Observed Numbers of Species

	Primary Forest	Rubber Agroforest	Damar Agroforest	Durian Agroforest	Rubber Plantation
Birds[a]	179	105	92	69	—
Collembola[b]					
Leaf litter	20.6	22.8	—	—	11.6
Soil	13.7	16.0	—	—	8.3
Mammals[c]	—	39	46	33	—
Trees[d]	171	92	—	—	1
Total plants[d]	382	266	—	—	6

[a] Thiollay, 1995.
[b] Deharveng, 1992.
[c] Sibuea and Herdimansyah, 1993.
[d] Michon and de Foresta, 1995.

of tropical forest biodiversity and profitable use of natural resources, since in addition to their biodiversity these multistrata damar agroforests in Sumatra are financially attractive.

Damar resins are utilized by industries in Indonesia or exported worldwide. In 1984, the export market represented one third of the harvested volume, a trade rising from 250 to 400 t/year between 1972 and 1983 (Michon et al., 1998). In 1994, the damar production was expected to reach 10,000 t (Dupain, 1994), at a value of U.S. $300 to 400/t. Of this trade, 80% is met by the damar agroforests. The economic value of the damar trade and its associated activities is of major significance to the villages around Krui. In 1993, the profits from damar production were U.S. $7.2 million from sales, U.S. $2.6 million from added value, and U.S. $1.4 million from wages. To this is added U.S. $0.3 million in profits made by Krui traders (Michon et al., 1998). This analysis excludes the locally consumed products from these agroforests, e.g., fruits, vegetables, spices, fuelwood, timber, palm thatching, rattan, bamboo, fibers, as well as paddy rice.

With the exception of plantation crops, many farming systems in the tropics, including traditional subsistence swidden farming, are based on mixtures and are frequently haphazard in their configuration and spacing. In contrast, monocultures are particularly prevalent in countries with temperate climates. It is not clear whether or not the tendency to complexity and random distribution of the components of farming systems in the tropics is a deliberate attempt by farmers in the tropics to mimic the diversity of natural ecosystems in order to minimize risk.

In contrast to traditional agroforestry, the recent development of agroforestry as a science by agronomists and foresters has tended to adopt the temperate model and to plant the tree component in lines, regular patterns, or along the contour of sloping land (see review by Cooper et al., 1996). This is especially the case in countries where farm size is large (e.g., Australia), where large areas of countryside are planted in geometric patterns.

Modern agroforestry has also tended to be a set of stand-alone technologies, that together form various land-use systems in which trees are sequentially or simultaneously integrated with crops and/or livestock (Nair, 1989). Recently, however, it has been suggested that agroforestry practices should be successional phases in the development of a productive and complex agroecosystem, akin to the succession of natural ecosystems (Leakey, 1996). In this way, trees producing different products can be used to fill niches in a mosaic of patches in the landscape, making the system ecologically more stable and biologically more diverse. It is anticipated that this diversity would increase with each phase of the agroecological succession.

Toward this end, current activities at ICRAF are focusing on the development of agroforestry as "a dynamic, ecologically-based, natural resource management system that, through the integration of trees on farms and in the landscape, diversifies and sustains production for increased social, economic and environmental benefits." One aspect of this is to determine, through the use of models, the best land-use options for agricultural productivity and biodiversity conservation: the choice between integration or segregation (van Noordwijk et al., 1995b).

In parallel with these developments in agroforestry there has also been a move to promote the domestication of indigenous trees, the "Cinderella" trees overlooked by science (Leakey and Newton, 1994a; Leakey and Jaenicke, 1995; Leakey and Izac, 1996).

Bringing the new ideas about agroforestry and about domestication together provides one with a new paradigm for sustainable land-use development that focuses on two aspects of biodiversity:

1. Diversifying agroecosystems
2. Capturing and enhancing intraspecific diversity

AGROFORESTRY AND THE DIVERSIFICATION OF AGROECOSYSTEMS

From past experience, domesticated trees are frequently grown in monocultures, but they could play an important role in species-rich multistrata agroforests (Leakey, 1996b). The development of multistrata systems that include cultivars of domesticated trees could increase the profitability of these agroforests. Thus this approach could, it seems, go a long way toward the establishment of land uses that will fulfill the needs of rural and urban populations for food and income, while maintaining much of the biological diversity of forests or rehabilitating degraded ecosystems.

Much research will be needed, however, to achieve this and to demonstrate that productivity and profitability are not necessarily environmentally damaging. Evidence already emerging from studies to develop viable alternatives to slash-and-burn agriculture suggests that the greenhouse gas emissions, especially methane, from areas where sources such as paddy fields are juxtaposed with perennial vegetation are lower than from areas monocropped with rice (van Noordwijk et al., 1995a). However, the successful establishment of trees on cleared sites is known to suffer from changes in the populations and species diversity of symbiotic microflora associated with land clearance (Leakey et al., 1993; Mason and Wilson, 1994), and similar changes probably occur in the beneficial micro- and mesofauna above- and belowground. Evidence exists for the negative effects of site clearance on soil fauna populations (Eggleton et al., 1995) and for the need to restore them to ensure soil fertility.

A challenge for agroforestry research is to develop economically and socially acceptable land-use systems that function like undisturbed ecosystems and maintain biodiversity. Could complex multistrata agroforests, like those of Sumatra, be developed in humid West Africa and in Latin America? The answer is almost certainly yes. Indeed, simple indigenous multistrata systems already exist, such as the cocoa farm, and the compound gardens of West Africa (Okafor and Fernandes, 1987), while in the Peruvian Amazon, multistrata agroforests have been found to be an economically attractive system (Table 2).

DEVELOPMENT OF MULTISTRATA AGROFORESTS

There are plenty of tree species that have traditionally provided local people with their daily needs for the full range of nontimber forest products, which could

Table 2 Comparison of Net Present Values and Internal Rates of Returns of Production Systems in 1985 and 1991 Prices Using a 15-Year Time Horizon in Yurimaguas, Peru

	1985 Prices		1991 Prices	
Production Option	Net Present Value (U.S.$ ha^{-1})	Internal Rate of Return (%)	Net Present Value (U.S.$ ha^{-1})	Internal Rate of Return (%)
Multistrata system	6444	219	6727	831
Peach palm	979	35	2061	64
Shifting cultivation	79	7	218	19

Note: Low input and high input continuous cultivation both had negative net present values.

From ICRAF, *1994 Annual Report,* Nairobi, Kenya, 1995.

now be incorporated into agroforestry systems. Table 3 illustrates candidate species for West Africa (see also Okafor and Lamb, 1994; Abbiw, 1990). How can these complex agroforests be further developed in the tropics? Probably three things are required. First, there is the need to identify the species of commercial importance of relevance to local markets. Second, there is a need to develop an entrepreneurial mentality among the community, who have traditionally been subsistence farmers and hunter–gatherers. Third, there is a need to determine how best the tree species may be combined to develop agroforests. The growth of the urban market and the absence of jobs in the urban areas may provide the commercial incentive required. What is probably missing is the demonstration of what is possible, particularly in the areas near urban markets.

With such a wide choice of species for the middle and upper strata of multistrata agroforests, clearly research to determine the best combinations and configurations is much needed, but also extremely difficult. There are, therefore, three research approaches that can be taken:

1. The testing of prototype systems (i.e., best-guess combinations), perhaps aimed at market needs, and developed with the help of farmers with some experience of compound gardens;
2. Research to test specific hypotheses aimed at the development of some principles regarding the optimal combinations and/or densities of trees in the different strata, which involves both complementarity of species biologically and in terms of labor demands and market opportunities;
3. Use of random mixtures of the species in unstructured combinations, as would probably be developed by farmers.

Research is needed to determine whether or not the apparently random distribution of trees in many existing examples of multistrata systems is indeed random, or whether farmers from experience grow species in certain combinations. Understanding this process would be of benefit in assisting attempts to know why multistrata systems evolve differently under different social and ecological conditions and would therefore help to transfer these systems more effectively to new areas.

Table 3 A Sample of the West African Tree/Shrub/Liane Species Appropriate for Growth in Multistrata Agroforests and for Domestication

	Common Names	Mature Height (m)
Anthocleista schweinfurthii	Ayinda	15–20
Antrocaryon micraster	Aprokuma/onzabili	40–50
Baillonella toxisperma	Moabi	45–55
Calamus spp.	Rattan	35–45
Canarium schweinfurthii	Aiele/Africa canarium/incense tree	45–55
Chrysophyllum albidum	Star apple	30–40
Cola acuminata	Kola nut	15–25
Cola lepidota	Monkey kola	10–20
Cola nitida	Kola nut	20–30
Coula edulis	Coula nut/African walnut	25–35
Dacryodes edulis	African plum/Safoutier	15–25
Entandrophragma spp.	Sapele/tiama/utile/sipo	50–60
Garcinia kola	Bitter kola	20–30
Gnetum africanum	Ero	0–10
Irvingia gabonensis	Bush mango/andok	20–30
Khaya spp.	African mahogany	50–60
Lovoa trichiloides	Bibolo/African walnut	40–50
Milicia excelsa	Iroko/mvule/odum	45–55
Nauclea diderichii	Opepe/kusia/bilinga	35–45
Pentaclethra macrophylla	Oil bean tree/Mubala/Ebé	20–30
Raphia hookeri and other spp.	Raphia palm	5–15
Ricinodendron heudelotii	Groundnut tree/nyangsang/essessang	40–50
Terminalia ivorensis	Framiré/Idigbo	45–55
Terminalia superba	Fraké/afara/limba	45–55
Tetrapleura tetraptera	Prekese/Akpa	20–30
Treculia africana	African breadfruit/etoup	20–30
Trichoscypha arborea	Anaku	15–25
Triplochiton scleroxylon	Ayous/obeche/wawa	55–65
Vernonia amydalina	Bitter leaf	0–10
Xylopia aethiopica	Spice tree	15–25

From Leakey, R. R. B., *Agroforestry Systems,* 1998. With permission.

There are a number of options on how to apply these approaches to the development of multistrata systems:

1. Enrichment planting within logged or degraded forest;
2. Planting in cleared forest land as a perennial tree alternative to slash-and-burn agriculture;
3. Planting under a plantation of either an upper or middle strata species.

Of these options, 1 and 3 have merit environmentally in that the system will be developed more quickly as a demonstration and will not leave the land bare at the establishment phase. On the other hand, option 2, the currently most practical from the perspective of smallholders, is probably the most relevant in terms of developing multistrata systems. Such systems are both relevant on good soils of land cleared at the forest margin and on already degraded land, perhaps with proximity to urban markets. Because this option is the most relevant practically, it is also probably the one on which the greatest research effort should be concentrated.

In all these options, research is especially required to determine the functional groups of species which will do well in the lower strata, where light will be a limiting factor. Currently in Southeast Asia, staple food crops are grown beside the multistrata agroforests, because they are light demanding. This may be the best arrangement, but it is also possible that new crops, or new cultivars of existing crops, could be integrated into multistrata agroforests.

BIODIVERSE AGROECOSYSTEMS

From the biodiversity viewpoint, there is a difference in agroecosystems between the *planned* biodiversity and the *unplanned,* or associated, biodiversity. The latter are all those organisms, above- and belowground, that have found niches to fill among the planted trees and crops. The extent to which unplanned biodiversity occurs in different agroecosystems is not well known or understood, although studies have started to address the effects of a broad spectrum of agricultural practices on wildlife populations (McLaughlin and Mineau, 1995; Perfecto and Snelling, 1995). Swift et al. (1996) have, however, drawn four very different scenarios for the relationships between agricultural intensification and biodiversity (Figure 3), although these may be very scale dependent (e.g., from farm to landscape) and probably also vary depending on the level of biodiversity at the time of planting. Thus, the biodiversity associated with an agroforest planted on recently cleared land at the forest margin, as an alternative to slash-and-burn agriculture, would almost certainly be very different from the same tree/crop mixture planted to rehabilitate already degraded land. There is a need for controlled experiments to determine these relationships between intensification and biodiversity. In addition there is a need to determine the patterns of diversity in different agroforestry systems and their implications for ecological functioning at different scales. There is currently unresolved debate about the functional role of species diversity in ecosystems (Johnson et al., 1996), which makes it difficult to suggest best practices. However, it seems that planned biodiversity should aim at maximizing ecosystem processes (nutrient cycling, production of different products, light requirements, etc.) and structural complexity, rather than increasing the number of species per se. However, species numbers may also impact on function. The few studies in which the diversity of agricultural ecosystems has been manipulated suggest that increases in diversity from 0 to 10 plant species alters ecosystem function, but that there is little effect beyond that point (Schulze and Mooney, 1993).

Species also vary in their importance in different food webs, where, once again, scale is important. For example, for a fruit tree species to support a population of monkeys will require a large habitat. A small area of fruit trees may therefore avoid monkey damage and so be preferable for production, but without a large area there may not be a market for the crop. Thus, ecological and economic factors affecting

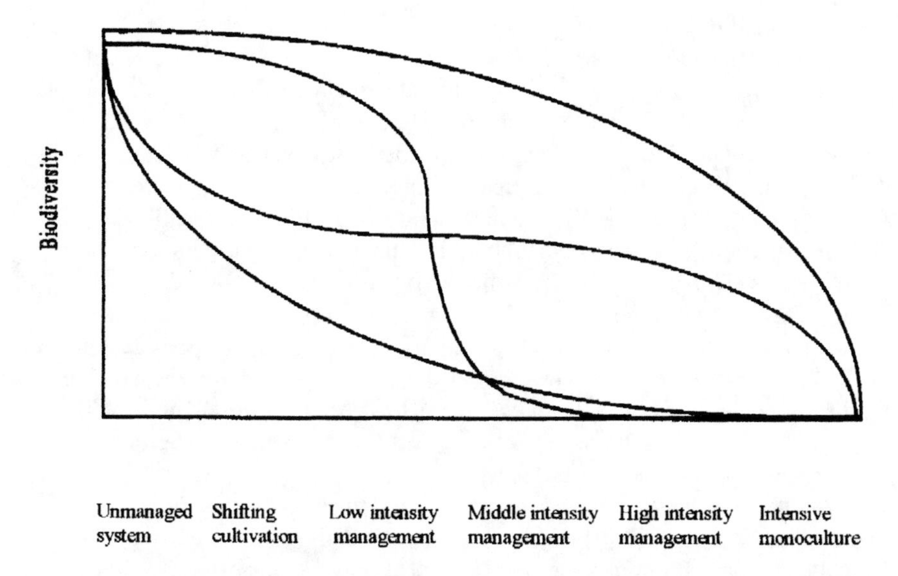

Figure 3 Hypothetical relationships between agricultural intensification and agroecosystem biodiversity. (From Swift, M. J. et al., 1996. *Functional Roles of Biodiversity: A Global Perspective,* Mooney, H. A. et al., Eds., SCOPE 55, John Wiley & Sons, New York. With permission.)

the viability of production systems can be antagonistic. Typically, predators require larger range than their prey; thus for the sustainable production of these fruits the area may need to be big enough to support leopards. Top predators are, however, often unwelcome in farmland while their absence creates pest management issues for farmers. In this context, a land-use mosaic with corridors between forest patches may be the most appropriate means of acquiring the scale needed to achieve some level of ecological equilibrium; however, the likely benefits of corridors between forest patches are not fully understood (Hobbs, 1992). This concept thus returns to issues of scale, land-use design, and the "integrate vs. segregate" debate (van Noordwijk et al., 1995b) in developing optimal land-use strategies.

FOREST PATCHES, BIOGEOGRAPHICAL ISLANDS, AND AGROFORESTRY

In the tropics, conservationists have focused their attention on the protection of natural forests and woodlands and, until recently (Schelas and Greenberg, 1996), have not given much attention to the widely dispersed forest patches throughout human-occupied landscapes. These patches are often critical components of a farmer's environment, being a source of products and environmental services of importance to the farmer's livelihood and welfare. As biogeographical islands, their role in maintaining biological diversity is also crucial. Issues of scale are central to this role, and thus, in a landscape mosaic, forest patches and areas of agroforestry are potentially complementary, especially when considering the need for ecological equilibrium and population size *vis à vis* genetic diversity. There is thus a need for ecologists and geneticists to become more involved in agroforestry research and the process of farmer reforestation. Biological and genetic diversity are, for example, becoming an issue in the densely populated, high-potential areas of Kenya. These are areas where the situation of "more people, less erosion" (Tiffen et al., 1994) is also becoming "more people, more trees" (Holmgren et al., 1994), as tree biomass is increasing at 4.7%/year, while the high human population is expanding at 3.0%. However, of concern is the fact that about half the trees planted on a farm are of a single species (*Grevillia robusta*), while the other half consists of about 10 to 20 species. This raises the question at what point the tree population reaches a size and distribution such that the controls that prevent isolated outbreaks of pests and disease break down. Not enough is known about the patch dynamics of such outbreaks relative to species numbers and configurations to steer agroforesters in the development of risk-averse strategies. The use of only a few tree species in agroforestry to provide corridors between forest patches could obviously allow their pests and diseases the opportunity to spread, while reducing the effectiveness of corridors for the ecological and genetic stability of other species. This reaffirms the need for both diversity and the need to consider the ecological implications of landscape design. In the case of *G. robusta*, which was introduced into East Africa from Australia, the risks of its dominance in the landscape could be exacerbated by its limited genetic base. Fortunately, new and more diverse accessions of *G. robusta* have now been introduced in East Africa, but it will take some time for these to be disseminated around the countryside. This example illustrates the need to ensure that agroforestry promotes the importance of diversity and not just the planting of trees by farmers.

Strictly, biogeographical islands are completely isolated populations, in which island size is positively related to biodiversity. The concept has, however, also been applied to shrinking habitats such as woodlands in farmland. Although Simberloff (1988) concluded that, in a wildlife conservation context, treating woodlands as islands in this way provided contradictory or inconclusive evidence, there does seem to be some merit in considering (1) how agroforestry could perhaps benefit biodiversity through increasing island sizes and (2) what might be the possible impacts of such action. It is not known, however, if the species–area relationships apply when habitats are being expanded. Can agroforestry, as a form of restoration ecology, reverse the 18 factors (e.g., low population density, inbreeding, catastrophe, competition, habitat destruction, disease, etc.) associated with species extinction listed by Soulé (1983)? Agroforestry can probably reverse some of them, but which of the 18 factors are the most crucial and which should these be the prime targets for the development of a healthy agroecosystem? At present, it is only genetic diversity that is getting any attention. The decrease in size of biogeographical islands lowers genetic diversity, increases the likelihood of inbreeding, and lowers the chances of future allopatric speciation (Soulé, 1986). Agroforestry, because of its production orientation, is promoting the retention and introduction of genetic diversity among the tree species planted by farmers. This aspect of tree domestication at least is one positive step toward a healthier and more productive environment. It is hoped that, as agroforestry is increasingly seen as applied ecology, other such steps will follow, such as the geographic expansion of the multistrata agroforestry systems already described.

DOMESTICATION OF TREES FOR TIMBER AND NONTIMBER FOREST PRODUCTS

Throughout the tropics there are numerous perennial woody species that have provided indigenous peoples with many of their daily needs for millennia. Many of these people have now left the land for urban life, but they still demand traditional food, medicines, and other natural products. These traditionally important woody plants are virtually undomesticated. These neglected "Cinderella" species have great genetic diversity and also play a key role in biological, chemical, and hydrological cycles, protecting soils and providing ecological niches. The food-producing species are also important for food security, especially in the dry season, as well as a source of vitamins and minerals critical for the health and nutrition of children and pregnant women.

There are four groups of wild trees, shrubs, and lianes which could be rapidly domesticated for agroforestry and which can be viewed as potentially important sources of income for farmers. These trees produce

1. The traditionally important wild foods, mostly fruits, nuts and leaves for vegetables;
2. The traditionally important fibers;
3. Locally and industrially important pharmaceuticals and other extractives such as gums and resins;
4. Commercially important quality timbers and woods.

The domestication of tree species is a dynamic process which develops from deciding which species to domesticate and proceeds through background socioeconomic studies, the collection of germplasm, genetic selection and improvement to the integration of domesticated species in land-use systems (Leakey and Newton, 1994b; Leakey and Jaenicke, 1995). Domestication is an ongoing process in which genetic and cultivation improvements are continuously refined. In genetic terms, domestication is accelerated and human-induced evolution. Domestication, however, is not only about selection. Simons et al. (1998) contend that it integrates the four key processes of the identification, production, management, and adoption of tree genetic resources.

Strategies for tree domestication will vary depending on the value of the products, the extent of intraspecific variation, and many other factors, but for high-value species such as those producing marketable forest products the vegetative propagation of superior genotypes identified from within the existing wild populations will be appropriate (Leakey and Simons, 1998a). This is the approach generally followed in horticulture. Thus, an individual with superior yield, fruit flavor, stem form, or wood quality can be mass produced by vegetative propagation. In this way it is possible to select those clones likely to develop above-average characteristics in any given trait. By a series of ongoing selections and an ever-increasing intensity of selection, it is also possible to achieve rapid and substantial genetic improvements. In situations where desirable traits are not easily identified, it is possible to multiply a number of copies of each plant and to establish these in field trials and then to observe the development of distinct genetic differences between each clone (Longman and Jeník, 1987).

When identifying and selecting trees for cloning by vegetative propagation, there are two important criteria. It is important, first, to ensure that those that are highly superior for the desired traits are chosen and, second, to ensure that the selected individuals are unrelated and as genetically diverse in other traits as possible. This calls for a risk-aversion strategy (see Leakey, 1991) that allows intensive selection (e.g., 1 out of 100 to 10,000 trees) from among trees from different populations, ideally from throughout the range of the species. Furthermore, this should be an ongoing rolling program of multiple-trait selection, in which more and more intensive selection is imposed through the addition of selection criteria for new traits while, at the same time, new sources of genetic stock are continually added as new accessions enter the program. These new accessions should come from further germ plasm exploration and from breeding programs and so continually broaden the genetic base of the planted trees. Molecular genetics techniques can be used to ensure that genetic diversity is maintained in the clones used for commercial production.

Selection procedures will vary depending on the product and the situation. For example, for indigenous fruits, rapid progress will be made if indigenous knowledge can be used. Usually rural people know which are the best individual trees in their area for yield, fruit size, or flavor. Thus, as with temperate apples, pears, etc., people can be asked to report the existence of superior trees, so reducing the task of screening large numbers of trees. On the other hand, for medicinal trees, it is more likely that a chemical screening process will be required, but the magnitude of this task can probably be reduced by starting on a population basis, since it is likely that trees from certain environments will be richer in the required metabolites. Meanwhile, for timber trees, log size and straightness are the first selection criteria. Various forms of "plus-tree" (i.e., elite genotype) provenance and progeny selection are well known. To these have recently been added some "predictive tests" which can be applied in the nursery as a procedure for mass screening from genetically diverse seedling populations.

COMMERCIALIZATION

For the success of a domestication program, there must be markets for the products. Many of the products from tropical trees are already sold on local and regional markets, and a few have broken through into the international marketplace. For example, while the pulp of the bush mango (*Irvingia gabonensis*) is eaten fresh, its extracted kernels and those of the related species, *I. wombolu*, are traded regionally throughout the year. From Cameroon, this trade extends to Nigeria, Equatorial Guinea, and Gabon (Falconer, 1990; Ndoye, 1995). Similarly, the trade of kola nuts extends from humid zone countries of West Africa up into the dry zone where there is a big demand by the Muslim community (Falconer, 1990). Chewing sticks likewise are traded northwards in West Africa, with a street value put on the trade from Kumasi market in Ghana of about U.S. $9 million/year (Falconer, 1992). Again, in Cameroon the bark of *Prunus africana* (Pygeum) is exported to Europe where pharmaceutical products estimated to be worth $150 million/year are produced for worldwide trade and treatment of prostate gland disorders (Cunningham and Mbenkum, 1993). In southern Africa, some indigenous fruits are marketed locally as wines and jams, with a liqueur from *Sclerocarya birrea* fruits (Amarula) now on the international market. In Amazonia, products from the peach palm (*Bactris gasipaes*) are also being exported. The palm heart trade has been estimated at around $50 million/year (Clement and Villachica, 1994), with the fruits also having a similar value.

Commercialization is both necessary and potentially harmful to farmers. Without it the market for products is small. On the other hand, it is potentially harmful to rural people if it expands to the point that outsiders with capital to invest come in and develop large-scale monocultural plantations for export markets.

Leakey and Izac (1996) have examined some of the issues requiring consideration to ensure the economic viability of small-scale production and under what conditions small-scale production can be competitive with large-scale production.

TREE DOMESTICATION IN PROGRESS

Since 1993, ICRAF has initiated programs to identify priority trees for domestication (Franzel et al., 1996) and started germplasm collection in the humid lowlands of West Africa, the semiarid lowlands of West Africa, the Miombo Woodlands of the southern Africa plateau, and in western Amazonia (Leakey and Simons, 1998b). Through these programs ICRAF is developing the concepts, strategies, and policies associated with capturing genetic diversity of a wide range of indigenous trees for growth in agroforestry systems. These will be developed as the programs evolve. In this way, it is hoped the promises of agroforestry to alleviate poverty and to mitigate environmental degradation and the loss of biodiversity will to be fulfilled.

CONCLUSIONS

The vision of agroforestry presented here is as an integrated land use that, through the capture of intraspecific diversity and the diversification of species on the farm, combines increases in productivity and income generation with environmental rehabilitation and the creation of biodiverse agroecosystems. In most places this is just a vision, but there are increasing numbers of examples where the vision is already a reality. The body of ecological data from agroforestry research is growing, and the research agenda is changing toward systems thinking. There is much to do to encourage these developments and the socioeconomic/policy conditions that promote them. There is, however, an urgent need for the reasons for hope to be heard above the shouts of doom and gloom about tropical forests, if the needs of growing human populations are to be met without sacrificing the biological diversity that keeps our ecosystems functional.

REFERENCES

Abbiw, D., 1990. *Useful Plants of Ghana: West African Uses of Wild and Cultivated Plants,* Intermediate Technology Publications, Royal Botanic Gardens, Kew, London.

Clement, C. R. and Villachica, H., 1994. Amazonian fruits and nuts: potential for domestication in various agroecosystems, in *Tropical Trees: Potential for Domestication and the Rebuilding of Forest Resources,* R. R. B. Leakey and A. C. Newton, Eds., HMSO, London, 230–238.

Cooper, P. J. M., Leakey, R. R. B., Rao, M. R., and Reynolds, L., 1996. Agroforestry and the mitigation of land degradation in the humid and sub-humid tropics of Africa, *Exp. Agric.,* 32:235–290.

Cunningham, A. B. and Mbenkum, F.T., 1993. Sustainability of harvesting *Prunus africana* bark in Cameroon: a medicinal plant in international trade, Report of WWF/UNESCO/Kew, People and Plants Programme, WWF, Godalming, Surrey, England.

Deharveng, L., 1992. Conservation of biodiversity in Indonesian agroforests, unpublished Report of SOFT-ORSTOM-BIOTROP quoted by Michon and de Foresta, 1995.

Dupain, D., 1994. *Une Regione Traditionallment Agroforestiere en Mutation: Le Pesisir,* CNGARC, Montpellier, France.

Eggleton, P., Bignell, D. E., Sands, W. I., Waite, B., and Wood, T., 1995. The species richness of termites (Isoptera) under differing levels of forest disturbance in Mbalmayo Forest Reserve, Southern Cameroon, *J. Trop. Ecol.,* 11:85–98.

Falconer, J., 1990. The major significance of "minor" forest products. The local use and value of forests in the West African humid forest zone, Community Forestry Note, No. 6, FAO, Rome.

Falconer, J., 1992. Non-timber forest products in Southern Ghana, ODA Forestry Series, No. 2, U.K. Overseas Development Administration, London.

Fernandes, E. C. M., O'Kting'ati, A., and Maghembe, J. A., 1984. The Chagga homegardens: a multistoreyed agroforestry cropping system on Mount Kilimanjaro (northern Tanzania), *Agrofor. Syst.,* 2:73–86.

Franzel, S., Jaenicke, H., and Janssen, W., 1996. Choosing the Right Trees. Setting Priorities for Multipurpose Tree Improvement, Research Report 8, ISNAR, The Hague.

Hobbs, R., 1992. The role of corridors in conservation: solution or bandwagon? *Trends Ecol. Evol.,* 7:389–392.

Holmgren, P., Masakha, E. J., and Sjöholm, H., 1994. Not all African land is being degraded: a recent survey of trees on farms in Kenya reveals rapidly increasing forest resources, *Ambio,* 23:390–395.

ICRAF, 1995. *1994 Annual Report,* ICRAF, Nairobi, Kenya.

ICRAF, 1997. *ICRAF Medium-Term Plan 1998-2000,* ICRAF, Nairobi, Kenya.

Ingram, J., 1990. The role of trees in maintaining and improving soil productivity — a review of the literature, in *Agroforestry for Sustainable Production: Economic Implications,* R. T. Prinsley, Ed., Commonwealth Science Council, London, 243–303.

Izac, A.-M. N. and Swift, M. J., 1994. On agricultural sustainability and its measurement in small-scale farming in sub-Saharan Africa, *Ecol. Econ.,* 11:105–125.

Jacob, V. J. and Alles, W. S., 1987. The Kandyan gardens of Sri Lanka, *Agrofor. Syst.*, 5:123–137.

Johnson, K. H., Vogt, K. A., Clark, H. J., Schmitz, O. J., and Vogt, D. J., 1996. Biodiversity and the productivity and stability of ecosystems, *Tree*, 11:372–377.

Leakey, R. E. and Lewin, R., 1996. *The Sixth Extinction: Biodiversity and Its Survival*, Weidenfeld and Nicholson, London.

Leakey, R. R. B., 1991. Towards a strategy for clonal forestry: some guidelines based on experience with tropical trees, in *Tree Breeding and Improvement*, J. E. Jackson, Ed., Royal Forestry Society of England, Wales and Northern Ireland, Tring, Herts, U.K., 27–42.

Leakey, R. R. B., 1996. Definition of agroforestry revisited, *Agrofor. Today*, 8:5–7.

Leakey, R. R. B., 1998. Agroforestry in the humid lowlands of West Africa: some reflections on future directions, *Agroforestry Systems*, (In press).

Leakey, R. R. B. and Izac, A.-M.N., 1996. Linkages between domestication and commercialization of non-timber forest products: implications for agroforestry, in *Domestication and Commercialization of Non-Timber Forest Products for Agroforestry, Non-Wood Forest Products*, No. 9, R. R. B. Leakey, A. B. Temu, M. Melnyk, and P. Vantomme, Eds., FAO, Rome, Italy, 1–7.

Leakey, R. R. B. and Jaenicke, H., 1995. The domestication of indigenous fruit trees: opportunities and challenges for agroforestry, in *Proceedings of 4th International BIO-REFOR Workshop*, Tampere, Finland, K. Suzuki, S. Sukurai, K. Ishii, and M. Norisada, Eds., BIO-REFOR, Tokyo, Japan, 15–26.

Leakey, R. R. B. and Newton, A. C., 1994a. Domestication of "Cinderella" species as a start of a woody-plant revolution, in *Tropical Trees: Potential for Domestication and the Rebuilding of Forest Resources*, R. R. B. Leakey and A. C. Newton, Eds., HMSO, London, 3–5.

Leakey, R. R. B. and Newton, A. C., 1994b. *Domestication of Tropical Trees for Timber and Non-timber Products*, MAB Digest 17, UNESCO, Paris.

Leakey, R. R. B. and Simons, A. J., 1998a. When does vegetative propagation provide a viable alternative to propagation by seed in forestry and agroforestry in the tropics and subtropics?, in *Problems of Forestry in Tropical and Sub-tropical Countries – The Procurement of Forestry Seed – The Example of Kenya*, 21 pp. (In press).

Leakey, R. R. B. and Simons, A. J., 1998b. The domestication and commercialization of indigenous trees in agroforestry for the alleviation of poverty, *Agrofor. Syst.*, 38:165–176.

Leakey, R. R. B. and Tomich, T. P., 1998. Domestication of tropical trees: a policy paradigm for sustainable development through smallholder agroforestry, in *Agroforestry in Sustainable Ecosystems*, L. E. Buck, J. P. Lassoie, and E. C. M. Fernandes, Eds., CRC Press/Lewis Publishers, Boca Raton, FL.

Leakey, R. R. B., Wilson, J., Newton, A. C., Mason, P. A., Dick, J. M., and Watt, A. D., 1993. The role of vegetative propagation, genetic selection, mycorrhizas and integrated pest management in the domestication of tropical trees, in *Proceedings of International BIO-REFOR Workshop*, Yogyakarta, Indonesia, 31–36.

Longman, K. A. and Jeník, J., 1987. *Tropical Forest and Its Environment*, 2nd ed., Longman Scientific Technical, Harlow, England.

Mason, P. A. and Wilson, J., 1994. Harnessing symbiotic associations: vesicular-arbuscular mycorrhizas, in *Tropical Trees: The Potential for Domestication and the Rebuilding of Forest Resources*, R. R. B. Leakey and A. C. Newton, Eds., HMSO, London, 165–175.

McLaughlin, A. and Mineau, P., 1995. The impact of agricultural practices on biodiversity, *Agric. Ecosys. Environ.*, 55:201–212.

Mellor, J. W., Delgado, C. L., and Blackie, M. J., 1987. Priorities for accelerating food production growth in Sub-Saharan Africa, in *Accelerating Food Production in Sub-Saharan Africa,* J. W. Mellor, C. L. Delgado, and M. J. Blackie, Eds., Johns Hopkins University Press, Baltimore, MD, 353–375.

Michon, G. and de Foresta, H., 1995. The Indonesian agroforest model. Forest resource management and biodiversity conservation, in *Conserving Biodiversity Outside Protected Areas: The Role of Traditional Agro-Ecosystems,* P. Halliday and D. A. Gilmour, Eds., IUCN, Gland, 90–106.

Michon, G. and de Foresta, H. 1996., The agroforest model as an alternative to the pure plantation model for domestication and commercialization of NTFPs, in *Domestication and Commercialization of Non-Timber Forest Products in Agroforestry Systems, Non-Wood Forest Products,* R. R. B. Leakey, A. B. Temu, and M. Melnyk, Eds., FAO, Rome, 160–175.

Michon, G., de Foresta, H., and Aliadi, H., 1998. Reinventing the forest: damar extraction and cultivation in Sumatra, Indonesia, in *People, Plants and Justice,* C. Zerner, Ed., Rainforest Alliance. (In press).

Nair, P. K., 1989. Agroforestry defined, in *Agroforestry Systems in the Tropics,* P. K. R. Nair, Ed., Kluwer Academic Publishers, Dordrecht, The Netherlands, 13–18.

Ndoye, O., 1995. *The Markets for Non-Timber Forest Products in the Humid Forest Zone of Cameroon and Its Borders: Structure, Conduct, Performance and Policy Implications,* CIFOR, Bogor, Indonesia.

Okafor, J. C. and Fernandes, E. C. M., 1987. Compound farms of southeastern Nigeria: a predominant agroforestry homegarden system with crops and small livestock, *Agrofor. Syst.,* 5:153–168.

Okafor, J. C. and Lamb, A., 1994. Fruit trees: diversity and conservation strategies, in *Tropical Trees: The Potential for Domestication and the Rebuilding of Forest Resources,* R. R. B. Leakey and A. C. Newton, Eds., HMSO, London, 34–41.

Perfecto, I. and Snelling, R., 1995. Biodiversity and the transformation of a tropical agroecosystem — ants in coffee plantations, *Ecol. Appl.,* 5:1084–1097.

Sanchez, P. A. and Leakey, R. R. B., 1998. Land-use transformation in Africa: three determinants for balancing food security with natural resources utilization, *Eur. J. Agron.,* (in press).

Sanchez, P. A., Buresh, R. J., and Leakey, R. R. B., 1998. Trees, soils and food security, *Philos. Trans. R. Soc. London,* (in press).

Schelas, J. and Greenberg, R., 1996. *Forest Patches in Tropical Landscapes,* Island Press, Washington, D.C.

Schulze, E.-D. and Mooney, H. A., 1993. Ecosystem function and biodiversity: a summary, in *Biodiversity and Ecosystem Function,* Ecological Series 99, E.-D., Schulze and H. A. Mooney, Eds., Springer Verlag, Berlin, 497–505.

Sibuea, T. and Herdimansyah, T. D., 1993. *The Variety of Mammal Species in the Agroforest Areas of Krui (Lampung), Muara Bungo (Jambi) and Maninjau (West Sumatra),* ORSTOM/HIMBIO, Paris.

Simberloff, D. S., 1988. The contribution of population and community biology to conservation science, *Annu. Rev. Ecol. Syst.,* 19:473–511.

Simons, A. J., Weber, J., and Maghembe, J. A., 1998. Genetic improvement of agroforestry trees, *Agrofor. Syst.,* (in press).

Soulé, M. E., 1983. What do we really know about extinction?, in *Genetics and Conservation,* C. M. Schonewald-Cox, S. M. Chambers, B. MacBryde, and W. L. Thomas, Eds., Benjamin/Cummings, Menlo Park, CA.

Soulé, M. E., 1986. *Conservation Biology: The Science of Scarcity and Diversity,* Sinaner Associates, Sunderland, MA.

Swift, M. J. and Anderson, J. M., 1993. Biodiversity and ecosystem function in agricultural systems, in *Biodiversity and Ecosystem Function,* E.-D. Schulze and H. A. Mooney, Eds., Springer Verlag, Berlin, 15–42.

Swift, M. J., Vandermeer, J., Ramakrishnan, P. S., Anderson, J. M., Ong, C. K., and Hawkins, B. A., 1996. Biodiversity and agroecosystem function, in *Functional Roles of Biodiversity: A Global Perspective,* H. A. Mooney, J. H. Cushman, E. Medina, O. E. Sala, and E.-D. Schulze, Eds., John Wiley and Sons, New York, 261–298.

Thiollay, J.-M., 1995. The role of traditional agroforests in the conservation of rain forest bird diversity in Sumatra, *Conserv. Biol.,* 9:335–353.

Tiffen, M., Mortimore, M., and Gichuki, F., 1994. *More People, Less Erosion: Environmental Recovery in Kenya,* ACTS Press, Naorobi.

van Noordwijk, M., Tomich, T. P., Winahyu, R., Murdiyarso, D., Suyanto, Partoharjono, S., and Fagi, A. M., 1995a. *Alternatives to Slash and Burn in Indonesia: Summary Report of Phase 1,* ICRAF, Nairobi, Kenya.

van Noordwijk, M., van Schaik, C. P., de Foresta, H., and Tomich, T. P., 1995b. Segregate or integrate nature and agriculture for biodiversity conservation, in *Proceedings of Biodiversity Forum,* 4–5 Nov. 1995, Jakarta, Indonesia, Convention on Biological Diversity, Montreal.

CHAPTER 9

The Role of Agroecosystems in Wildlife Biodiversity

Thomas E. Lacher, Jr., R. Douglas Slack, Lara M. Coburn, and Michael I. Goldstein

CONTENTS

INTRODUCTION: THE INTERACTION BETWEEN WILDLIFE AND AGROECOSYSTEMS

Agriculture is among the most important of all human enterprises. A small number of species of crops, domesticated by a variety of early civilizations, now

1-56670-290-9/99/$0.00+$.50

provides the basis of most of our food consumption. Fifteen species of plants, primarily grains, provide over 90% of all human energy needs, and over 98% of all human food is produced in terrestrial habitats (Paoletti et al., 1992). Agriculture, forestry, and human settlements occupy 95% of all terrestrial environments, whereas nondeveloped areas such as national parks account for only 3.2% worldwide (Pimental et al., 1992). The balance between protected areas and modified landscapes has shifted strongly toward the latter, and there is little doubt that agricultural diversification and expansion has decreased biodiversity over the past two centuries (Dahlberg, 1992). Thus, concerns over the effects of agriculture on wildlife have increased in recent years.

The major impacts on wildlife are caused by habitat conversion and habitat fragmentation. The U.S. provides a good example of this process in a developed country. About 70% of the U.S. (excluding Alaska) is held in private ownership by millions of individuals, although 50% of the land is in the hands of only 2% of the population. About 50% of the country is either cropland, pasture land, or rangeland, owned by approximately 4.7 million individuals (U.S. Department of Agriculture, 1996). Over 200 different species of crops are produced on this land; however, 80% of this total production is accounted for by four species: hay, wheat, corn, and soybeans (U.S. Department of Agriculture, 1996). When forests or grasslands are converted to an agroecosystem, virtually all native species of plants and many of the animals are lost. There is often some degree of utilization of the agricultural fields by vertebrates and invertebrates, but when these species cause losses of crops, they are controlled, usually by chemical means. This often results in the elimination of nontarget organisms as well. Some natural ecosystems, for example, wetlands, have been particularly severely impacted by agricultural expansion. Up until the 1950s, approximately 87% of all wetland conversion was attributable to agriculture, though recent legislation has reduced that percentage. In fact, between 1982 and 1992, 57% of wetland losses were due to urban expansion, and only 20% to agriculture (U.S. Department of Agriculture, 1996).

Economic incentives frequently contribute to habitat conversion. Increased economic pressures and new technological innovations can cause losses of biological diversity in the early stages of development (Howitt, 1995). This is especially a problem in the tropics. Developed country policies often determine the agricultural practices of developing countries; developing countries generally function as "price takers" and are largely exporters of primary products to the developed world (McNeely and Norgaard, 1992). Agricultural development projects financed through international aid agencies have neglected environmental issues in the past. The impact on wildlife in developing tropical nations has been substantial; however, there are several innovative proposals to link economic and ecological systems in agricultural development (McNeely and Norgaard, 1992).

We present a summary of the effects of agriculture on wildlife, both positive and negative. For each individual scenario, there are both benefits and costs. We present four different case studies that attempt to capture the complexity of issues and effects. Finally, we close with some recommendations.

EFFECTS OF AGROECOSYSTEMS ON WILDLIFE

Positive Effects of Agriculture on Wildlife

Several aspects of agroecosystems can positively affect wildlife populations. One aspect of the fragmentation of an agroecosystems/forest mosaic is the creation of edge habitat. This results in the edge effect, or the tendency for the variety and density of some species of plants and animals to increase at the border between different plant communities (Forman, 1997). Edges, or ecotones, contain species from both habitats as well as a subset of species considered to be edge specialists (Yahner, 1988). Some of these species are important game species (e.g., white-tailed deer in North America), which has resulted in a management practice among game biologists of creating edge habitat (Yoakum and Dasmann, 1971). The cost of edge creation is a reduction in the amount of forested habitat available and a decline in the abundance and richness of forest species if fragmentation becomes too severe. There are, however, some species that exist and even thrive in altered or fragmented habitats, especially those that have small area requirements or that are mobile and can easily move among habitat patches (Merriam, 1991; Noss and Cooperrider, 1994).

Grain agriculture often leaves residual seeds on the ground after harvest that serve as a valuable resource for many species of wildlife. Rice plantations provide rice grains as food as well as surrogate wetlands for many species of waterfowl (see below for a detailed case study). In the U.S., concern over the environmental impact of agriculture has also led to the passage of legislation geared toward the enhancement of wildlife habitat. Although not specifically tied to agriculture, the Endangered Species Act of 1973 has led to changes in agricultural practices when species of concern were potentially impacted. For example, concerns over the impact of irrigation on salmon fisheries led to the restriction of access to federally supplied irrigation water in California (Day, 1996). The Farm Act of 1996 created several valuable programs for the protection of wildlife (Table 1) and has included the conservation

Table 1 Programs and Provisions of the 1996 Federal Agriculture Improvement Act (Farm Act)

1. Environmental Quality Incentives Program
2. Wetlands Reserve Program and Conservation Reserve Program
3. Farmland Protection Program
4. Swampbuster and Wetland Provisions
5. Wildlife Habitat Incentives Program
6. Flood Risk Reduction Program
7. Emergency Watershed Protection Program
8. Conservation of Private Grazing Land
9. National Natural Resources Conservation Foundation
10. Conservation Farm Option
11. State Technical Committees

Source: U.S. Department of Agriculture, *America's Private Land: A Geography of Hope,* U.S. Department of Agriculture, Washington, D.C., 1996.

of wildlife habitat as a goal of agricultural programs. For example, the Wildlife Incentives Program was the first agricultural program developed exclusively for the creation and protection of wildlife habitat. The Conservation Reserve Program was originally presented in the 1985 Farm Act and reauthorized, in 1996, the setting aside or converting of as much as 36.4 million acres of environmentally sensitive farmland, through 2001. The Wildlife Habitat Incentives Program of 1996 set aside $200 million for restoration programs in the Everglades Agricultural Area provision (Day, 1996). All of this legislation, explicitly linked to agriculture, has benefited wildlife.

Negative Effects of Agriculture on Wildlife

The greatest negative effect of agriculture on wildlife is the conversion of natural vegetation to an agroecosystem. Habitat loss directly reduces biodiversity. At least 71 species and subspecies of vertebrates[1] and at least 217 species of plants[2] have gone extinct in North America (north of Mexico) since the arrival of Europeans. Over 95% of our original virgin forests are now gone from the lower 48 states (Postel and Ryan, 1991). The decline in large mammalian predators as a result of habitat loss to agriculture has resulted in an increase in the population of deer, which have now become an agricultural pest in many regions (Day, 1996). These losses of species and habitats are the result of the gamut of human activities, of which agricultural activities form a major part.

Given that a certain amount of land will be dedicated to providing for human food and nutrition, the next most significant effect is fragmentation. Habitat fragmentation is defined as the subdivision of continuous habitat over time; the most important large-scale cause of habitat fragmentation is expansion and intensification of human land use (Burgess and Sharpe, 1981; Harris, 1984). Fragmentation is considered to be an important cause of local extinction (Wilcox and Murphy, 1986). Fragmentation results in a loss of original habitat, a reduction in the sizes of the patches of remaining habitat, and an increase in the degree of interpatch distances, all of which increase the rate of local extinction (Harris, 1984; Wilcove et al., 1986). Even certain aspects of the edge effect, positive for some species, are detrimental to others. Edges can serve as potential ecological traps for breeding birds by concentrating nests in a small area where the risk of predation is high (Rudnicky and Hunter, 1993). There is also an apparently high rate of nest parasitism of breeding birds by cowbirds in edge habitats (Brittingham and Temple, 1983).

Activities other than deforestation associated with agriculture also pose threats to wildlife. The use of agricultural chemicals expanded after the end of World War II and their impact on avian populations became a national issue after the publication of Rachel Carson's *Silent Spring* in 1962. Several recent review volumes have addressed the impacts of agricultural chemicals on wildlife and discussed the attempts to mitigate these impacts (Kendall and Lacher, 1994; Colborn et al., 1996).

[1] See The Nature Conservancy, 1992. *Extinct Vertebrate Species in North America,* unpublished draft list, March 4, 1992. The Nature Conservancy, Arlington, VA.

[2] See Russell, C. and Morse, L., 1992. *Extinct and Possibly Extinct Plant Species of the United States and Canada,* unpublished report, review draft, 13 March 1992, The Nature Conservancy, Arlington, VA.

Although persistent pesticides, like DDT, which was an issue in Carson's time, are rarely used, other more acutely toxic compounds now pose a threat of mortality to wildlife. Others, so-called endocrine disrupters, may cause long-term reproductive damage (Colborn et al., 1996). Some of the most significant victims of pesticides have been nontarget species of insects. The loss of these potential pollinators will have far-reaching effects, even on many agricultural crops (Buchmann and Nabhan, 1996). States like California, which was once the largest U.S. user of agricultural toxicants, now have ambitious programs to reduce chemical use (Anderson, 1995). This includes compounds recently suspected of acting as endocrine-disrupting chemicals (Fry, 1995).

Noss and Cooperrider (1994) present a series of summary tables in chapter 3 of their book on the impacts of a variety of land-use practices. Concern for increased rates of local extinction and the concomitant loss of biodiversity as a result of agricultural development is a growing issue in tropical regions as well as temperate zones (Holloway, 1991).

CASE STUDIES: THE USE OF AGROECOSYSTEMS BY WILDLIFE

Wildlife and Rice Cultivation

In the U.S., Texas is one of seven states that grow rice (Texas Rice Task Force, 1993). The Texas rice crop is grown in the gulf prairies and marshes of the upper Texas coast (Gould, 1975). The native tall grass prairies historically extended inland from extensive coastal marshes for approximately 20 to 150 km. The prairies were characterized by nearly level to gently sloping topography interspersed with small, rain-filled depressions. Prior to the 1900s, the prairies of the upper Texas coast were grazed by herds of bison (*Bison bison*) and wild horses (Robertson and Slack, 1995). As the land was settled, bison and wild horses were replaced with free-ranging cattle and later with agricultural crops (Craigmiles, 1975; Stutzenbaker and Weller, 1989). Rice was first introduced to the coastal prairies in the mid-1800s. By 1954, a peak of 254,000 ha of rice were harvested on the gulf prairies (Hobaugh et al., 1989). Currently, about 110,000 ha of rice are harvested in Texas producing an aggregate addition to the Texas economy of almost $1 billion (Texas Rice Task Force, 1993). In addition, the economy of the rice growing region of the state is enhanced by significant expenditures for recreational hunting.

Rice fields are prepared for planting in late winter with actual planting occurring in March or April. Fields are flooded shortly thereafter until immediately prior to harvest in August. These flooded fields provided large expanses of wetlands for some resident birds to use. In portions of the Texas rice belt a second crop ("ratoon crop") results from resprouting from the initial planting and is harvested in October. Harvested fields contain waste grain and are left to stand fallow for up to 2 years. During the subsequent seasons, the fallow fields are grazed by cattle. Therefore, a typical, 3-year, rice–pasture rotation system involves three fields; during early winter rice is harvested in one field, another field is plowed in preparation for planting rice the next spring, and the third field is being grazed (Hobaugh et al., 1989).

At a landscape scale, the tall grass prairies of the upper Texas coast were a relatively homogeneous matrix of tall prairie grasses with small, scattered, natural depressions (Figure 1a). At smaller scales within the matrix, the landscape was heterogeneous, with grasses, forbs, and scattered natural wetlands with associated aquatic vegetation. Because of the intensive rice-cropping system, the resulting landscape is a reversed image of the native tall grass prairie environment (Figure 1b) — a heterogeneous mosaic at the landscape scale, with homogenous field-sized stands of vegetation, prepared fields, or pastures.

The rice-cropping system in Texas lies adjacent to a heavily industrialized region with over 30% of the U.S. petroleum industry and more than 50% of the U.S. chemical production occurring in this region (Robertson and Slack, 1995). In addition, the Houston–Galveston metropolitan area is the fourth largest metropolitan area in the U.S. As a result of these economic pressures, the area of wetland habitats has declined dramatically in Texas, and especially in coastal regions of the state (Anderson, 1996; Moulton et al., 1997). Current estimates of losses show that >35% (84,000 ha) of Texas' coastal marshes have been destroyed since the 1950s (Anderson, 1996).

The net effect of the landscape mosaic produced by the Texas rice-cropping system has been a dramatic change in use by migratory birds since the advent of rice agriculture. Lesser snow geese (*Chen caerulescen*), greater white-fronted geese (*Anser albifrons*), and Canada geese (*Branta canadensis*) only began to use the prairie after rice agriculture became established on the upper Texas coast (Hobaugh et al., 1989). Waterfowl were commonly associated with the small natural depressions in the native prairies (McIlhenny, 1932). However, it wasn't until the 1940s and 1950s with mechanization of rice farming, extensive irrigation, and the 3-year rice rotation system that geese and waterfowl began to exploit the system fully (Hobaugh et al., 1989; Robertson and Slack, 1995). Waterfowl and geese are attracted to the mosaic of habitats because of the availability of waste grain after harvest in the fall and the extensive areas of standing and impounded water associated with roost ponds. Gawlik (1994) documented as much as 87 kg/ha in stubble fields immediately after harvest. Waste rice is an important source of food for wintering ducks, geese, and numerous granivorous passerines (Terry, 1996). Similarly, the importance of waste rice to wintering waterfowl has been documented for the Central Valley of California (Alisauskas et al., 1988; Brouder and Hill, 1995; Gawlik, 1994). In addition to rice grains, green vegetation emerging during the winter in harvested rice fields and in fields that had been prepared for the rice crop the following season, becomes an important source of food for geese (Hobaugh, 1985; Gawlik, 1994). Well over 2 million waterfowl and geese winter on the upper Texas coast with the bulk of these birds found using freshwater wetlands associated with rice agriculture (Haskins, 1996). The extensive use of rice-cultivated land by wintering lesser snow geese has been identified as a significant component of the observed high population growth rates. These high population densities have resulted in significant alteration to Arctic coastal salt marsh plant communities (Abraham and Jefferies, 1997).

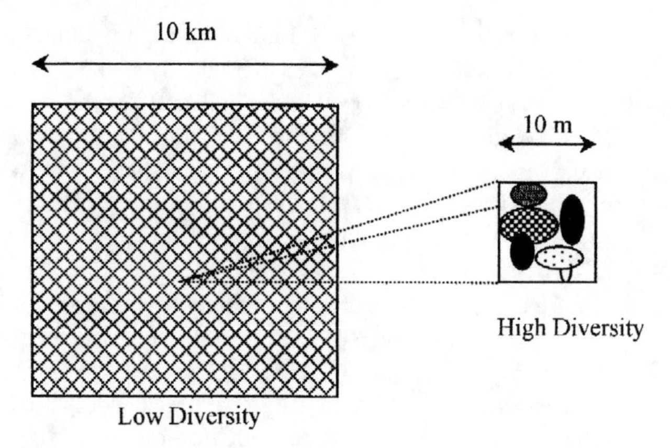

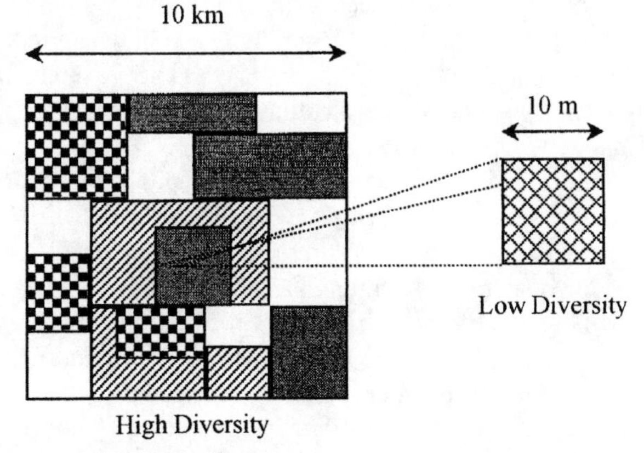

Figure 1 Schematic representation of the Texas coastal tall grass prairie (A) before the advent of rice cultivation (mid-1980s) and (B) after rice production. Rice production has fragmented the larger landscape into patches of homogeneous stands of vegetation, prepared fields, or pastures but has greatly reduced landscape heterogeneity at small spatial scales.

McFarlane (1994) and Terry (1996) have documented the use of the rice system by more than 70 species of birds during an annual cycle. Most species were associated with wetland habitats such as roost ponds, flooded rice fields, and natural depressions. Sheridan et al. (1989) have documented at least 22 species of colonial-nesting waterbirds nesting in 42 colonies located on the upper Texas coast including rice lands. In addition, over 35 species of migratory shorebirds were documented for the upper Texas coast, with 16 species found by Terry (1996). Migratory shore birds take advantage of the wetland habitats associated with rice agriculture, as well as moist, open fields prepared for next year's rice crop.

Migratory Birds, Agroecosystems, and Agricultural Chemicals

Swainson's hawks (*Buteo swainsoni*) are long-distance migrants whose habits in North America have been well documented (England et al., 1997). Breeding habitat in North America consists of open grassland and shrub steppe semiarid ecosystems from Mexico to the prairie provinces of Canada. Birds nest in trees adjacent to large fields, often utilizing agricultural grassland habitat for locating food and other daily activities (Bloom, 1980). Hawks hunt on the ground or in midair, opportunistically eating insects, small mammals, reptiles, and birds (Bednarz, 1986). Adults primarily feed on mammals and birds to supply the nutritional requirements to growing nestlings. Fledglings, aggregates of nonbreeding hawks, and aggregates of premigratory mixed-age hawks forage primarily for insects such as grasshoppers and dragonflies (Woffinden, 1986; Johnson et al., 1987).

Breeding in western North America during the boreal summer, Swainson's hawks migrate to the nonbreeding grounds in South America with the advancing austral summer, maintaining both climatic and habitat similarities (Figure 2). The journey of up to 10,000 km in each direction takes less than 2 months, and, once settled in southern South America, hawks generally reside in the agricultural grasslands of the Argentine pampas (White et al., 1989; Woodbridge et al., 1995; Goldstein, 1997). This habitat is similar to agricultural prairies found throughout their North American range.

Utilizing agricultural areas west and north of the capital city of Buenos Aires during the nonbreeding season, Swainson's hawks were found in the Argentine provinces of La Pampa, Cordoba, Buenos Aires, Santa Fe, and San Luis (Goldstein, 1997). Hawks were encountered sunbathing and foraging in freshly tilled fields or in fields whose crop height was less than 40 cm. Crops used included alfalfa, corn, sorghum, soybean, and sunflower. Flocks of hawks followed insect outbreaks, traveling over a small region as stages in crop growth changed throughout the season. Other insectivorous species, such as the Chimango caracara (*Milvago chimango*), burrowing owl (*Athene cunicularia*), Franklin's gull (*Larus pipixcan*), and southern lapwing (*Vanellus chilensis*) followed these insect outbreaks as well. Aplomado falcons (*Falco femoralis*) also feed on agricultural sites in the pampas, predating insectivorous songbirds.

Increasing monoculture and more intensively managing alfalfa in these regions have also resulted in heavy reliance on agrochemicals for crop protection from insect pests. Subsequent hot and dry conditions of the pampas during the austral summers of 1994–95 and 1995–96 led to severe grasshopper outbreaks, exacerbating the problem of reliance on chemical controls. Typically, the inexpensive organophosphate insecticide monocrotophos (MCP) was used for grasshopper controls. During this time, when insect outbreaks and chemical controls were at their maximum, the largest flocks of Swainson's hawks, up to 12,000 birds, were seen.

Agrochemical controls during the austral summers from 1994 through 1996 led to 19 documented hawk mortality incidents, accounting for approximately 6000 dead Swainson's hawks over two seasons (Goldstein et al., 1996; Goldstein, 1997). Hawks died in fields while foraging for grasshoppers, in roosts after returning from foraging bouts, and along the trajectory from fields to roosting trees. The agrochemical MCP

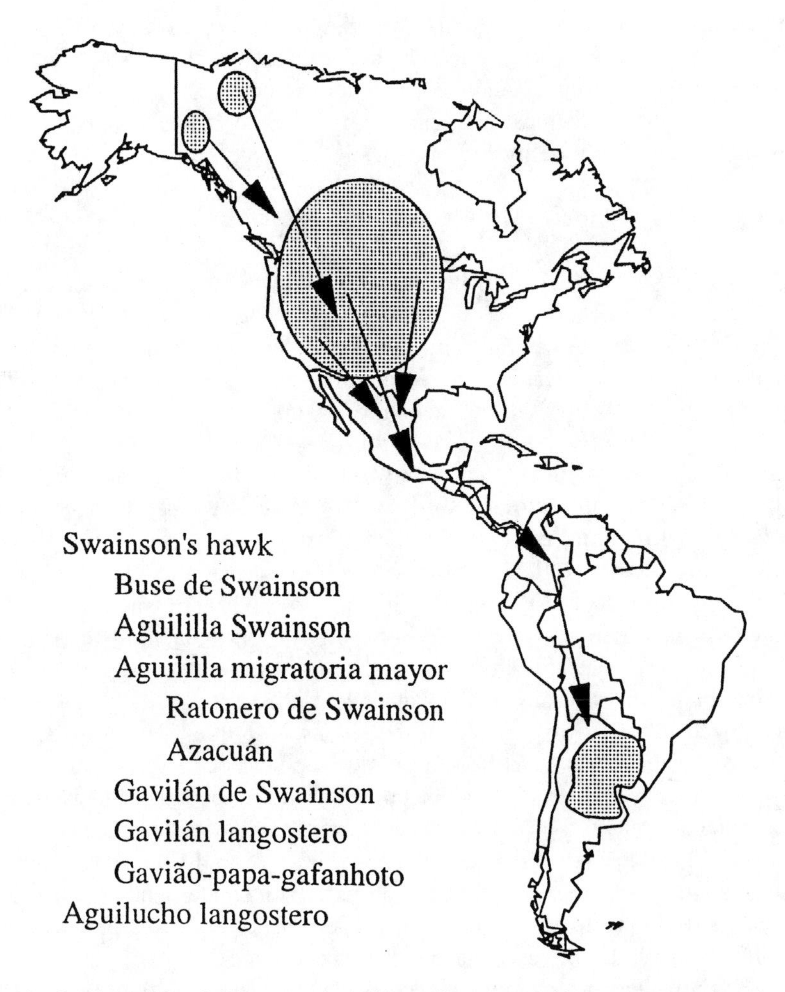

Figure 2 Swainson's hawk (*Buteo swainsoni*) breeding range, migratory route, and nonbreeding range, with a list of the common names for the Swainson's hawk used across the Americas. (Courtesy of M. Fuller, unpublished data.)

was determined responsible for mortality in birds from all 6 sites sampled and in 17 of 19 sites overall, based on forensic analysis and farmer testimony (Goldstein, 1997). The mortality incidents were highly publicized in the scientific and lay news media, resulting in the establishment of an international working group whose function it was to resolve potential future conflicts between agricultural production and wildlife habitat use prior to the 1996–97 austral summer season. University scientists, agrochemical representatives, conservation activists, and government personnel from Argentina, the U.S., and Canada joined together to designate an MCP-free zone in the area of previous Swainson's hawk mortality.

During 1996, use of MCP in alfalfa or as a grasshopper control agent was made illegal in Argentina. An ecotoxicology program was initiated, with field and laboratory training for government agents, students, and veterinarians living in the pampas. Grassroots campaigns described grasshopper-eating hawks as allies to farmers during the time when they were required to transition from MCP to another chemical. The OP dimethoate and the synthetic pyrethroid cypermethrin were most frequently chosen as chemical alternatives. With the successful removal of MCP from this zone, hawk mortality was completely eliminated.

Agricultural Practices in Coffee Agroecosystems

Coffee (*Coffea arabica*) originated in Africa and was introduced to Latin America in the early 18th century by the Dutch. Nearly one third of the world coffee now comes from Latin America where it is the leading agricultural commodity for many countries and the leading source of foreign exchange. In all, 44% of the permanent cropland is now coffee, including 750,000 ha in Central America (Perfecto et al., 1996).

Coffee is a shade-tolerant species and was traditionally grown under the canopy of taller trees, often native species. Coffee in the traditional system was allowed to grow fairly tall (3 to 5 m) under a 60 to 90% cover of shade. Plants were grown at a relatively low density (1000 to 2000/ha), took 4 to 6 years to first harvest, and had a life span of over 30 years. Soil erosion was low and there was little need for agrochemical use (Perfecto et al., 1996). Several factors influenced the shift to a more-intensified approach to cultivation, called sun coffee. First, the spread of coffee leaf rust to Latin America caused phytopathologists to reason that the problem would be minimized if coffee were grown in the sun as, therefore, less moisture would accumulate on the leaves. This led to the development of more densely planted, high-yield varieties that would produce up to four times the kilograms per hectare of traditional plantations. Sun coffee is kept shorter (2 to 3 m) and planted at densities of 3000 to 10,000/ha. Time to first harvest is shorter (3 to 4 years), but plantation life span is less (12 to 15 years). In addition, there is a greater input of agricultural chemicals and a higher likelihood of erosion. The high cost of inputs, however, made sun coffee nearly 50% more expensive than shade coffee (Perfecto et al., 1996). This does not include the environmental cost of sun coffee production. Nevertheless, sun coffee has spread throughout the region and now is the most common practice in Colombia (60% of all production; Perfecto et al., 1996).

Concern over neotropical migratory birds (NTMBs) has refocused attention on shade coffee. Wunderle and Waide (1993) surveyed overwintering neotropical migrants in the Bahamas and the Greater Antilles and observed that shade coffee plantations provided habitat for species normally restricted to forests. Russell Greenberg of the Smithsonian Migratory Bird Center and colleagues have conducted several studies on the levels of biodiversity, including neotropical migrants, that are supported in sun vs. shade plantations. As a generalization, shade coffee supports more biodiversity than sun coffee; however, there is a great deal of variation among types of shade coffee that merits examination.

Greenberg et al. (1997) did a comparison of bird species composition in fragments of forest, matorral (second-growth shrub land), and three types of coffee plantations: sun coffee, shade coffee with a *Gliricidia sepium* overstory, and shade coffee with a canopy of several species of the genus *Inga*. Both *Gliricidia* and *Inga* and nitrogen-fixing legumes are commonly planted in shade coffee plantations in Latin America. Forest remnants had the highest richness, followed by the matorral and *Inga* sites. Sun coffee and *Gliricidia* had the lowest species richness, this in spite of the fact that the *Gliricidia* shade plantations were at a lower elevation with a potentially larger pool of species to draw from. This suggests that shade coffee alone can support somewhat higher richness than sun coffee, but the canopy species is important. When the canopy is composed of a single species, the gain in richness approaches early second growth, but is still less than forest fragments and far less than undisturbed forest. Greenberg recommends that plantations should be as structurally diverse as possible and a mixed agroforestry system could provide this kind of habitat.

There are data to suggest that shade coffee provides habitat for other species as well. Estrada and colleagues (1993; 1994) found that bat species richness in shade coffee plantations in Veracruz, Mexico was higher than in adjacent agricultural fields but lower than in forest; the same was true for terrestrial mammals. Biodiversity of other groups of vertebrates and invertebrates is higher as well (Perfecto et al., 1996).

Several conservation organizations are now promoting shade coffee and mixed agroforestry systems of production as more sustainable and environmentally benign. The Rainforest Alliance has launched an ECO-OK certification program for coffee to support more sustainable practices (Wille, 1994). Conservation International has a similar Sustainable Coffee Initiative with sites in Mexico, Guatemala, Colombia, and Peru which promotes shade coffee and organic techniques (Greenberg, 1996).

Trees as Row Crops: Plantation Forestry and Wildlife

Currently in the U.S., approximately 490 million acres of land are used for commercial timber production (American Forest and Paper Association, AF&PA, 1996). The majority is held by private individuals (59%), with the remainder held by the forestry industry, national forests, and other public agencies (Powell et al., 1992). Timber is important economically; however, the impacts of timber production on wildlife vary and are often extensive. The process of clear-cutting and reforestation creates an artificial cycle of disturbance, resulting in truncated succession, a loss of species richness, and a loss in structural and functional diversity. Native stock is often replaced by nonnative or genetically "improved" species which results in a loss of genetic purity or genotypes of native stock. Pesticides and herbicides are also used with unknown, secondary consequences on the native flora and fauna (Noss and Cooperrider, 1994).

One of the most obvious and frequently cited impacts of timber production on biodiversity is an increase in habitat fragmentation. Habitat fragmentation simply refers to the subdividing of a continuous habitat over time (Pickett and Thompson, 1978; Foster, 1980) and processes that affect an intact forest may be exaggerated when the forest community is fragmented (Noss, 1983). Fragmentation of forests also increases the proportion of edge to interior habitat as the size of the forest decreases (Ranney et al., 1981). Numerous studies on NTMBs have cited increased competition, nest predation, and cowbird parasitism associated with fragmentation and increased edge habitat (Brittingham and Temple, 1983; Small and Hunter, 1988; Yahner and Scott, 1988; Wilcove and Robinson, 1990; Rudnicky and Hunter, 1993).

Another major impact of timber production on biodiversity is the loss of the habitat itself. Of the virgin forests in the U.S., 85% had been destroyed by 1980, with losses of approximately 95 to 98% in the lower 48 states (Postel and Ryan, 1991). Species dependent on these forest habitats, such as the endangered red-cockaded woodpecker (*Picoides borealis*) suffer when these habitats are destroyed or altered (Yahner, 1995). The red-cockaded woodpecker has almost been eradicated from its former range, in part, due to the destruction of its primary habitat, longleaf pine forests. Longleaf pine forests have been reduced by approximately 98% and are the most endangered forest type (Noss, 1989). This reduction is due to logging, fire suppression, and replacement with faster growing, more economical loblolly pine (Dickson et al., 1995). Other species that have experienced declines, or have been extirpated at least partially because of logging practices, include the spotted owl (*Strix occidentalis*), the red wolf (*Canis niger*), the wood duck (*Aix sponsa*), the ivory-billed woodpecker (*Campephilus principalis*), and the passenger pigeon (*Ectopistes migratorius*) (Bellrose, 1976; Paradiso and Nowak, 1982; Robinson and Bolen, 1984; Gill, 1990; Block et al., 1995).

Although the majority of forest management practices appear to have a detrimental impact on biodiversity, some practices do benefit some species. For examples, small clear-cuts (<10 ha) appear to benefit some songbirds and mammals which require dense, brushy, vegetation for cover and food (Scott and Yahner, 1989; Hughes and Fahey, 1991; Yahner, 1993). Edge does benefit some game animals, such as the white-tailed deer (*Odocoileus virginianus*) (Yoakum and Dasmann, 1971), and many game managers emphasize the creation and maintenance of edge despite its potential negative impacts on other species.

Timber management practices have recently come under fire from various organizations and this increased concern over the impacts of timber management has led in part to the implementation of the Sustainable Forestry Initiative (SFI) by the AF&PA. Several priorities outlined by SFI directly emphasize biodiversity (AF&PA, 1996). As advances, innovations, and technologies are developed and implemented, these negative effects on biodiversity may be ameliorated in the future.

CONCLUSIONS — NET EFFECTS OF AGROECOSYSTEMS

Loss of Biodiversity

There is little question that, over time, agriculture has resulted in a loss of biodiversity. The losses have been particularly serious in areas where little natural habitat was left among agroecosystems, such as in the central plains of the U.S. The prairie fauna once included grizzly bears, wolves, mountain lions, elk, deer, and millions of bison, all of which are now extinct throughout most of their former grassland ranges. The eastern deciduous forest of the U.S. was nearly completely deforested at the turn of the century, although remnant patches of forest remained on hilltops and in valleys. As agriculture shifted westward, many species were able to recolonize, regenerating forests and reducing the long-term severity of the agricultural impact.

If lands are properly managed in a mosaic, losses of biodiversity need not be permanent. Studies indicate that extensive areas of tropical forest, once thought to be virgin, were at one time under intensive cultivation. Pristine forests in the Darien gap in Panama were apparently subjected to over 4000 years of human disturbance, and are probably no older than 350 years (Bush and Colinvaux, 1994). Diversity is extremely high with no indication that this history of agricultural activity depleted the biodiversity of the region. These forests never were subjected to extensive deforestation, however, and this likely prevented large-scale local extinctions. Indeed, Mellink (1991) observed that the presence of isolated farms in the San Luis Potosi plateau of Mexico actually increased regional bird species richness.

The implementation of policies recommended under the 1996 Farm Act will help to preserve patches of natural habitat and will facilitate the protection of residual populations of native species. If patches are kept small and isolated, however, they may not protect viable populations over long periods of time. Areas of native vegetation should be as large as possible and interconnected via corridors of habitat to maximize their effect. The extensive nature of modern agriculture is much different than the practices of the precolonial people of the Darien region in Panama, and the long-term consequences on wildlife populations under current management practices will be negative.

Change in Community Structure

Agroecosystems simplify the environment. Generally, they contain fewer species than the native flora, and they contain less diversity in foliage structure than native ecosystems. They also contain far fewer species of invertebrates and vertebrates than natural ecosystems. Thus, the communities of organisms associated directly with agroecosystems represent but a tiny subset of the total biodiversity of the region, and the community structure is also simplified. These changes affect not only species diversity, but also the functional diversity of agroecosystems. When agroecosystems are extensive and remaining habitat is fragmented, local extinctions that result from fragmentation will also reduce the species and functional diversity of the region as a whole. All of these effects have been observed and documented (McNeely et al., 1990; Day, 1996).

Changes in community structure can result in changes in ecosystem function. A common consequence of human activities is the extirpation of large predators. The elimination of predator populations can allow herbivore populations to increase, thus increasing the impact of herbivores on plant community structure and function. This can lead to large-scale changes in community structure and function (Rasmussen, 1941; Paine, 1966). Changes in species composition can also affect community processes in unexpected ways by altering the functional diversity of communities (Tilman et al., 1997; Hooper and Vitousek, 1997). Thus, the effects of altered species diversity can extend far beyond the boundaries of agroecosystems and can result in major shifts in ecosystem function as well as composition.

Recommendations for the Mitigation of Impacts

The joint production of crops and wildlife is a relatively new concept (Howitt, 1995). Paoletti et al. (1992) present a table of choices among farming systems that either reduce or enhance biodiversity in the regional agroecosystem. Indeed, given the preponderance of lands under some form of human management, we must begin to assign a larger role to preserving biodiversity in agroecosystems (Pimental et al., 1992). In the U.S., the realization that most biodiversity exists on private lands has led to the development of new initiatives to make conservation more attractive to private landowners. California is a habitat mosaic of federal and private lands and an extensive region of intensive agricultural development. California also is the most biologically diverse state with the largest number of federally listed candidates or endangered species in the country (Scott et al., 1995). Two thirds of these listed species are on private lands. A federally sponsored program called Habitat Conservation Plans (HCP) provides a mechanism where landowners agree to an overall plan to protect an endangered species and its habitat in exchange for a permit to alter some portion of the habitat in the planning area (Scott et al., 1995). HCPs can vary in their geographic scope from a single parcel or landowner to larger areas and multiple landowners. This mechanism was instituted first in California over a controversy concerning several threatened species of butterflies. The success of this experience led to the 1982 amendment to the Endangered Species Act, allowing HCPs. Although the majority of these HCPs are in California, they will be implemented nationwide, with approximately 40 plans approved and at least another 150 in progress (Beatley, 1995).

There are numerous efforts to reduce the use and impact of agricultural chemicals on agroecosystems (Kendall and Lacher, 1994). The U.S. Environmental Protection Agency has supported the creation of dialogue groups to assist in the resolution of conflicts over agricultural chemical use and the protection of wildlife (Avian Effects Dialogue Group, 1994) and has developed a new paradigm for the assessment of environmental risk (U.S. Environmental Protection Agency, 1992). California has become a world leader in the development of methods to reduce pesticide use (Anderson, 1995). The sustainable agriculture movement also emphasizes the use of native biota for pest control (Miller and Rossman, 1995).

The impact of agricultural activities on tropical wildlife is of growing concern. Economic factors strongly influence agricultural practices in the tropics (McNeely and Norgaard, 1992). Frequently, pressures for colonization result from economic hardship in urban areas, and the lack of technical expertise of colonists results in a sequence of poor land-use practices from deforestation through inappropriate agriculture to abandoned pasture land. This land-use succession has been referred to as "nutrient mining" (Southgate and Clark, 1993). Southgate and Clark (1993) make the point that farmers and ranchers in countries where crop and livestock yields have improved seldom encroach on natural habitats. In countries with poor yields and increasing populations, new areas are continually being cleared. Programs that increase yields effectively buy time for the implementation of population control. However, most donor and foreign aid agencies are currently reducing support of agricultural development programs. Protectionism in developed countries also inhibits the transfer of technologies to developing countries, inhibiting their competitiveness. Some environmental groups unwittingly contribute to this process by ignoring or downplaying agricultural development and emphasizing conservation efforts only, which can exacerbate rather than mitigate pressures on protected areas. The dynamics of land use on a mosaic landscape are complex and interrelated; it is not possible to view wildlife conservation and agricultural development as independent processes. Agricultural development projects funded by international aid agencies must address conservation and the mitigation of environmental impacts, and conservation projects must look more closely at the interrelationship between agricultural productivity and deforestation.

Public awareness of the impacts of agroecosystems on biodiversity in general, and wildlife in particular, has led to the development of new laws and regulations in countries throughout the world. There is a new appreciation of the concept of the management of the whole landscape for multiple uses, and a better understanding of the interrelations among landscape units. Agriculturists and conservationists alike are coming to an agreement that the old way of doing things is no longer the best way, considering the diverse expectations of a more-educated, globally connected populace.

REFERENCES

Abraham, K. F. and Jefferies, R. L., 1997. High goose populations: causes, impacts and implications, in Arctic Ecosystems in Peril: Report of the Arctic Goose Habitat Working Group, B. D. J. Batt, Ed., Arctic Goose Joint Venture Special Publication. U.S. Fish and Wildlife Service, Washington, D.C., and Canadian Wildlife Service, Ottawa, Ontario, 7–72.

Alisauskas, R. T., Ankney, C. D., and Klaas, E. E., 1988. Winter diets and nutrition of midcontinental lesser snow geese, *J. Wildlife Manage.,* 52:403–414.

American Forest and Paper Association, 1996. Sustainable Forestry for Tomorrow's World, First annual progress report on the American Forest and Paper Association's Sustainable Forestry Initiative, AF&PA, Inc., Washington, D.C.

Anderson, D. W., 1995. Society responds to contamination — changes in pest control practices reduce toll on wildlife, *Calif. Agric.,* 49:65–72.

Anderson, J., 1996. Texas Wetlands Conservation Plan (Draft), Texas Parks and Wildlife Department, Austin.

Avian Effects Dialogue Group, 1994. *Assessing Pesticide Impacts on Birds,* Resolve, Inc., Washington, D.C.

Beatley, T., 1995. Habitat conservation plans: a new tool to resolve land use conflicts, *Land Lines,* 7:1–6.

Bednarz, J. C., 1986. Swainson's hawk, in *Proceedings of the Southwest Raptor Management Symposium,* R. L. Glinski, B. G. Pendelton, M. B. Moss, M. N. LeFranc, Jr., B. A. Millsap, and S. W. Hoffman, Eds., National Wildlife Foundation Sci. Tech. Ser. 11, Washington, D.C., 87–96.

Bellrose, F. C., 1976. *Ducks, Geese, and Swans of North America,* Stackpole Books, Harrisburg, PA.

Block, W. M., Finchand, D. M., and Brennan, L. A., 1995. Single-species versus multiple-species approaches for management, in *Ecology and Management of Neotropical Migratory Birds,* T. E. Martin and D. M. Finch, Eds., Oxford University Press, New York, 461–476.

Bloom, P. H., 1980. The Status of the Swainson's Hawk in California, Project W-54-R-12, U.S. Dept. of the Interior, Bureau of Land Management, Dept. of Fish and Game, Sacramento, CA.

Brittingham, M. C. and Temple, S. A., 1983. Have cowbirds caused forest songbirds to decline?, *BioScience,* 33:31–35.

Brouder, S. M. and Hill, J. E., 1995. Winter flooding of ricelands provides waterfowl habitat, *Calif. Agric.,* 49:58–60, 62–64.

Buchmann, S. L. and Nabhan, G. P., 1996. *The Forgotten Pollinators,* Island Press, Washington, D.C.

Burgess, R. L. and Sharpe, D. M., 1981. *Forest Island Dynamics in Man-Dominated Landscapes,* Springer-Verlag, New York.

Bush, M. B. and Colinvaux, P. A., 1994. Tropical forest disturbance: paleoecological records from Darien, Panama, *Ecology,* 75:1761–1768.

Colborn, T., Dumanoski, D., and Myers, J. P., 1996. *Our Stolen Future,* Plume/Penguin, New York.

Craigmiles, J. P., 1975. Advances in rice — through research and application, in *Six Decades of Rice Research in Texas,* J. E. Miller, Ed., Texas Agricultural Experiment Station Research Monographs, No. 4, College Station, TX, 1–8.

Dahlberg, K. A., 1992. The conservation of biological diversity and U.S. agriculture: goals, institutions, and policies, *Agric. Ecosyst. Environ.,* 42:177–193.

Day, K., 1996. Agriculture's links to biodiversity, *Agric. Outlook,* December:32–37.

Dickson, J. G., Thompson, F. R., III, Conner, R. N., and Franzreb, K. F., 1995. Silviculture in central and southeastern oak-pine forests, in *Ecology and Management of Neotropical Migratory Birds,* T. E. Martin and D. M. Finch, Eds., Oxford University Press, New York, 245–266.

England, A. S., Bechard, M. J., and Houston, C. S., 1997. Swainson's hawk: *Buteo swainsoni,* in *The Birds of North America,* Vol. 265, A. Poole and F. Gill, Eds., The Academy of Natural Sciences, Philadelphia, PA, and the American Ornithologists Union, Washington, D.C., 1–27.

Estrada, A., Coates-Estrada, R., and Merrit, D., Jr., 1993. Bat species richness and abundance in tropical rain forest fragments and in agricultural habitats at Los Tuxtlas, Mexico, *Ecography,* 16:309–318.

Estrada, A., Coates-Estrada, R., and Merrit, D., Jr., 1994. Non-flying mammals and landscape changes in the tropical rain forest region of Los Tuxtlas, *Ecography,* 17:229–241.

Forman, R. T. T., 1997. *Land Mosaics,* Cambridge University Press, Cambridge, U.K.

Foster, R. B., 1980. Heterogeneity and disturbance in tropical vegetation, in *Conservation Biology. An Evolutionary Ecological Perspective,* M. E. Soule and B. A. Wilcox, Eds., Sinauer, Sunderland, MA, 75–96.

Fry, D. M., 1995. Unexpected side effects of chemicals acting as hormone mimics, *Calif. Agric.,* 49:67.

Gawlik, D. E., 1994. Competition and Predation as Processes Affecting Community Patterns of Geese, M.S. thesis, Texas A&M University, College Station, TX.

Gill, F. B., 1990. *Ornithology,* W. F. Freeman and Co., New York.

Goldstein, M. I., 1997. Toxicological Assessment of a Neotropical Migrant on Its Non-Breeding Grounds: Case Study of the Swainson's Hawk in Argentina, M.S. thesis, Clemson University, Clemson, SC.

Goldstein, M. I., Woodbridge, B., Zaccagnini, M. E., Canavelli, S. B., and Lanusse, A., 1996. An assessment of mortality of Swainson's hawks on wintering grounds in Argentina, *J. Raptor Res.,* 30:106–107.

Gould, F. W., 1975. *Texas Plants: A Checklist and Ecological Summary,* Texas A&M University, College Station, TX.

Greenberg, R., 1996. Birds in the tropics: the coffee connection, *Birding*, December:472–481.

Greenberg, R., Bichier, P., Angon, A. C., and Reitsma, R., 1997. Bird populations in shade and sun coffee plantations in central Guatemala, *Conserv. Biol.,* 11:448–459.

Harris, L. D., 1984. *The Fragmented Forest: Island Biogeography Theory and the Preservation of Biotic Diversity,* University of Chicago Press, Chicago.

Haskins, J., 1996. *Analyses of Selected Migratory Game Bird Survey Data (1955–1995). Region 2,* U.S. Fish and Wildlife Service, Albuquerque, NM.

Hobaugh, W. C., 1985. Body condition and nutrition of snow geese wintering in southeastern Texas, *J. Wildlife Manage.,* 49:1028–1037.

Hobaugh, W. C., Stutzenbaker, C. D., and Flickinger, E. L., 1989. The rice prairies, in *Habitat Management for Migrating and Wintering Waterfowl in North America,* L. M. Smith, R. L. Pederson, and R. M. Kaminski, Eds., Texas Tech University Press, Lubbock, 367–383.

Holloway, J., 1991. Biodiversity and tropical agriculture: a biogeographic view, *Outlook Agric.,* 20:9–13.

Hooper, D. U. and Vitousek, P. M., 1997. The effects of plant composition and diversity on ecosystem processes, *Science*, 277:1302–1305.

Howitt, R. E., 1995. How economic incentives for growers can benefit biological diversity, *Cal. Agric.,* 49:28–33.

Hughes, J. W. and Fahey, T. J., 1991. Availability, quantity, and selection of browse by white-tailed deer after clearcutting, *J. For.,* 89:31–36.

Johnson, C. G., Nickerson, L. A., and Bechard, M. J., 1987. Grasshopper consumption and summer flocks of non-breeding Swainson's hawks, *Condor,* 89:676–678.

Kendall, R. J. and Lacher, T. E., Jr., 1994. *Wildlife Toxicology and Population Modeling: Integrated Studies of Agroecosystems* (Special Publication Series, Society of Environmental Toxicology and Chemistry), Lewis Publishers, Chelsea, MI.

McFarlane, R. W., 1994. Birdlife on the Katy Prairie — Section 4, in *Balancing Growth and Conservation: Proceedings Katy Prairie Conference,* Texas Parks and Wildlife Department, Austin, TX.

McIlhenny, E. A., 1932. The blue goose in its winter home, *Auk*, 49:279–306.

McNeely, J. A. and Norgaard, R. B., 1992. Developed country policies and biological diversity in developing countries, *Agric. Ecosyst. Environ.,* 42:194–204.

McNeely, J. A., Miller, K. R., Reid, W. V., Mittermeier, R. A., and Werner, T. B., 1990. *Conserving the World's Biological Diversity,* IUCN, Gland, Switzerland; WCI, CI, WWF-US, and the World Bank, Washington, D.C.

Mellink, E., 1991. Bird communities associated with three traditional agroecosystems in the San Luis Potosi Plateau, Mexico, *Agric. Ecosyst. Environ.,* 36:37–50.

Merriam, G., 1991. Corridors and connectivity: animal populations in heterogeneous environments, in *Nature Conservation 2: The Role of Corridors,* D. A. Saunders and R. J. Hobbs, Eds., Surrey Beatty and Sons, Chipping Norton, NSW, Australia, 133–142.

Miller, D. R. and Rossman, A. Y., 1995. Systematics, biodiversity, and agriculture, *BioScience,* 45:680–686.

Moulton, D. W., Dahl, T. E., and Dal, D. M., 1997. *Texas Coastal Wetlands; Status and Trends, Mid-1950s to Early 1990s,* U.S. Fish and Wildlife Service, Albuquerque, NM.

Noss, R. F., 1983. A regional landscape approach to maintain biodiversity, *BioScience,* 33:700–706.

Noss, R. F., 1989. Longleaf pine and wiregrass: keystone components of an endangered ecosystem, *Nat. Areas J.,* 9:211–213.

Noss, R. F. and Cooperrider, A. Y., 1994. *Saving Nature's Legacy,* Island Press, Washington, D.C.

Paine, R. T., 1966. Food web complexity and species diversity, *Am. Nat.,* 100:65–75.

Paoletti, M. G., Pimentel, D., Stinner, B. R., and Stinner, D., 1992. Agroecosystem biodiversity: matching production and conservation biology, *Agric. Ecosyst. Environ.,* 40:3–23.

Paradiso, J. L. and Nowak, R. M., 1982. Wolves, in *Wild Animals of North America,* J. A. Chapman and G. H. Feldhammer, Eds., Johns Hopkins Press, Baltimore, MD, 460–474.

Perfecto, I., Rice, R. A., Greenberg, R., and Van der Voort, M., 1996. Shade coffee: a disappearing refuge for biodiversity, *BioScience,* 46:598–608.

Pickett, S. T. A. and Thompson, J. H., 1978. Patch dynamics and design of nature reserves, *Biol. Conserv.,* 13:27–37.

Pimental, D., Stachow, U., Takacs, D. A., Brubaker, H. W., Dumas, A. R., Meaney, J. J., O'Neil, J. A. S., Onsi, D. E., and Corzilius, D. B., 1992. Conserving biological diversity in agricultural/forestry systems, *BioScience*, 42:354–362.

Postel, S. and Ryan, J. C., 1991. Reforming forestry, in *State of the World 1991: A Worldwatch Institute Report on Progress Toward a Sustainable Society,* L. Starke, Ed., W. W. Norton, New York, 74–92.

Powell, D. S., Faulkner, J. L., Darr, D. D., Zhu, Z., and MacCleery, D., 1992. Forest Resources of the United States, Gen. Tech. Report RM-234, U.S. Department of Agriculture Forest Service.

Ranney, J. W., Bruner, M. C., and Levenson, J. B., 1981. The importance of edge in the structure and dynamics of forest islands, in *Forest Island Dynamics in Man-Dominated Landscapes,* R. L. Burgess and D. M. Sharpe, Eds., Springer-Verlag, New York, 67–95.

Rasmussen, D. I., 1941. Biotic communities of Kaibab Plateau, Arizona, *Ecol. Monogr.,* 11:230–275.

Robertson, D. R. and Slack, R. D., 1995. Landscape change and its effects on the wintering range of a lesser snow goose *Chen caerulescens caerulescens* population: a review, *Biol. Conserv.,* 71:179–185.

Robinson, W. L. and Bolen, E. G., 1984. *Wildlife Ecology and Management,* Macmillan, New York.

Rudnicky, T. C. and Hunter, M. L., Jr., 1993. Avian nest predation in clearcuts, forests, and edges in a forest-dominated landscape, *J. Wildlife Manag.,* 57:358–364.

Scott, D. P. and Yahner, R. H., 1989. Winter habitat and browse use by snowshoe hares, *Lepus americanus*, in a marginal habitat in Pennsylvania, *Can. Field Nat.,* 103:560–563.

Scott, T., Standiford, R., and Pratini, N., 1995. Private landowners critical to saving California biodiversity, *Calif. Agric.,* 49:50–54, 57.

Sheridan, P. F., Slack, R. D., Ray, S. M., McKinney, L. W., Klima, E. F., and Calnan, T. R., 1989. Biological components of Galveston Bay, in *Galveston Bay: Issues, Resources, and Management – Series 13,* National Oceanic and Atmospheric Administration, Washington, D.C.

Small, M. F. and Hunter, M. L., 1988. Forest fragmentation and avian nest predation in forested landscape, *Oecologia,* 76:62–64.

Southgate, D. and Clark, H. L., 1993. Can conservation projects save biodiversity in South America?, *Ambio,* 22:163–166.

Stutzenbaker, C. D. and Weller, M. W., 1989. The Texas coast, in *Habitat Management for Migrating and Wintering Waterfowl in North America,* L. M. Smith, R. L. Pederson, and R. M. Kaminski, Eds., Texas Tech University Press, Lubbock, 385–404.

Terry, K. L., 1996. An Evaluation of Avian Use of a Rice-Wetland System, M.S. thesis, Texas A&M University, College Station.

Texas Rice Task Force, 1993. *Future of the Texas Rice Industry – Executive Summary,* Texas Agricultural Experiment Station, Texas A&M University, College Station.

Tilman, D., Knops, J., Wedlin, D., Reich, P., Ritchie, M., and Siemann, E., 1997. The influence of functional diversity and composition of ecosystem processes, *Science,* 277:1300–1302.

U.S. Department of Agriculture, 1996. *America's Private Land: A Geography of Hope,* U.S. Department of Agriculture, Washington, D.C.

U.S. Environmental Protection Agency, 1992. Framework for Ecological Risk Assessment, EPA/630/R-92/001, Washington, D.C.

White, C. M., Boyce, D. A., and Straneck, R., 1989. Observations on *Buteo swainsoni* in Argentina, 1984, with comments on food, habitat alteration, and agricultural chemicals, in *Raptors in the Modern World,* B. U. Meyburg and R. D. Chancellor, Eds., World Working Group on Birds of Prey and Owls, Berlin, 79–87.

Wilcove, D. S. and Robinson, S. K., 1990. The impact of forest fragmentation on bird communities in Eastern North America, in *Biogeography and Ecology of Forest Bird Communities,* A. Keast, Ed., SPB Academic Publishing, The Hague, The Netherlands, 319–331.

Wilcove, D. S., McLellan, C. H., and Dobson, A. P., 1986. Habitat fragmentation in the temperate zone, in *Conservation Biology: The Science of Scarcity and Diversity,* M. E. Soule, Ed., Sinauer, Sunderland, MA, 237–256.

Wilcox, B. S. and Murphy, D. D., 1986. Conservation strategy: the effects of fragmentation on extinction, *Am. Nat.,* 125:879–887.

Wille, C., 1994. The birds and the beans, *Audubon Mag.,* 96:58–64.

Woffinden, N. D., 1986. Notes on the Swainson's hawk (*Buteo swainsoni*) in central Utah, USA: insectivory, premigratory aggregations, and kleptoparasitism, *Great Basin Nat.,* 46:302–304.

Woodbridge, B., Finley, K. K., and Seager, S. T., 1995. An investigation of the Swainson's hawk in Argentina, *J. Raptor Res.,* 29:202–204.

Wunderle, J. M. and Waide, R. B., 1993. Distribution of overwintering Nearctic migrants in the Bahamas and Greater Antilles, *Condor,* 95:904–933.

Yahner, R. H., 1988. Changes in wildlife communities near edges, *Conserv. Biol.,* 2:233–239.

Yahner, R. H., 1993. Effects of long-term forest clear-cutting on wintering and breeding birds, *Wilson Bull.,* 105:239–255.

Yahner, R. H., 1995. *Eastern Deciduous Forest: Ecology and Wildlife Conservation,* University of Minnesota Press, Minneapolis.

Yahner, R. H. and Scott, D. P., 1988. Effects of forest fragmentation on depredation of artificial nests, *J. Wildlife Manage.,* 52:158–161.

Yoakum, J. and Dasmann, W. P., 1971. Habitat management practices, in *Wildlife Management Techniques,* R. H. Giles, Ed., The Wildlife Society, Washington, D.C., 173–231.

CHAPTER 10

Natural Systems Agriculture

Jon K. Piper

CONTENTS

1-56670-290-9/99/$0.00+$.50

INTRODUCTION: ENVIRONMENTAL PROBLEMS ASSOCIATED WITH MODERN AGRICULTURE

"One Kansas Farmer Feeds 101 People and You," proclaims a billboard alongside Interstate 135, near Salina, KS. Modern agriculture has been overwhelmingly successful in terms of output per farmer, acre, or hour worked. Agricultural productivity has steadily increased as a result of technological advances in machinery, fertilizer, and pesticides coupled with the intensive use of plant genetic diversity to improve yield through plant breeding. For example, yields of corn and sorghum increased severalfold in the U.S. between the 1930s and the 1980s (Jordan et al., 1986).

In terms of return on labor, industrial agriculture based on monocultures of annual grains is unquestionably a highly productive form of food and feed production. This productivity has arisen largely through simplifying agroecosystems into monocultures and tailoring them to maximize yield. In the process, however, many of the links between organisms and the soil that serve to regulate natural communities are ignored or disrupted. The following account surveys some of the more severe environmental effects deriving from large-scale monocultures.

With the publication of *Silent Spring* 35 years ago (Carson, 1962), the public began to become aware of unforeseen environmental consequences of modern agriculture and to question or not whether increasing agricultural production alone was a worthy goal. Three of the most obvious environmental consequences of high-production agriculture are fossil fuel dependency (and its consequent contribution to global warming), contamination of soil and water with toxic chemical residues, and rates of topsoil loss that exceed the natural rates of soil formation. Additional important consequences include the net depletion of aquifer water for irrigation and the loss of biodiversity from crops, land races, and crop wild relatives.

A general consequence of our modern agricultural system is dependency on fossil fuel–based energy. Pimentel et al. (1995) estimate that 10% of all energy used in U.S. agriculture is expended to offset the losses of soil nutrients and water caused by erosion. Over the last few decades, it has taken increasingly more fossil fuel energy to produce a unit of grain in the U.S. (Pimentel, 1984; Cleveland, 1995), with a recent ratio of total energy expended in agriculture (including transportion and processing) to food energy consumed in the U.S. of about 10:1 (Lovins et al., 1995).

Another consequence resulting from decades of chemical application on agricultural soils is contamination of surface waters and groundwaters by toxic chemicals. Particularly troublesome are unsafe levels of nitrate derived from applied fertilizer and residues from pesticides aimed at harmful insects, weeds, and pathogenic fungi.

Nitrate concentrations in groundwater are strongly correlated with overlying land use (Singh and Sekhon, 1979; Hallberg, 1986). Crops often do not take up all nitrogen applied before it leaches below the zone of biological activity in soil. This excess nitrogen in agricultural soils can slowly leach into deep aquifers, even years after fertilizer application ceases.

By the early 1960s, the properties of such long-lived, low-toxicity, and bioaccumulating substances as DDT began to emerge (Carson, 1962). In temperate regions, DDT has a half-life of 59 years. Once bioconcentrated in such top predators as carnivorous fish and bald eagles, DDT sharply reduces the reproductive potential of these species. Long-term exposure has been associated with mutagenesis and carcinogenesis. DDT was banned in 1969; other long-lived pesticides were banned in the 1970s. They were largely replaced with several types of short-lived, acutely toxic compounds (e.g., organophosphates). Residues of many of these subtances are still present all over the planet. Concentration of DDE (a DDT metabolite) is increasing in some Great Lakes and Arctic species (Hileman, 1994).

It was not until 1979 that routine agricultural use of the new generation of pesticides was linked to groundwater contamination. Researchers discovered that some of these apparently short-lived, unstable compounds can become extremely persistent once below the soil biologically active zone, where the usual biological degradation does not occur (Zaki et al., 1982; Cohen et al., 1995). Just as with nitrate, complete cessation of pesticide use would not immediately halt the increasing presence of pesticides in groundwater.

Many of these chemicals are threats to human health, especially among farm-workers who are exposed directly and rural families dependent upon drinking water from wells. Accumulating epidemiological evidence suggests that agricultural chemicals are associated with increased risks of many types of cancers (Blair et al., 1992; Zahm and Blair, 1992).

An additional unintended consequence of widespread and constant pesticide application is evolution of pesticide resistance in target organisms. The result is a need for higher application rates of some pesticides as well as continuous research to develop new substances to control the targeted pests. This phenomenon has been termed the *pesticide treadmill;* we work harder and harder to stay in the same place but with ever-increasing costs to environmental quality.

Despite the enormous problems presented by chemical contamination and fossil fuel dependency, the most serious problem for the long-term sustainability of agriculture is soil loss. Soil erosion is the primary conservation problem on much of U.S. cultivated cropland, and occurs mostly during short intervals of heavy rain or high wind and when the surface is not protected by a mulch or crop canopy (Larson et al., 1997). During the last few decades, about one third of the world arable land area has been lost through soil erosion, and this loss continues at an estimated annual rate exceeding 10 million ha/year (Pimentel et al., 1995). On average, soil on about 90% of U.S. cropland is being lost faster than it is being formed. Because the effects of erosion on some soil physical attributes are irreversible, erosion rates alone may not be good indicators of soil degradation and, consequently, soil quality can decline faster than the erosion rate. Once virgin soil is cultivated, organic carbon rapidly decreases (Campbell and Souster, 1982), large pores crucial for soil function are destroyed, changes in some physical properties increase the rate of erosion, rates of nutrient leaching can increase (Blank and Fosberg, 1989), and populations of such beneficial invertebrates as earthworms decline (Edwards and Lofty, 1975). Prairie soils can lose 30 to 60% of their organic carbon, 30 to 40% of nitrogen, and up to 25% of phosphorus from the A horizon after only a few decades of cultivation (Anderson and Coleman, 1985; Schoenau et al., 1989; Woods, 1989). Many of the consequences of soil degradation, such as reduced crop productivity, have been offset to by improvements in fertilizer and irrigation technology and the development of new, higher-yielding varieties.

The industrialization of agriculture, typified by widespread annual monocultures, has led to such profound problems as soil loss, loss of genetic diversity in cultivars, fossil fuel dependency, depletion and contamination of water supplies, pesticide poisoning of farmworkers and nontarget wild species, and development of pesticide resistance in pests. Modern agricultural methods, while highly productive in the short run, are sustainable only as long as topsoil is intact, fossil fuel supplies are affordable, and effective pesticides are available. This may be justified when fossil fuels are cheap and environmental costs can be ignored, but such practices make us vulnerable over the long run.

Definitions of sustainability abound. In view of the issues listed above, a sustainable agriculture for the Great Plains should address simultaneously several key environmental problems of modern agriculture. It should feature reduced or eliminated soil erosion, efficient use of land area and soil nutrients, improved water use efficiency, reduced reliance on synthetic nitrogen fertilizer, decreased risk of pest and disease epidemics, effective chemical-free weed management, reduced fossil energy requirements, reduced chemical contamination of soil and water, and the opportunity for farmers to hedge their bets among several agricultural products. A good working definition of sustainable agriculture is one that includes grain production with (1) no chemical contamination of the environment (via pesticides or fertilizers), (2) no dependence on nonrenewables (e.g., fossil fuel, fossil water), and (3) no net soil loss. This working definition is limited in that it leaves out such important considerations as sociology, economy, and justice. But it provides a beginning point for a biological research agenda.

By using this definition of sustainability, what type of system could simultaneously satisfy all three criteria? Natural grassland ecosystems provide appropriate models of long-term sustainability because they run on sunlight and rainfall, resist pests, weeds, and disease epidemics, and most importantly because they do not lose soil beyond the natural rate of formation. Some of these aspects are explored in depth in the following sections.

THE PRAIRIE MODEL

Natural grassland ecosystems may represent our best benchmarks for sustainability. Prairies (1) protect the soil from erosion, (2) provide their own nitrogen fertility requirements through the activities of both free-living and symbiotic nitrogen-fixing organisms, (3) avoid devastation by weedy invaders, insect pests, and plant diseases, and (4) run on sunlight and available precipitation (Table 1).

The vegetation structure of prairies has two important general characteristics that contribute to sustainability: the perennial plant growth habit and diversity. Prairies are composed primarily of herbaceous perennial plants growing in diverse arrays.

The perennial roots and canopies of prairie plants provide many benefits. These include (1) topsoil protection from wind and water erosion, (2) improved soil quality

Table 1 Comparison between Conventional, Industrial Agricultural Systems and Native Prairie Ecosystems for Some Factors that Contribute to Sustainability

Factor	Industrial Agriculture	Native Prairie
Fragility	High	Low
Resilience	Low	High
Biodiversity	Low	High
Potential for nutrient loss	High	Low
Connectance (biotic interdependence)	Low	High
Energy sources	Solar, fossil fuel	Solar
Nutrient sources	From fertilizers	Locally derived, recycled

Adapted from J. D. Soule and J. K. Piper, *Farming in Nature's Image,* Island Press, Washington, D.C., 1992. With permission.

with time, (3) restoration of original soil structure and function following disturbance, (4) biodiversity of soil-dwelling organisms, (5) resistance to weed establishment, and (6) stable populations of beneficial insects.

Several studies have demonstrated that the reestablishment of perennial cover on retired cropland can reduce soil erosion while restoring soil quality. The greater root biomass associated with perennial grasses (Richter et al., 1990) gives carbon inputs into the soil that can be several times greater than those into cultivated soils (Anderson and Coleman, 1985; Buyanovsky et al., 1987; McConnell and Quinn, 1988) and reduces rates of nutrient leaching relative to annual crops (Paustian et al., 1990). Active soil organic matter, available nutrients, water-stable aggregates, and polysaccharide content may recover under perennial grasses fairly quickly (Jastrow, 1987; McConnell and Quinn, 1988; Gebhart et al., 1994; Burke et al., 1995).

The benefits of a perennial cover were recognized by the authors of the U.S. Conservation Reserve Program (CRP), authorized by Title XII of the 1985 Food Security Act. This program redirected monetary resources and human efforts toward soil conservation and indirectly toward control of agricultural non-point-source pollution. It was designed to protect the most vulnerable U.S. cropland, with a goal to shift 16 to 18 million ha of highly erodible land from annual crop production to perennial vegetation for 10 years. Overall, the program keeps about 595 million t of soil from eroding into U.S. streams and rivers annually, equivalent to a 21% reduction in erosion on cropland (Bjerke, 1991).

The second characteristic of prairies that contributes to sustainability is plant biodiversity. Benefits of plant biodiversity include (1) nitrogen supplied by legumes, (2) management of herbivorous insects and some plant diseases, (3) soil biodiversity, and (4) ecosystem stability.

Legumes play a critical role in supplying nitrogen to most natural ecosystems. Over periods of several years, perennial legumes can increase the concentrations of both carbon and nitrogen in the soil, as well as influence the size and activity of the microbial community (Berg, 1990; Halvorson et al., 1991). Similarly, legumes are important in providing nitrogen within many pastures as well as multiple cropping systems (Davis et al., 1986; Mallarino et al., 1990). Studies have consistently shown higher dry matter yields in grass/legume mixtures than in grass monocultures (e.g., Barnett and Posler, 1983; Posler et al., 1993).

Biodiversity also plays an important role in pest regulation. The presence of non-host plant species can reduce insect density by interfering chemically or visually with host-finding behavior and thus colonization, feeding efficiency, movement among host individuals, and mate finding (Bach, 1980; Risch, 1981; Andow, 1990; Bottenberg and Irwin, 1992a; Coll and Bottrell, 1994). Moreover, reduced suitability of the microhabitat can reduce insect tenure time, oviposition, and larval survival, and can increase emigration rate (Tukahira and Coaker, 1982; Kareiva, 1985; Elmstrom et al., 1988). In some cases, diverse stands provide a more favorable habitat for parasitoids and predators, leading to reduced levels of insect herbivores (Letourneau and Altieri, 1983; Letourneau, 1987). The weight of published evidence suggests that reduced resource concentration, rather than increased numbers of natural enemies, accounts for most of the observed herbivore reductions within polyculture (Andow, 1991).

Similarly, numerous studies have shown benefits of plant species diversity in minimizing certain plant diseases (Burdon, 1987), particularly those diseases vectored by insects (Zitter and Simons, 1980). Establishing host plants within diversified stands can reduce insect landing rate, and thus initial colonization of the plot, by interfering chemically or visually with host-finding behavior (Irwin and Kampmeier, 1989; Bottenberg and Irwin, 1992a,b). This can, in turn, lead to reduced levels of disease in mixtures relative to monoculture (Power, 1987; Allen, 1989; Bottenberg and Irwin, 1992c). Plant species diversity can thus lower the rate of pathogen transmission among individual host plants.

Besides these benefits, biodiversity can beget biodiversity. For example, Miller and Jastrow (1993) noted a significant relationship between underground fungal and floristic species richness in prairie restorations. Diversity of soil organisms can have profound effects on plant mycorrhizal associations, nutrient-uptake ability, and nutrient retention and cycling.

Biodiversity can have ecosystem-level benefits, too. Several experimental studies have demonstrated that species-rich communities are more resilient and more efficient at using resources than species-poor communities (Naeem et al., 1994). McNaughton (1977; 1985) conducted experiments involving the grasslands of Serengeti National Park, Tanzania. Areas of roughly 16 m^2 with different diversities were marked; then exclosures were fenced to prevent grazing by migrant herds of zebra, gazelles, and wildebeest. Grazers reduced the biomass of diverse areas by only about 25%, whereas the less diverse areas lost about 75% of their biomass. Nitrogen limits the number of plant species that coexist in Minnesota grasslands; thus Tilman and Downing (1994) manipulated the number present by varying the amount of nitrogen applied on 207 plots each of 16 m^2. They started measuring biomass in 1987. The year 1988 featured a drought which reduced biomass differently on the different plots. After the drought, the species-diverse plots produced half their predrought biomass whereas the species-poor plots produced only about an eighth or less. In these studies, the more diverse communities were more resistant to change. Such studies indicate that plant biodiversity contributes directly to the resilience of grassland communities. Tilman et al. (1996) found that, in experimental plots of perennial grassland plant species, community productivity increased with plant biodiversity. Moreover, soil available nitrogen was used more completely in the more diverse plots, leading to less leaching potential.

Finally, resistance to invasion is another collective attribute of complex communities (Case, 1990; Drake, 1991). This property is important for the ability of a community to resist weeds and other exotic organisms.

AN AGRICULTURE MODELED ON THE PRAIRIE ECOSYSTEM

Elements of the Prairie Model

Remnant plant communities of the North American prairie are persistent biotic assemblages in which complex webs of interdependent plants, animals, and microbes garner, retain, and recycle critical nutrients, and protect the soil. As such they provide our best models of the types of communities needed to restore sustainable diversity to compromised ecosystems. Such diverse species assemblages, whose composition varies across soil types, tend to retain species, resist invasion by exotics, and are resilient during short-term climatic variation. Agricultural systems modeled on natural grassland ecosystems would comprise diverse plantings of perennial species that would prevent soil loss, provide much of their own nitrogen requirement via symbiotic nitrogen fixation, and resist invasion by weeds as well as outbreaks of insect pests and plant pathogens. They would be structural and functional analogs of prairie plant communities, composed predominantly of representatives from four major plant guilds: perennial C_3 and C_4 grasses, nitrogen-fixing species (primarily legumes), and composites (Asteraceae). These functional groups include the majority of prairie vegetation (Kindscher and Wells, 1995; Piper, 1995).

A major objective of research toward a sustainable agriculture is to develop innovative methods of production that minimize negative environmental impacts. The CRP, although very successful in terms of soil preservation, is expensive ($1.8 billion/year) (Osborn, 1993) and provides no edible product from the idled land. Hence, the goal of the Land Institute is to develop polycultures of perennial grains that protect the soil and provide the restorative properties of a perennial cover while yielding significant amounts of edible grain. Grain-producing mixtures of perennial grasses, legumes, and composites (e.g., sunflower species) would mimic the vegetation structure and sustainable function of native grasslands in some fundamental ways. Species composition of such perennial mixtures would vary geographically and with soil type. Several promising candidates for a perennial grain agriculture have been identified and evaluated (Wagoner, 1990; Soule and Piper, 1992). Potentially, the sustainable features of such an agriculture include improved soil retention and health, more efficient use of land area, lower fossil fuel dependence, diversity within and between crops to reduce vulnerability to pest and pathogen outbreaks, and greater on-farm predictability and flexibility. Because approximately 20% of U.S. on-farm energy usage is associated with traction (Lovins et al., 1995), perennial grain agriculture, by reducing seedbed preparation and cultivation, application of synthetic chemicals, and irrigation will also translate into savings in energy and materials costs for farmers.

Hence, elements of the prairie model to mimic a natural systems agriculture are (1) herbaceous perennials as grains and (2) species grown in diverse fields. The working model, then, comprises several perennial grain species representing four functional groups (i.e., C_4 grasses, C_3 grasses, nitrogen-fixing legumes, and composites) planted together.

To design persistent and diverse prairie-like grain fields, one must be cognizant of the broad similarity of grassland communities as well as the details of their differences. Surveys of locally and regionally occurring species and their local distributions (e.g., Piper, 1995) suggest the types of species likely to participate. First, there are some broad rules that hold across locations (e.g., representation by each of the four major guilds). Second, although one cannot predict perfectly the composition of the final successful community, the general structure of natural communities supported on different soil types gives clues to the types of species to emphasize in the mix (e.g., nitrogen fixers on poor soils). Third, because occasional extreme years can limit diversity considerably (Tilman and El Haddi, 1992), high biodiversity should enable a community to weather better the wide precipitation swings that characterize continental climates.

Perennials as Grains

The development of perennial seed crops for agronomic mixtures consists of two interrelated efforts. The first is the breeding of new perennial grains. This can involve the domestication of currently wild species as well as the improvement of wide crosses between annual grain crops and their perennial relatives. The second area comprises long-term studies of intercrop compatibility within diversified plantings. This work involves studies to discern beneficial and inhibitory crop interactions, growth and seed yield patterns, and effects on soil quality.

Examples of the first approach, the domestication of wild perennials, are experiments with the cool-season grasses mammoth wildrye and intermediate wheatgrass, the warm-season grass eastern gamagrass, the legume Illinois bundleflower, and Maximilian sunflower. Promising examples of the second approach, the development of perennial grains via wide crosses between annual grains and wild congeners, include studies with hybrid grain sorghum (*Sorghum bicolor* × *S. halepense*) and "Permontra" hybrid perennial rye (*Secale cereale* × *S. montanum*).

An obvious consideration before any new crop is adopted is yield. Seed yields of any new perennial grains need to be sufficiently high and stable to make their adoption by farmers compelling. A second consideration is the possible loss of long-term viability as a species is selected for higher seed yield. Studies of several perennial species (Reekie and Bazzaz, 1987; Horvitz and Schemske, 1988; Piper, 1992; Piper and Kulakow, 1994), however, have indicated that, within the ranges investigated, there are no strict "trade-offs" between increased seed yield, vegetative growth, likelihood of future reproduction, or survivorship. Jackson and Dewald (1994) demonstrated conclusively that a severalfold seed yield increase in eastern gamagrass was not associated with a decline in plant growth or survivorship. Similarly, there was no apparent trade-off between higher seed yield and rhizome production in crosses between annual and perennial sorghum species (Piper and Kulakow, 1994). Such results hold promise for research to increase seed yield without losing the perennial nature of a species.

Species Diversity

Intercropping, the simultaneous raising of different crops in the same place, makes use of species that complement one another spatially or seasonally. Relative to monocultures, intercropped systems can display more efficient use of land, labor, or resources, increased yield, and reduced loss to insects, diseases, and weeds (Francis, 1986; Vandermeer, 1989). Overyielding, a yield advantage in mixture relative to monoculture, can occur when interspecific competition in a mixture is less intense than intraspecific competition or where plant species enhance the growth of one another. Many factors can lead to overyielding. Crops may be released from competition for light by having different light requirements or differences in architecture that minimize shading. Roots of different species may explore different soil layers, or crop species may have complementary nutrient requirements or uptake abilities. Differences in seasonal period of nutrient uptake among crops can also promote overyielding. Intercrops may be more productive than monocultures crops for improving soil fertility, controlling soil erosion, lowering risk of total crop failure, and decreasing crop losses to insects and diseases. Thus, intercropping may satisfy several crop production goals simultaneously.

One difficulty facing the plant breeder is that it may be difficult to predict from its performance in monoculture how a crop will behave in polyculture. For example, some plants change their root architecture or patterns of nutrient uptake when grown in association with different species (Goodman and Collison, 1982; Jastrow and Miller, 1993). Shorter plants that are vigorous in monoculture may be shaded out by taller neighbors in polyculture. Thus, selection of varieties for use in perennial polyculture is inherently more complex than selection of varieties for monocultures due to interactions between variety and cropping system that may be unpredictable. So, in addition to the traits needed by any viable crop (e.g., adaptation to the growing environment, tolerance of or resistance to prevailing insect or disease organisms, and reasonably high and stable yield potential), scientists breeding crops for polyculture must also select for or against competitive ability, shade tolerance, and modifications to plant architecture that allow coexistence (Smith and Francis, 1986). Hence, evaluation of species interactions is critical in designing a breeding program for polycultures. Moreover, in perennial systems the outcomes of species interactions may differ in different years (e.g., Barker and Piper, 1995).

Mechanisms that reduce overlap in resource demand among coexisting species usually involve differences in location and timing of resource use. Roots of neighboring species may explore different soil layers or a requirement might be met by different resources. Mixtures of legumes with nonlegumes frequently demonstrate yield advantages over monocultures because the two species are tapping different N sources which minimizes competition for this nutrient. Similarly, intercrops may be released from competition for light, and show greater overall productivity, if canopies of component crops occupy different vertical layers (Davis et al., 1984; Clark and Francis, 1985). Differences in length of the growing period or in the seasonal periods of nutrient uptake among crops (e.g., Piper, 1993a) can also reduce direct competition and thus promote overyielding (Francis et al., 1982; Smith and Francis, 1986).

Alternatively, a plant may benefit its neighbors by providing cover (Vandermeer, 1980), nitrogen (Wagmare and Singh, 1984), pest or disease protection (Risch et al., 1983; Burdon, 1987), protection from desiccating winds (Radke and Hagstrom, 1976), physical support against lodging (references in Trenbath, 1976), enhancement of mycorrhizal associations (Jastrow and Miller, 1993), or by attracting pollinators (Rathcke, 1984). Indirect facilitation can occur where one species "traps" nutrients that would otherwise leach or be lost from the system, and which later become available to other species (e.g., Agamathu and Broughton, 1985).

The Land Institute's ongoing research agenda in Natural Systems Agriculture revolves around four basic agronomic questions:

1. Can a perennial grain yield as well as an annual grain crop?
2. Can a perennial grain polyculture overyield?
3. Can a perennial grain polyculture provide its own nitrogen fertility?
4. Can a perennial grain polyculture manage weeds, insect pests, and plant pathogens?

RESEARCH AGENDA AND FINDINGS THAT SUPPORT THE MODEL

Question 1: Can a Perennial Grain Yield As Well As an Annual Grain?

Work at the Land Institute to domesticate perennial grains began in 1978 with an inventory of nearly 300 herbaceous perennial species for their suitability to the environment of central Kansas and their promise as seed crops. A second inventory examined the agronomic potential in 4300 accessions of perennial grass species within the C_3 genera *Bromus, Festuca, Lolium, Agropyron,* and *Elymus* (*Leymus*). From these inventories, a handful of perennial species was chosen for exploring the principles of perennial grain agriculture.

Eastern gamagrass (*Tripsacum dactyloides* [L.] L.) is a large C_4 bunchgrass native from the southeastern U.S. and Great Plains southward to Bolivia and Paraguay (Great Plains Flora Association, 1986). A relative of maize (de Wet and Harlan, 1978), eastern gamagrass has long been recognized as a nutritious and highly productive forage. Because of its high-quality seed (27 to 30% protein and 7% fat, Bargman et al., 1989) and large seed size, however, gamagrass shows much promise also as a grain crop for consumption by people, livestock, or both. Ground seed has baking properties similar to those of cornmeal. The major hurdle facing eastern gamagrass as a grain crop is low seed yield (typically around 100 kg/ha, but as high as 250 to 300 kg/ha in some material at the Land Institute, Piper, unpublished data).

Mammoth wildrye (*Leymus racemosus* (Lam.) Tsveler) is a rhizomatous C_3 grass native to Bulgaria, Romania, Turkey, and western parts of the former Soviet Union. Grain of this and closely related species was reportedly eaten by Asian and European people historically, especially in drought years when annual grain crops faltered (Komarov, 1934). As is typical of cool-season grasses, wild rye displays most of its growth in late autumn and early spring. Seed yield in Land Institute trials has been as high 830 kg/ha (Piper, 1993b).

Illinois bundleflower (*Desmanthus illinoensis* (Michx.) MacM.) is a nitrogen-fixing legume that forms a deep taproot in its first year. It is native to the Great Plains, with a range extending north to Minnesota, east to Florida, and as far west as New Mexico (Great Plains Flora Association, 1986). It grows best during warm weather, flowering from late June onward. Small lenticular seeds are borne within clusters of brown pods beginning in late July. Highest yields have approached 2000 kg/ha (Piper, 1993b; unpublished data). The nutritional quality of the seeds (38% protein, 34% carbohydrate, Piper et al., 1988) suggests great potential as a grain legume.

Wild or Maryland senna (*Cassia marilandica*) is a legume native to the south-eastern region of the Great Plains. Flowering in Kansas takes place from late August to early September, producing racemes of insect-pollinated yellow flowers that become brown-black pods later in the fall. *C. marilandica* produces thick, deep roots, but does not appear to form symbioses with nitrogen-fixing *Rhizobium* bacteria. The Land Institute has examined its year-to-year patterns of seed yield to address the biological question of whether a herbaceous perennial can produce a sustained, high yield. It provides a good model for studying the population dynamics of a high-seed-yielding perennial (Piper, 1992).

Maximilian sunflower (*Helianthus maximilianii* Schrad.) is native to dry to moist open prairies throughout the Great Plains. Its range extends eastward to Maine and North Carolina, and westward to the Rocky Mountains from southern Canada to Texas (Great Plains Flora Association, 1986). Maximum seed yields ranged from 1460 to 1840 kg/ha in Land Institute trial plots in 1996 (J. K. Piper, unpublished data). In addition to its potential value as a food or oil seed crop (seed is 21% oil, Thompson et al., 1981), Maximilian sunflower appears to inhibit weed growth allelopathically, and may therefore be especially important during the establishment phase of a perennial grain field.

Grain sorghum (*Sorghum bicolor*), a native of the African continent, is grown extensively in the southern Great Plains as a seed crop for animal feed. It is weakly perennial in tropical regions, but is killed by frost at higher latitudes. Johnsongrass (*Sorghum halepense*), a weedy relative of cultivated sorghum, is in the U.S. a troublesome weed that overwinters by production of rhizomes, fleshy underground stems capable of winter survival. It may be feasible to convert a tetraploid variety of grain sorghum from an annual to a perennial growth habit by combining in hybrids good grain quality with the ability to produce winter-hardy rhizomes (Piper and Kulakow, 1994). The ease of making this transfer will depend on the number of genes controlling the production of rhizomes and whether or not overwintering ability is genetically associated with poor agronomic characteristics.

Under favorable growing conditions, high yields of some perennial grasses can range from 1500 to 2000 kg/ha (Ahring, 1964; Ensign et al., 1983; Mueller-Warrant et al., 1994). Extrapolated seed yields ranging from 1720 to 2090 kg/ha have been recorded for some perennial grasses and legumes in rain-fed, unfertilized plots at the Land Institute (Piper, 1992; 1993b; Piper and Kulakow, 1994). These experimental yields compare favorably with the benchmark yield for Kansas winter wheat of 1800 lb/ac (1960 kg/ha).

Germplasm evaluations at the Land Institute indicate that genetic variability for improvement of such traits as seed yield, loss of seed dormancy, shatter resistance, resistance to viral diseases, and overall vigor is available. Moreover, including in a gamagrass breeding program such favorable types as a gynomonoecious form prolifica (a mutant sex form in which all florets are female and thus seed producing) may increase seed yield severalfold (Dewald et al., 1987). Currently, the Land Institute's gamagrass breeding program is focusing on 23 high-yielding, disease-resistant families to create a synthetic variety for further improvement.

Question 2: Can a Perennial Polyculture Overyield?

Competition between individual plants should be stronger in monocultures than in mixtures because conspecific neighbors are most similar morphologically, phenologically, and in nutritional requirements. If intraspecific competition is stronger than interspecific competition or if facilitation is occurring, plants should yield relatively better in mixture than in monoculture, resulting in overyielding. Conversely, if the effects of interspecific competition are greater than those of intraspecific competition (as when species differ greatly in size, for example), then plants should perform relatively better in monocultures. If the relative effects of different species are neutral, then there is no yield advantage to polyculture.

Overyielding is common in traditional polyculture systems of Latin America, Africa, and Asia (Francis, 1986; Vandermeer, 1989). Whether overyielding can occur in perennial grain systems, and whether it can persist for more than 1 year, is less explored. A few examples from experiments performed at the Land Institute support the model.

In a study of 28 accessions of eastern gamagrass and Illinois bundleflower, a relative yield total (RYT) of 1.19 was obtained (Muto, 1990). This translates into a 19% yield advantage in mixture relative to the monocultures. A three-species study using eastern gamagrass, Illinois bundleflower, and mammoth wildrye produced an RYT of 1.26 (Barker and Piper, 1995). In both of these cases, overyielding occurred in more than 1 year.

Question 3: Can a Perennial Polyculture Provide Its Own Nitrogen Fertility?

The growing environment of perennial systems changes annually. Typical initial conditions include high levels of sunlight and available soil nutrients. With time, concentrations of some available nutrients decline while root mass and shading increase. With canopy closure, species interactions that began as competition for soil nutrients may end as competition primarily for light, with plants exhibiting a trade-off between tolerance of low soil resources and tolerance of shade (Tilman, 1985). Various factors, including root architecture, physiology (C_3 vs. C_4 photosynthetic pathway), and ability to fix atmospheric nitrogen can result in differences among species in net soil resource levels. Understanding the net effects of perennial grains on soil nutrient status is an important consideration for long-term fertility management of perennial stands to be grown with few or no inputs.

The ability of legumes to fix atmospheric nitrogen via symbioses with root nodule–forming bacteria has important implications for agricultural sustainability. It can release a plant from competition with neighbors for soil nitrogen and promote the growth of neighbors if fixed nitrogen subsequently becomes available to them. Nitrogen transfer from legumes to grasses may occur via leakage and excretion from roots (Simpson, 1965), following decay of nodules and roots (Haynes, 1980), and by direct mycorrhizal exchange (van Kessel et al., 1985; Eissenstat, 1990). Perennial grasses may receive from 46 to 80% of their nitrogen directly from companion legumes (Brophy et al., 1987). In agricultural systems, legume nitrogen can prove both energy efficient and cost-effective (Mallarino et al., 1990; Posler et al., 1993).

Ideally, in a sustainable agriculture based on perennial grain mixtures, a significant portion of the available nitrogen should arise from one or more leguminous companion crops. Therefore, an important question is whether or not a perennial polyculture can provide sufficient nitrogen fertility via legume nitrogen fixation to compensate for removal of this nutrient in harvested seed.

Work at the Land Institute has provided some indirect evidence of a benefit of a leguminous grain to companion species. A study monitored over several years showed that Illinois bundleflower grows and yields equally well on high and low nitrogen soil in years with adequate precipitation (Barker and Piper, 1995). This suggests that this legume can compensate for low soil nitrogen without a measurable penalty of lowered growth or seed yield. A piece of evidence is that eastern gamagrass yields better, and its yield does not decline with time, in mixtures with bundleflower relative to gamagrass monoculture (Piper, 1998). Finally, plots containing Illinois bundleflower have higher soil nitrate concentrations 3 to 5 years after establishment than plots containing only grasses (Piper, unpublished data).

Question 4: Can a Perennial Polyculture Manage Weeds, Herbivorous Insects, and Plant Pathogens?

Weeds

Weeds are a major source of competition with crops for light, water, and nutrients. Because most agricultural weeds are adapted to disturbed habitats, and can oftentimes take up soil water and nutrients faster than crops, they are a chronic problem where soils are repeatedly tilled. Thus, it is not surprising that herbicides account for 69% (by weight) of all pesticide use in the U.S. Nearly 90 million ha, more than half of all U.S. cropland, are treated with herbicide (Pimentel et al., 1991). Perennial grain polycultures are likely to compete well against weeds. Two important aspects of the crop/weed relationship in perennial polyculture are the possible synergistic effects of polycultures on weeds and the direct competitive relationship between perennials and weeds.

In contrast to rotational sequences of monocultures, which can manage weeds over time, intercropping combines the weed-suppressing effects of different crops within a single season. Crop polycultures may intercept more light, water, and nutrients than monocultures. For example, mixtures of C_3 and C_4 species display active crop growth during a greater proportion of the year, potentially eliminating the need for herbicide application (Evers, 1983). If weed mass in polyculture rows is less than what is predicted by weed mass in the respective monocultures, this may indicate a synergistic effect of polycultures on weeds.

An intercrop advantage may accrue in two ways (Liebman and Dyck, 1993). Greater crop yield may be coupled with lower weed growth in polyculture. This may occur if there is greater resource preemption by crops in the mixture (less available to weeds) or where there is allelopathic suppression by one or more crops. Alternatively, the intercrop may display a yield advantage while failing to suppress weed growth below levels observed within monocultures of the component crops. This may result if the intercrops use resources not exploitable by weeds or where intercrop use of resources is more efficient than in monoculture.

Studies done primarily in traditional cropping systems in the tropics have shown that internal weed management in polycultures is often greater than in monocultures of the respective crops (Liebman, 1988). Similarly, several temperate zone intercropping systems have been shown to improve weed management in important ways. Liebman and Dyck (1993) reviewed 51 published intercropping studies where the main crop was intersown with a "smother" crop. In 47 cases (92.2%), weed biomass was lower in mixture than where the main crop was grown alone. When intercrops were composed of two or more main crops, weed biomass of the intercrop was lower than in all of the component sole crops in 12 cases (50%), intermediate between component crops in 10 cases (41.7%), and higher than all sole crops in only 2 cases (8.3%).

For certain combinations, increasing crop species diversity per se may suppress weeds. In an ongoing experiment at the Land Institute using experimental mixtures of 4, 8, 12, and 16 perennial species (see section on Community Assembly below), percentage cover by annual weeds in the second year was inversely related to the diversity of perennials sown, an effect which was stronger in the third year (see Figure 1B, later).

Second, because of their permanent canopy, deep and extensive root systems, and vigorous regrowth in spring perennial plants should in general compete well against weeds. Several factors affect the relative ability of crop species to suppress weeds, however. Differences in height, canopy thickness, rooting zone, and phenology are likely to influence crop/weed interactions. Studies of prairie restoration typically show an initial period of dominance by weedy annuals, followed by increasing dominance by an array of herbaceous perennials (e.g., Holt et al., 1995).

Effective weed control has occurred in two separate experiments at the Land Institute. In one study, a plot containing rows planted with five densities of Maximilian sunflower and a control (no sunflowers), weed biomass was significantly reduced in the sunflower plots relative to the control (discussed in Piper, 1993b). By the second year, a sunflower density of 3.6 plants/m^2 reduced weed biomass between rows to levels only 25 to 50% of the control. In May of the third year, weed biomass in sunflower rows was 44% of weed biomass in the control. Here, effective weed control was maintained across years despite changes in the weed community from predominantly annuals in the first year to perennials by the third year.

In a second experiment that examined weed growth, a triculture comprising wildrye, eastern gamagrass, and Illinois bundleflower at equal densities, species combinations differed in their ability to control weeds (Piper, 1993c). Weed biomass was consistently lowest in rows with eastern gamagrass as a component, despite seasonal and yearly changes in species composition of the weed community. These results point to eastern gamagrass as the primary weed controller among the three species, probably via shading, although unmeasured underground interactions were also likely important throughout the 3-year study.

Insect Pests

For many vegetable and grain crops, pest insect density and level of feeding damage on host species are lower within diversified stands than in monoculture (Risch et al., 1983; Andow, 1986; 1991; Coll and Bottrell, 1994). In an extensive review of the literature, Andow (1986) found that 131 of 203 (64.5%) monophagous species were reported to be less abundant in polyculture cropping systems relative to monoculture. This phenomenon, in which host plants associated with other, non-host plant species suffer less herbivore attack than host plants in monoculture, is known as associational resistance (Tahvanainen and Root, 1972).

Although many of the pest management benefits obtained within diversified annual systems may transfer to perennial polycultures, typically perennial systems are characterized by less soil disturbance and greater year-to-year continuity of the host species. Pests therefore have the opportunity to maintain and even increase population density in host plant patches that are relatively stable and predictable. Unfortunately, the long-term dynamics of insect populations within perennial monocultures vs. polycultures remains relatively unexplored.

The *Desmanthus illinoensis*/*Anomoea flavokansiensis* association represents one system for exploring whether perennial grain polycultures can reduce insect density relative to monoculture. The leaf beetle *A. flavokansiensis* Moldenke (Coleoptera: Chrysomelidae: Clytrinae) specializes locally on Illinois bundleflower. From mid-June to early August adults feed on young bundleflower leaves and inflorescences. At high densities, *A. flavokansiensis* can reduce seed yield and is thus an important consideration for long-term stands that are to be grown without insecticides.

The potential to manage *A. flavokansiensis* via intercropping its host species with other, non-host perennial species has been examined (Piper, 1996). Replicated plots, comprising monocultures of Illinois bundleflower, and two- and three-species mixtures of bundleflower with the eastern gamagrass and mammoth wildrye, were established at two sites. Over a 5-year period, the way that the plant species diversity affects *A. flavokansiensis* density on individual host plants and how beetle density changes with time in perennial stands were examined.

Insects were counted frequently from mid-June to early August. In the first 3 years, beetle density was generally low (<1 per plant), and did not differ among treatments. In the fourth year, however, beetle density peaked at 15 and 25 insects per plant at the two sites, and was highest within bundleflower monoculture for most dates. In year 5, density was again low, but tended to remain higher in monoculture at one site. The results show that beetle density on Illinois bundleflower can be reduced in polyculture and hold promise for the management of this type of insect herbivore within perennial grain polycultures. Differences in beetle density were probably due more to differences in resource concentration, combined with physical barriers provided by the grasses, than an increased presence of natural enemies in polycultures.

It is encouraging that density of this monophagous insect on its perennial host was reduced in some instances in polyculture even 4 and 5 years following establishment. Monitoring of the year-to-year population dynamics of this insect may predict how similar specialist herbivores will react to plant species diversity in perennial grain mixtures.

Plant Disease

A potential hazard confronting stands of perennial grains is the establishment and long-term residency of systemic diseases. Perennials may be capable of storing inoculum from year to year, enabling spread of the disease throughout the field in later years. One possibility for the management of such diseases is to grow host plants within polycultures.

Eastern gamagrass is infected by the pathogens sugarcane mosaic virus strain maize dwarf mosaic virus B and maize dwarf mosaic virus. These are aphid-borne, nonpersistently transmitted polyviruses that infect and then overwinter in gamagrass (Seifers et al., 1993). Such diseases have the potential to increase in severity with time. Symptoms of infection vary from a general mosaic to chlorotic and necrotic spots dispersed throughout the leaf. At high levels of severity, these diseases reduce growth and seed yield (Piper et al., 1996).

To examine the potential for polyculture to manage levels of aphid-borne disease in a perennial system, plots comprising monocultures of eastern gamagrass as well as mixtures of gamagrass with two nonhost species, mammoth wildrye and Illinois bundleflower, were established and monitored for 5 years (Piper et al., 1996). For the first 2 years at one of two sites, disease incidence (percent of plants infected) and severity (intensity of symptoms) were generally lower in plots in which bundleflower was a component. Because there were two biculture treatments in this study, the main factor seemed to be the presence of bundleflower within a plot, not merely lower host plant density relative to monoculture. Illinois bundleflower appears sufficiently different, in terms of architecture and biochemistry, from gamagrass to provide an effective barrier to vector movement. The results suggest that incidence and severity of these viral diseases can be delayed in some mixes of perennial grains, and that the effect can persist for more than 1 year.

This study showed that manipulations of host plant density and diversity may be useful in designing cultural methods for minimizing disease intensity and spread in the early years. Intercropping eastern gamagrass with such nonhost species as Illinois bundleflower appears to represent a disease management strategy complementary to selecting for viral resistance.

COMMUNITY ASSEMBLY

Theoretical and experimental studies of community organization have revealed some intriguing phenomena with implications for the development of stable perennial grain polycultures. Such studies have shown that complexity is often the consequence of specific events that occur during community assembly — events that are generally undetectable from an examination of extant community patterns alone (Post and Pimm, 1983; Gilpin, 1987; Robinson and Dickerson, 1987; Drake, 1991; Drake et al., 1993). Complex stable communities in nature are not created as is; they assemble by the addition of species through invasion and by the loss of species through local extinction. Species that form important intermediate states must be included within the appropriate time frame. The historical sequence in which species are added to and removed from a community determines its final structure.

Community stability changes as species are progressively added to or subtracted from model food webs (Tregonning and Roberts, 1979; Roberts and Tregonning, 1981). A principal result of this work is that most "new" complex systems are unstable; stability comes with age. Drake (1990a) found that with unlimited invasion opportunities communities eventually reach persistent species compositions where none of the species "outside" the community can enter them. In addition, there may be so-called Humpty-Dumpty effects (Drake, 1990b; Luh and Pimm, 1993), meaning it may be impossible to reassemble a community solely from the species present in the final, persistent state. This is because species present early in assembly that were important in determining the final structure may not be present in the final composition.

Such results suggest that mimicking the history of a natural community is a powerful tool for constructing stable perennial polycultures. Persistent, diverse grain fields may be created more efficiently by starting with many more than the desired number of species, and allowing the community to "shake down" or "collapse," than by experimenting with many mixes of the desired numbers of species. If this prediction is correct, it will confirm an important role for history in constructing stable, complex communities.

In 1994, the Land Institute began a study to explore whether plant biodiversity and the sequence of species introductions can contribute to the assembly of perennial grain polycultures. Treatments consist of four incrementally diverse mixtures of herbaceous perennial species that represent the C_4 grasses, C_3 grasses, nitrogen-fixing legumes, and composites. Several of the species are potential perennial grains. The initial seed mixtures comprise 4, 8, 12, and 16 grassland species (Table 2). None of the species was present at the site prior to the experiment. Each lower diversity treatment was nested within its higher diversity counterpart (see Naeem et al., 1994). The treatments vary the size of the species pool, while keeping guild representation constant.

Table 2 Species Composition of Four Diversity Treatments at the Land Institute

	C_4 Grasses	C_3 Grasses	Legumes	Composites
I	*Tripsacum dactyloides*	*Leymus racemosus*	*Desmanthus illinoensis*	*Helianthus maximilianii*
II	*Tripsacum dactyloides*	*Leymus racemosus*	*Desmanthus illinoensis*	*Helianthus maximilianii*
	Sorghum bicolor × *S. halepense*	*Elymus glaucus*	*Dalea purpurea*	*Helianthus mollis*
III	*Tripsacum dactyloides*	*Leymus racemosus*	*Desmanthus illinoensis*	*Helianthus maximilianii*
	Sorghum bicolor × *S. halepense*	*Elymus glaucus*	*Dalea purpurea*	*Helianthus mollis*
	Panicum virgatum	*Agropyron intermedium*	*Lotus corniculatus*	*Ratibida pinnata*
IV	*Tripsacum dactyloides*	*Leymus racemosus*	*Desmanthus illinoensis*	*Helianthus maximilianii*
	Sorghum bicolor × *S. halepense*	*Elymus glaucus*	*Dalea purpurea*	*Helianthus mollis*
	Panicum virgatum	*Agropyron intermedium*	*Lotus corniculatus*	*Ratabida pinnata*
	Eragrostis trichodes	*Agropyron smithii*	*Amorpha canescens*	*Liatris pycnostachya*

Once seeded, half the plots within each treatment are left alone to assemble without further intervention. The other half of the plots are resown with any species that fail to establish, or that disappear after having established initially. The first treatment broadly corresponds to a theoretical recipe of letting excessively species-rich systems collapse to stable end points (Tregonning and Roberts, 1979; Roberts and Tregonning, 1981), the second to methods of allowing both invasion and local extinction during assembly (Post and Pimm, 1983; Drake, 1990a).

During the first 4 years, plot diversity varied considerably in the 12- and 16-species treatments, but remained steadier in the 4- and 8-species treatments (Figure 1A). Lower diversity in the fourth year for the higher diversity treatments was due to declining numbers of weeds in these plots (Figure 1B). Plant communities of

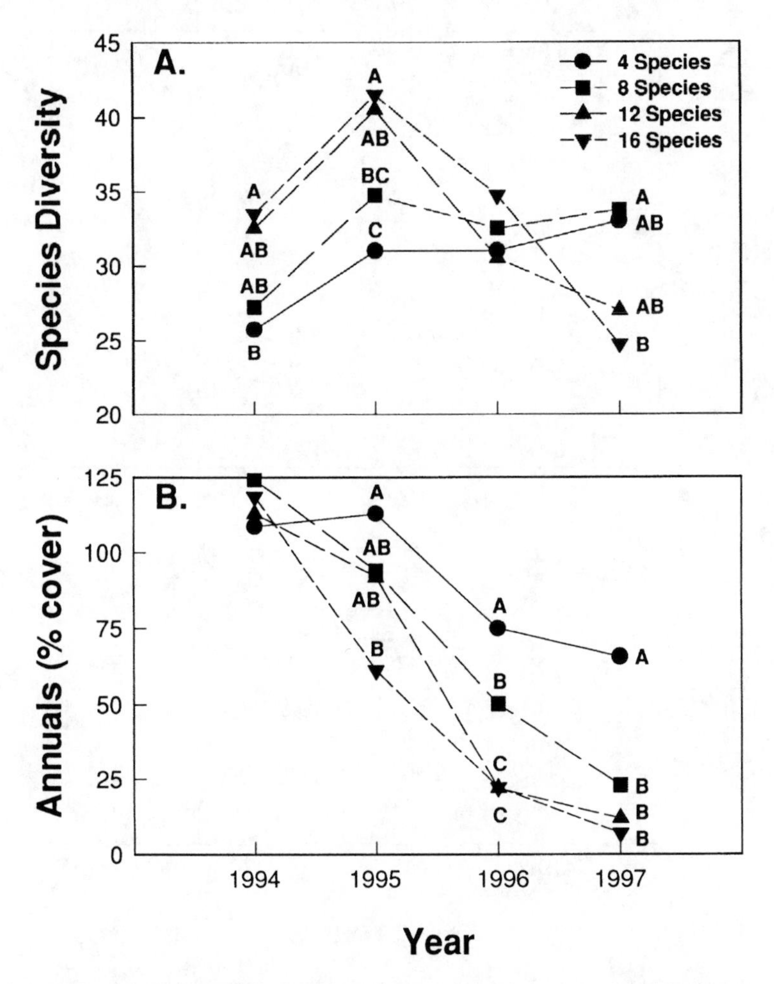

Figure 1 (A) Species diversity and (B) percentage cover by annual species over 4 years for four diversity treatments seeded in 1994 ($n = 16$ plots). For each year, symbols with the same letter do not differ at $p < 0.05$.

healthy prairie are over 99% perennial species, whereas annuals constituted most of the cover within the community assembly plots. By the second year, 1995, however, the percentage of cover by annuals was already decreasing as treatment diversity increased. In the fourth year, weedy cover in the most diverse treatment was ~1/10 that of the least diverse treatment (7 vs. 65%). Even field bindweed (*Convolvulus arvensis*), a pernicious perennial weed, declined as treatment diversity increased (data not shown). Of the four major plant guilds, percentage cover by perennial C_3 grasses and composites increased significantly with diversity treatment, and legumes showed a trend in this direction (Figures 2 and 3). Percentage cover by perennial grasses and composites is on a steady upward trajectory, with overall cover by perennial composites, in particular, increasing more than 12-fold from 1994 to 1997.

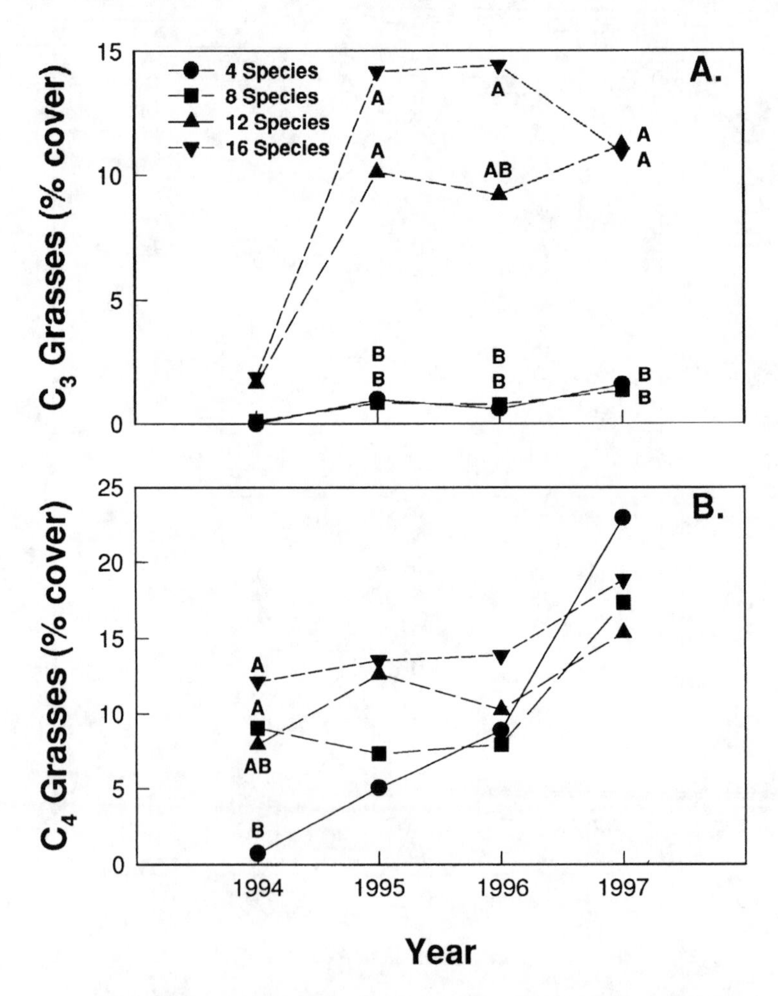

Figure 2 Percentage cover by (A) perennial C3 grasses and (B) perennial C4 grasses over 4 years for four diversity treatments seeded in 1994 (n = 16 plots). For each year, symbols with the same letter do not differ at $p < 0.05$.

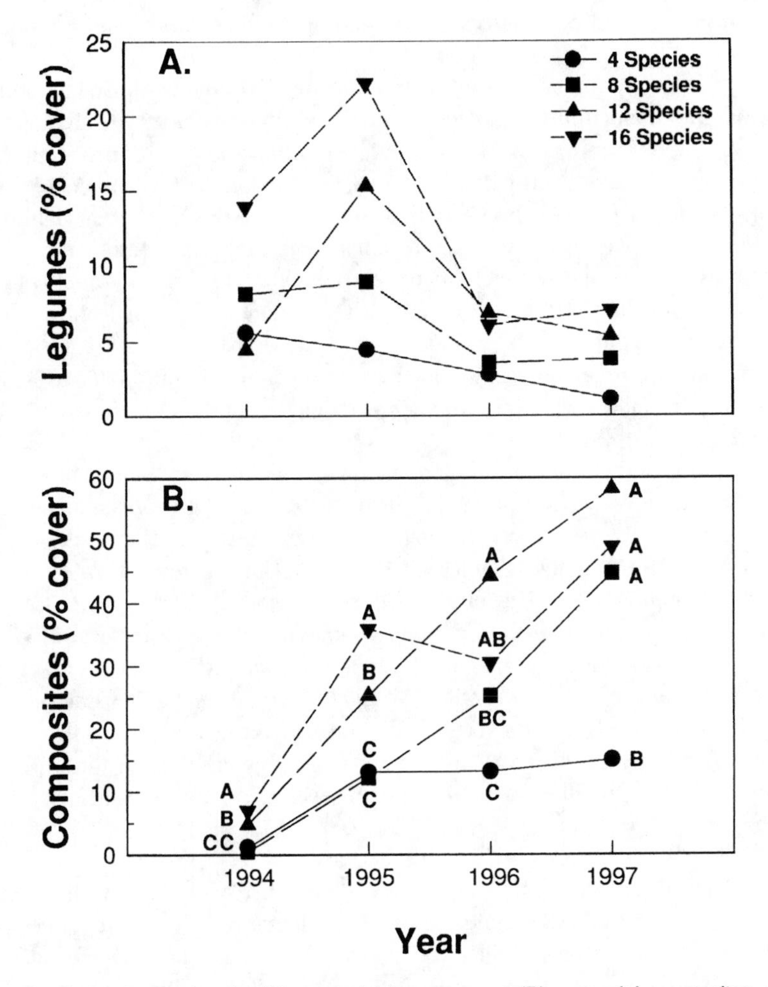

Figure 3 Percentage cover by (A) perennial legumes and (B) perennial composites over 4 years for four diversity treatments seeded in 1994 (n = 16 plots). For each year, symbols with the same letter do not differ at $p < 0.05$.

The results show that, after just 2 or 3 years, assembly of these plots toward diverse, perennial communities becomes apparent.

CONCLUDING REMARKS

For several years, biodiversity has been the major buzzword among conservation biologists as measures of diversity are commonly used as indicators of the condition of ecological systems. Such books as Olson et al. (1995) demonstrate that the concept of diversity is beginning to find significant applications in U.S. agriculture, too. Diversity is widely thought to be a beneficial, even essential, property of healthy biological, social, and economic systems, and agriculture intersects the realms of all three.

Many of the environmental problems associated with industrialized agriculture arise from the monoculture mind-set. Perennial grain polyculture, as a component of a diversified farming operation, could simultaneously produce significant amounts of grain while providing the benefits of soil and water conservation currently achieved through the CRP. It would enhance the long-term viability of agriculture by providing food without exhausting or contaminating the natural resource base upon which productivity depends. Perennial grain fields could provide an alternative to the CRP, increase year-to-year predictability, and allow greater within-season flexibility (i.e., provide grain, hay, or pasture). In exploring this paradigm shift from annual grain monoculture to perennial grain fields as analogs of natural grassland ecosystems, we acknowledge that the work is inherently long term and transdisciplinary.

The grain varieties that would be grown in perennial polyculture do not yet exist. Thus, a long-term domestication process is required. Because we may not see great progress for a few years, such an endeavor requires great patience on the parts of both scientists and funding agencies. Moreover, because perennial grains will be required to maintain themselves in the field for a period of years instead of months as with annual crops, it is necessary to incorporate multiple-year patterns of growth and seed yield into crop development research. For example, will yields be predictable year to year, or will they oscillate wildly? How will interactions among crop species in polyculture change over years? Unlike the grower of annual crops, who has some flexibility in changing crops after each growing season, the grower of perennial polyculture will have to project over several possibly very different growing seasons.

The work is also necessarily interdisciplinary and cooperative, not only within the plant sciences but among the plant sciences and other sciences, as well as among the "hard" sciences and the social sciences. It is, therefore, going to require a great deal of cooperation among people who previously may not have interacted much outside their areas of expertise. A research team to explore natural systems agriculture would include one or more ecologists, including those who specialize in crop–weed interactions, soils, insects, and plant pathogens. Several plant breeders would be needed to focus on perennial grasses, legumes, and composites. The crop development and ecological research would be aided by scientists with expertise in biotechnology and computer modeling. An environmental historian would put the work in context. This list is by no means exhaustive, and several additions are likely.

The work at this point remains somewhat risky both in terms of satisfying funders seeking short-term payoffs as well as for professional advancement (Soule and Piper, 1992). As we continue to deplete our soils and contaminate the environment, however, the long-term riskiest path for agriculture may well be the one we are on at present. Perennial grain agriculture represents a largely unexplored approach that will provide new avenues and insights for agricultural researchers and other scientists. Innovative and creative alternatives, however, are likely to provide the greatest insights and discoveries within science.

Humanity's utter dependence upon a thin layer of active soil is summed up in "Albrecht's dilemma," named after Missouri soil scientist William A. Albrecht. To feed ourselves, we appropriate nature's bounty by plowing up the prairie soil, planting crops which produce more useful food than could be obtained from the wild, but in the process accelerate the loss of nutrients from the soil. The movement of productive soil from the land to the sea is hastened, to our peril, by this necessary human activity. Opie (1993) has stated that Albrecht's dilemma is in essence an ethical paradox facing humanity and the environment. It is the conflict between the inevitability of soil destruction by farming and the imperative of soil protection to ensure a continued food supply. Most approaches toward a sustainable agriculture can at best reduce somewhat the present environmental problems created by modern agricultural production. Natural systems agriculture, in contrast, offers the possibility of achieving ecological restoration as a consequence of agricultural production. Studies of experimental grassland systems are showing that diversity, productivity, and sustainability can go together (Tilman et al., 1996). The question facing us is whether the tragedy of soil loss will be allowed to continue or whether the natural systems approach will be explored as a means of reconciling the two priorities of food production and preserving our ecological capital.

A perennial polyculture of grains is really a "domesticated prairie." It is admittedly a radical departure from current agriculture. It involves new crop species, selected under much more complex regimes than are used for crops destined for monoculture, and new designs for crop mixtures. Novel approaches toward a sustainable agriculture highlight the critical importance of maintaining biodiversity. Efforts to safeguard wild species, and the ecosystems that contain them, are necessary to provide humanity with the widest possible range of options for the domestication of perennial grains. The principles discovered along the way are likely to be applicable across an array of agroecosystem types. Thus, in the work to create a natural systems agriculture much new research ground must be broken, but in the process, it is hoped, the broken ground of the prairie will begin to be healed.

ACKNOWLEDGMENT

I thank Brian Donahue for critiquing an earlier version of this chapter.

REFERENCES

Agamathu, P. and Broughton, W. J., 1985. Nutrient cycling within the developing oil palm–legume ecosystem, *Agric. Ecosyst. Environ.*, 13:111–123.

Ahring, R. M., 1964. The Management of Buffalograss *Buchloe dactyloides* (Nutt.) Engelm. for Seed Production in Oklahoma, Tech. Bull. T-109, OK State Univ. Agric. Exp. Stn., Stillwater.

Allen, D. J., 1989. The influence of intercropping with cereals on disease development in legumes, presented at CIMMYT/CIAT Workshop on Research Methods for Cereal/Legume Intercropping in Eastern and Southern Africa, CIAT Regional Bean Program, Arusha, Tanzania.

Anderson, D. W. and Coleman, D. C., 1985. The dynamics of organic matter in grassland soils, *J. Soil Water Conserv.*, 40:211–216.

Andow, D. A., 1986. Plant diversification and insect population control in agroecosystems, in *Some Aspects of Integrated Pest Management,* D. Pimentel, Ed., Cornell University Press, Ithaca, NY, 277–368.

Andow, D. A., 1990. Population dynamics of an insect herbivore in simple and diverse habitats, *Ecology,* 71:1006–1017.

Andow, D. A., 1991. Vegetational diversity and arthropod population response, *Annu. Rev. Entomol.,* 36:561–586.

Bach, C. E., 1980. Effects of plant density and diversity on the population dynamics of a specialist herbivore, the striped cucumber beetle, *Acalymma vittata* (Fab.), *Ecology,* 61:1515–1530.

Bargman, T. J., Hanners, G. D., Becker, R., Saunders, R. M., and Rupnow, J. H., 1989. Compositional and nutritional evaluation of eastern gamagrass (*Tripsacum dactyloides* (L.) L), a perennial relative of maize (*Zea mays* L.), *Lebensm. Wiss. Technol.,* 22:208–212.

Barker, A. A. and Piper, J. K., 1995. Growth and seed yield of three grassland perennials in monocultures and mixtures, in *Prairie Biodiversity: Proceedings of the 14th North American Prairie Conference,* D. C. Hartnett, Ed., Kansas State University Press, Manhattan, 193–197.

Barnett, F. L. and Posler, G. L., 1983. Performance of cool-season perennial grasses in pure stand and in mixtures with legumes, *Agron. J.,* 75:582–586.

Berg, W. A., 1990. Herbage production and nitrogen accumulation by alfalfa and Cicer Milkvetch in the southern plains, *Agron. J.,* 82:224–229.

Bjerke, K., 1991. An overview of the agricultural resources conservation program, in *The Conservation Reserve — Yesterday, Today and Forever,* L. A. Joyce, J. E. Mitchell, and M. D. Skold, Eds., USDA Forest Service, Washington, D.C., 7–10.

Blair, A., Zahm, S. H., Pearce, N. E., Heineman, E. F., and Fraumeni, J. F., Jr., 1992. Clues to cancer etiology from studies of farmers, *Scand. J. Work Environ. Health,* 18:209–215.

Blank, R. R. and Fosberg, M. A., 1989. Cultivated and adjacent virgin soils in northcentral South Dakota: I. Chemical and physical comparisons, *Soil Sci. Soc. Am. J.,* 53:1484–1490.

Bottenberg, H. and Irwin, M. E., 1992a. Canopy structure in soybean monocultures and soybean-sorghum mixtures: impact on aphid (Homoptera: Aphididae) landing rates, *Environ. Entomol.,* 21:542–548.

Bottenberg, H. and Irwin, M. E., 1992b. Flight and landing activity of *Rhopalosiphum maidis* (Homoptera: Aphididae) in bean monocultures and bean-corn mixtures, *J. Entomol. Soc.,* 27:143–153.

Bottenberg, H. and Irwin, M. E., 1992c. Using mixed cropping to limit seed mottling induced by soybean mosaic virus, *Plant Dis.,* 76:304–306.

Brophy, L. S., Heichel, G. H., and Russelle, M. P., 1987. Nitrogen transfer from forage legumes to grass in a systematic planting design, *Crop Sci.,* 27:753–758.

Burdon, J. J., 1987. *Diseases and Plant Population Biology,* Cambridge University Press, Cambridge, U.K.

Burke, I. C., Lauenroth, W. K., and Coffin, D. P., 1995. Soil organic matter recovery in semiarid grasslands: implications for the Conservation Reserve Program, *Ecol. Applic.,* 5:793–801.

Buyanovsky, G. A., Kuceraand, C. L., and Wagner, G. H., 1987. Comparative analyses of carbon dynamics in native and cultivated ecosystems, *Ecology*, 68:2023–2031.

Campbell, C. A. and Souster, W., 1982. Loss of organic matter and potentially mineralizable nitrogen from Sasketchewan due to cropping, *Can. J. Soil Sci.*, 62:651–656.

Carson, R., 1962. *Silent Spring*, Houghton Mifflin, Boston, MA.

Case, T. J., 1990. Invasion resistance arises in strongly interacting species-rich model competition communities, *Proc. Natl. Acad. Sci. U.S.A.*, 87:9610–9614.

Clark, E. A. and Francis, C. A., 1985. Transgressive yielding in bean:maize intercrops; interference in time and space, *Field Crops Res.*, 11:37–53.

Cleveland, C. J., 1995. The direct and indirect use of fossil fuels and electricity in USA agriculture, 1910–1990, *Agric. Ecosyst. Environ.*, 55:111–121.

Cohen, B., Wiles, R., and Bondoc, E., 1995. *Weed Killers by the Glass*, Environmental Working Group, Washington, D.C.

Coll, M. and Bottrell, D. G., 1994. Effects of nonhost plants on an insect herbivore in diverse habitats, *Ecology*, 75: 723–731.

Davis, J. H. C., van Beuningen, L., Ortiz, W. V., and Pino, C., 1984. Effect of growth habit of beans (*Phaseolus vulgaris* L.) on tolerance to competition from maize when intercropped, *Crop Sci.*, 24:751–755.

Davis, J. H. C., Woolley, J. N., and Moreno, P. A., 1986. Multiple cropping with legumes and starchy roots, in *Multiple Cropping Systems*, C. A. Francis, Ed., Macmillan, New York, 133–160.

Dewald, C. L., Burson, B. L., de Wet, J. M. J., and Harlan, J. R., 1987. Morphology, inheritance and evolutionary significance of sex reversal in *Tripsacum dactyloides* (Poaceae), *Am. J. Bot.*, 74:1055–1059.

de Wet, J. M. J. and Harlan, J. R., 1978. *Tripsacum* and the origin of maize, in *Maize Breeding and Genetics*, D. B. B. Walden, Ed., John Wiley and Sons, New York, 129–141.

Drake, J. A., 1990a. The mechanics of community assembly and succession, *J. Theor. Biol.*, 147:1–28.

Drake, J. A., 1990b. Communities as assembled structures: do rules govern pattern?, *Trends Ecol.Evol.*, 5:159–164.

Drake, J. A., 1991. Community assembly mechanics and the structure of an experimental species ensemble, *Am. Nat.*, 131:1–26.

Drake, J. A., Flum, T. E., Witteman, G. J., Voskuil, T., Hoylman, A. M., Creson, C., Kenney, D. A., Huxel, G. R., LaRue, C. S., and Duncan, J. R., 1993. The Construction and assembly of an ecological landscape, *J. Anim. Ecol.*, 62:117–130.

Edwards, C. A. and Lofty, J. F., 1975. The influence of cultivation on animal populations, in *Progress in Soil Biology*, J. Vanek, Ed., Academia Publ., Prague, Czechoslovakia, 399–408.

Eissenstat, D. M., 1990. A comparison of phosphorus and nitrogen transfer between plants of different phosphorus status, *Oecologia*, 82:342–347.

Elmstrom, K. M., Andow, D. A., and Barclay, W. W., 1988. Flea beetle movement in a broccoli monoculture and diculture, *Environ. Entomol.*, 17:299–305.

Ensign, R. D., Hickey, V. G., and Bernardo, M. E., 1983. Seed yield of Kentucky bluegrass as affected by post-harvest residue removal, *Agron. J.*, 75:549–551.

Evers, G. W., 1983. Weed control on warm season perennial grass pastures with clovers, *Crop Sci.*, 23:170–171.

Francis, C. A., 1986. *Multiple Cropping Systems*, Macmillan, New York.

Francis, C. A., Prager, M., and Tejada, G., 1982. Effects of relative planting dates in bean (*Phaseolus vulgaris* L.) and maize (*Zea mays* L.) intercropping, *Field Crops Res.*, 5:45–54.

Gebhart, D. L., Johnson, H. B., Mayeaux, H. S., and Polley, H. W., 1994. The CRP increases soil organic matter, *J. Soil Water Conserv.*, 49:488–492.

Gilpin, M. E., 1987. Experimental community assembly: competition, community structure and the order of species introductions, in *Restoration Ecology: A Synthetic Approach to Ecological Research,* W. R. Jordan, M. Gilpin, and J. Aber, Eds., Cambridge University Press, Cambridge, U.K., 151–161.

Goodman, P. J. and Collison, M., 1982. Varietal differences in uptake of 32P labelled phosphate in clover plus ryegrass swards and monocultures, *Ann. Appl. Biol.,* 100:559–565.

Great Plains Flora Association, 1986. *Flora of the Great Plains,* University Press of Kansas, Lawrence.

Hallberg, G. R., 1986. From hoes to herbicides: agriculture and groundwater quality, *J. Soil Water Conserv.,* 41:357–364

Halvorson, J. J., Smith, J. L., and Franz, E. H., 1991. Lupine influence on soil carbon, nitrogen and microbial activity in developing ecosystems at Mount St. Helens, *Oecologia,* 87:162–170.

Haynes, R. J., 1980. Competitive aspects of the grass-legume association, *Adv. Agron.,* 33:227–261.

Hileman, B., 1994. Environmental estrogens linked to reproductive abnormalities, cancer, *Chem. Eng. News,* 72:19–23.

Holt, R. D., Robinson, G. R., and Gaines, M. S., 1995. Vegetation dynamics in an experimentally fragmented landscape, *Ecology*, 76:1610–1624.

Horvitz, C. C. and Schemske, D. W., 1988. Demographic cost of reproduction in a neotropical herb: an experimental fields study, *Ecology*, 69:1741–1745.

Irwin, M. E. and Kampmeier, G. E., 1989. Vector behavior, environmental stimuli, and the dynamics of plant virus epidemics, in *Spatial Components of Plant Disease Epidemics.,* M. J. Jeger, Ed., Prentice-Hall, Englewood Cliffs, 14–39.

Jackson, L. L. and Dewald, C. L., 1994. Predicting evolutionary consequences of greater reproductive effort in *Tripsacum dactyloides*, a perennial grass, *Ecology*, 75:627–641.

Jastrow, J. D., 1987. Changes in soil aggregation associated with tallgrass prairie restoration, *Am. J. Bot.,* 74:1656–1664

Jastrow, J. D. and Miller, R. M., 1993. Neighbor influences on root morphology and mycorrhizal fungus colonization in tallgrass prairie plants, *Ecology*, 74:561–569.

Jordan, J. P., O'Donnell, P. F., and Robinson, R. R., 1986. Historical evolution of the state agricultural experiment station system, in *New Directions for Agriculture and Agricultural Research,* K. A. Dahlberg, Ed., Rowman and Allanheld, Totowa, 147–162.

Kareiva, P. M., 1985. Finding and losing host plants by Phyllotreta: patch size and surrounding habitat, *Ecology*, 66:1809–1816.

Kindscher, K. and Wells, P. V., 1995. Prairie plant guilds: a multivariate analysis of prairie species based on ecological and morphological traits, *Vegetatio*, 117:29–50.

Komarov, V. L., 1934. *Flora of the U.S.S.R. II. Gramineae,* Bot. Inst. Acad. Sci. U.S.S.R., Leningrad. [Translation. Office of Technical Services, U.S. Department of Commerce, Washington, D.C., 1963.]

Larson, W. E., Lindstrom, M. J., and Schumacher, T. E., 1997. The role of ssevere storms in soil erosion: a problem needing consideration, *J. Soil Water Conserv.,* 52:90–95.

Letourneau, D. K., 1987. The enemies hypothesis: tritrophic interaction and vegetational diversity in tropical agroecosystems, *Ecology*, 68:1616–1622.

Letourneau, D. K. and Altieri, M. A., 1983. Abundance patterns of a predator, *Orius tristicolor* (Hemiptera: Anthocoridae), and its prey, *Frankliniella occidentalis* (Thysanoptera: Thripidae) in polycultures versus monocultures, *Environ. Entomol.,* 12: 1464–1469.

Liebman, M., 1988. Ecological suppression of weeds in intercropping systems: a review, in *Weed Management in Agroecosystems: Ecological Approaches,* M. A. Altieri and M. Liebman, Eds., CRC Press, Boca Raton, FL, 197–212.

Liebman, M. and Dyck, E., 1993. Crop rotation and intercropping strategies for weed management, *Ecol. Appl.,* 3:92–122.

Lovins, A. B., Lovins, L. H., and Bender, M., 1995. Agriculture and energy, *Enc. Energy Technol. Environ.,* 1:11–18.

Luh, H.-K. and Pimm, S. L., 1993. The assembly of ecological communities: a minimalist approach, *J. Anim. Ecol.,* 62:749–765.

Mallarino, A. P., Wedin, W. F., Perdomo, C. H., Goyenola, R. S., and West, C. P., 1990. Nitrogen transfer from white clover, red clover, and birdsfoot trefoil to associated grass, *Agron. J.,* 82:790–795.

McConnell, S. G. and Quinn, M.-L., 1988. Soil productivity of four land use systems in southeastern Montana, *Soil Sci. Soc. Am. J.,* 52:500–506.

McNaughton, S. J., 1977. Diversity and stability of ecological communities: a comment on the role of empiricism in ecology, *Am. Nat.,* 111:515–525.

McNaughton, S. J., 1985. Ecology of a grazing system: the Serengeti, *Ecol. Monogr.,* 55:259–294.

Miller, R. M. and Jastrow, J. D., 1993. Mycorrhizal fungal and floristic diversity in a successional tallgrass prairie, in *Abstracts of the 9th North American Conference on Mycorrhizae,* L. Peterson and M. Schelkle, Eds., Guelph, Ontario, 21.

Mueller-Warrant, G. W., Young, W. C., III, and Mellbye, M. E., 1994. Influence of residue removal method and herbicides on perennial ryegrass seed production: II. Crop tolerance, *Agron. J.,* 86:684–690.

Muto, P. J., 1990. Evaluation of Illinois bundleflower and eastern gamagrass germplasm in monocultures and bicultures, *Land Inst. Res. Rep.,* 7:25–28.

Naeem, S., Thompson, L. J., Lawlor, S. P., Lawton, J. H., and Woodfin, R. M., 1994. Declining biodiversity can alter the performance of ecosystems, *Nature*, 368:734–737.

Olson, R., Francis, C., and Kaffka, S., 1995. *Exploring the Role of Diversity in Sustainable Agriculture,* American Society of Agronomy, Crop Science Society of America, Soil Science Society of America, Madison.

Opie, J., 1993. *Ogallala: Water for a Dry Land,* University of Nebraska Press, Lincoln.

Osborn, T., 1993. The Conservation Reserve Program: status, future, and policy options, *J. Soil Water Conserv.,* 48:271–278.

Paustian, K., Andrén, O., Clarholm, M., Hansson, A.-C., Johansson, G., Lagerlöf, J., Lindberg, T., Pettersson, R., and Sohlenius, B., 1990. Carbon and nitrogen budgets of four agroecosystems with annual and perennial crops, with and without N fertilization, *J. Appl. Ecol.,* 27:60–84.

Pimentel, D., 1984. Energy flow in agroecosystems, in *Agricultural Ecosystems,* R. Lowrance, B. R. Stinner, and G. J. House, Eds., John Wiley and Sons, New York, 121–132.

Pimentel, D., McLaughlin, L., Zepp, A., Lakitan, B., Kraus, T., Kleinman, P., Vacini, F., Roach, W. J., Graap, E., Keeton, W. S., and Selig, G., 1991. Environmental and economic effects of reducing pesticide use, *BioScience*, 41:402–409.

Pimentel, D., Harvey, C., Resosudarmo, P., Sinclair, K., Kurz, D., McNair, M., Crist, S., Shpritz, L., Fitton, L., Saffouri, R., and Blair, R., 1995. Environmental and economic costs of soil erosion and conservation benefits, *Science,* 267:1117–1123.

Piper, J. K., 1992. Size structure and seed yield over 4 years in an experimental *Cassia marilandica* (Leguminosae) population, *Can. J. Bot.,* 70:1324–1330.

Piper, J. K., 1993a. Soil, water and nutrient change in stands of three perennial crops, *Soil Sci. Soc. Am. J.,* 57:497–505.

Piper, J. K., 1993b. A grain agriculture fashioned in nature's image: the work of The Land Institute, *Great Plains Res.,* 3:249–272.

Piper, J. K., 1993c. Neighborhood effects on growth, seed yield, and weed biomass for three perennial grains in polyculture, *J. Sustain. Agric.,* 4(2):11–31.

Piper, J. K., 1995. Composition of prairie plant communities on productive versus unproductive sites in wet and dry years, *Can. J. Bot.,* 73:1635–1644.

Piper, J. K., 1996. Density of *Anomoea flavokansiensis* on *Desmanthus illinoensis* in monoculture and polyculture, *Entomol. Exp. Appl.,* 81:105–111.

Piper, J. K., 1998. Growth and seed yield of three perennial grains within monocultures and mixed stands, *Agric. Ecosyst. Environ.,* 68:1–11.

Piper, J. K. and Kulakow, P. A., 1994. Seed yield and resource allocation in *Sorghum bicolor* and F1 and backcross generations of *S. bicolor x S. halepense* hybrids, *Can. J. Bot.,* 72:468–474.

Piper, J., Henson, J., Bruns, M., and Bender, M., 1988. Seed yield and quality comparison of herbaceous perennials and annual crops, in *Global Perspectives on Agroecology and Sustainable Agricultural Systems,* P. Allen and D. Van Dusen, Eds., University of California, Santa Cruz, 715–719.

Piper, J. K., Handley, M. K., and Kulakow, P. A., 1996. Incidence and severity of viral disease symptoms on eastern gamagrass within monoculture and polycultures, *Agric. Ecosyst. Environ.,* 59:139–147.

Posler, G. L., Lenssen, A. W., and Fine, G. L., 1993. Forage yield, quality, compatibility, and persistence of warm-season grass — legume mixtures, *Agron. J.,* 85:554–560.

Post, W. M. and Pimm, S. L., 1983. Community assembly and food web stability, *Math. Biosci.,* 64:169–192.

Power, A. G., 1987. Plant community diversity, herbivore movement, and an insect-transmitted disease of maize, *Ecology,* 68:1658–1669.

Radke, J. K. and Hagstrom, R. T., 1976. Strip intercropping for wind protection, in *Mutliple Cropping,* American Society for Agronomy Special Publication No. 27, R. I. Papendick, P. A. Sanchez, and G. B. Triplett, Eds., American Society of Agronomy, Madison, WI, 201–222.

Rathcke, B., 1984. Competition and facilitation among plants for pollination, in *Pollination Biology,* L. Real, Ed., Academic Press, Orlando, 305–329.

Reekie, E. G. and Bazzaz, F. A., 1987. Reproductive effort in plants — 3: Effect of reproduction on vegetative activity, *Am. Nat.,* 129:907–919.

Richter, D. D., Babbar, L. I., Huston, M. A., and Jaeger, M., 1990. Effects of annual tillage on organic carbon in a fine-textured Udalf: the importance of root dynamics to soil carbon storage, *Soil Sci.,* 149:78–83.

Risch, S. J., 1981. Insect herbivore abundance in tropical monocultures and polycultures: an experimental test of two hypotheses, *Ecology,* 62: 1325–1340.

Risch, S. J., Andow, D., and Altieri, M. A., 1983. Agroecosystem diversity and pest control: data, tentative conclusions, and new research directions, *Environ. Entomol.,* 12:625–629.

Roberts, A. and Tregonning, K., 1981. The robustness of natural systems, *Nature,* 288:265–266.

Robinson, J. V. and Dickerson, J. E., Jr., 1987. Does invasion sequence affect community structure?, *Ecology,* 68:587–595.

Schoenau, J. J., Stewart, J. W. B., and Bettany, J. R., 1989. Forms of cycling of phosphorus inprairie and boreal forest soils, *Biogeochemistry*, 8:223–237.

Seifers, D. L., Handley, M. K., and Bowden, R. L., 1993. Sugarcane mosaic virus strain maize dwarf mosaic virus B as a pathogen of eastern gamagrass, *Plant Dis.*, 77:335–339.

Simpson, J. R., 1965. The transference of nitrogen from pasture legumes associated with grass under several systems of management in pot culture, *Aust. J. Agric. Res.*, 16:915–926.

Singh, B. and Sekhon, G. S., 1979. Nitrate pollution of groundwater from farm use of nitrogen fertilizers — a review, *Agric. Environ.*, 4:207–225.

Smith, M. E. and Francis, C. A., 1986. Breeding for multiple cropping systems, in *Multiple Cropping Systems,* C. A. Francis, Ed., Macmillan, New York, 219–249.

Soule, J. D. and Piper, J. K., 1992. *Farming in Nature's Image: An Ecological Approach to Agriculture,* Island Press, Washington, D.C.

Tahvanainen, J. O. and Root, R., 1972. The Influence of vegetational diversity on the population ecology of a specialized herbivore, *Phyllotreta cruciferae* (Coleoptera: Chrysomelidae), *Oecologia*, 10:321–346.

Thompson, T., Zimmerman, D., and Robers, C., 1981. Wild *Helianthus* as a genetic resource, *Field Crops Res.*, 4:333–343.

Tilman, D., 1985. The resource-ratio hypothesis of plant succession, *Am. Nat.*, 125:827–852.

Tilman, D. and Downing, J. A., 1994. Biodiversity and stability in grasslands, *Nature*, 367:363–365.

Tilman, D. and El Haddi, A., 1992. Drought and biodiversity in grasslands, *Oecologia*, 89:257–264.

Tilman, D., Wedin, D., and Knops, J., 1996. Productivity and sustainability influenced by biodiversity in grassland ecosystems, *Nature*, 379:718–720.

Tregonning, K. and Roberts, A., 1979. Complex systems which evolve towards homeostasis, *Nature*, 281:563–564.

Trenbath, B. R., 1976. Plant interactions in mixed crop communities, in *Multiple Cropping,* American Society for Agronomy Special Publication No. 2, R. I. Papendick, P. A. Sanchez, and G. B. Triplett, Eds., American Society for Agronomy, Madison, 129–170.

Tukahira, E. M. and Coaker, T. H., 1982. Effect of mixed cropping on some insect pests of Brassicas; reduced *Brevicoryne brassicae* infestations and influences on epigeal predators and the disturbance of oviposition behavior in *Delia brassicae*, *Entomol. Exp. Appl.*, 32:129–140.

Vandermeer, J. H., 1980. Indirect mutualism: variations on a theme by Stephen Levine, *Am. Nat.*, 116:441–448.

Vandermeer, J. H., 1989. *The Ecology of Intercropping,* Cambridge University Press, Cambridge, U.K.

van Kessel, C., Singleton, P. W., and Hoben, H. J., 1985. Enhanced N-transfer from soybean to maize by vesicular arbuscular mycorrhizal (VAM) fungi, *Plant Physiol.*, 79:562–563.

Wagmare, A. B. and Singh, S. P., 1984. Sorghum-legume intercropping and its effects on nitrogen fertilization. I. Yield and nitrogen uptake by crops, *Exp. Agric.*, 20:251–259.

Wagoner, P., 1990. Perennial grain development — past efforts and potential for the future, *CRC Crit. Rev. Plant Sci.*, 9:381–408.

Woods, L. E., 1989. Active organic matter distribution in the surface 15 cm of undisturbed and cultivated soil, *Biol. Fertil. Soils,* 8:271–278.

Zahm, S. H. and Blair, A., 1992. Pesticides and non-Hodgkin's lymphoma, *Cancer Res.* (Suppl.), 52:5485s–5488s.

Zaki, M. H., Moran, D., and Harris, D., 1982. Pesticides in groundwater: the aldicarb story in Suffolk County, N.Y., *Am. J. Public Health,* 72:1391–1395.

Zitter, T. A. and Simons, J. N., 1980. Management of viruses by alteration of vector efficiency and by cultural practices, *Annu. Rev. Phytopathol.,* 18:289–310.

CHAPTER 11

Managing Agroecosystems as Agrolandscapes: Reconnecting Agricultural and Urban Landscapes

Gary W. Barrett, Terry A. Barrett, and John David Peles

CONTENTS

INTRODUCTION

During the past decade, several interface fields of study, including agroecosystem ecology and landscape ecology, have emerged that integrate ecological theory and management practices within the realm of applied ecology (Barrett, 1984; 1992). Agroecosystem ecology is based on the premise that natural ecosystems are models for the long-term management of agriculture and on the philosophy of working with nature rather than against it (Jackson and Piper, 1989; Barrett, 1990). Landscape ecology considers the development and dynamics of (1) spatial heterogeneity, (2) spatial and temporal interactions and exchanges across the landscape, (3) influences

1-56670-290-9/99/$0.00+$.50

of spatial heterogeneity on biotic and abiotic processes, and (4) the management of spatial heterogeneity for societal benefit (Risser et al., 1984).

The management of agricultural systems has traditionally focused on the agro-ecosystem (i.e., crop field or landscape patch), rather than on the total agrolandscape (i.e., watershed or region in which the crop fields are elements in the landscape matrix) level of resolution (Barrett, 1992). Increasing the crop yield has been the main management goal (National Research Council, 1989). Policies and practices to maximize crop yield have involved the use of increased subsidies such as for fossil fuels, fertilizers, and pesticides (National Research Council, 1989). These management strategies have resulted in (1) decreased crop and biotic diversity, (2) net energy loss, (3) profit loss for farmers, and (4) extensive nonpoint pollution of the environment (Altieri et al., 1983; National Research Council, 1989; Barrett and Peles, 1994).

In recent years, agricultural management strategies have begun to focus on increasing biotic (genetic, species, landscape) diversity (Barrett, 1992), on reducing energy inputs (Odum, 1989), and on increasing food safety (see National Research Council, 1996, for a review) rather than only on crop yield. It has become increasingly clear that we cannot sustain agricultural productivity by viewing agricultural systems independent from other landscape elements or ecological/urban systems (i.e., we must develop a holistic agrolandscape perspective in addition to an agro-ecosystem perspective). We argue that a landscape approach must be established in which landscape units, such as watersheds, are managed as functional systems based on the concept of holistic, long-term sustainability (Lowrance et al., 1986; Barrett, 1992). This holistic approach differs from a picture of the world according to Callicott (1989) which breaks a highly integrated functional system into separate, discrete, and functionally unrelated sets of particulars. A piecemeal or fragmented approach permits the radical rearrangement of parts of the landscape without concern for upsetting the functional integrity and organic unity of the whole. By definition, and by necessity, the agrolandscape approach must integrate aesthetic, biological, physical, and ecological factors; must couple urban (heterotrophic) with rural (autotrophic) systems; and must establish land-use policies based on sound ecological theory (Barrett, 1989; Elliott and Cole, 1989). Sustainability, a common theme of many recent paradigms, is defined here as the ability to keep a system in existence or to prevent it from falling below a given threshold of health (Barrett, 1989). Goodland (1995) defined sustainability, as it pertains to the environment, as "maintenance of natural capital."

The sustainable landscape approach, which considers agroecosystems as components of the total landscape (Jackson and Piper, 1989), encourages the integration of concepts such as sustainable agriculture, biotic diversity, and levels of organization (Barrett, 1992; Barrett et al., 1997). The focus of this approach is to manage for sustainability of the total landscape based on an understanding of how agroecosystem units function as an integrated whole (Figure 1).

In this chapter, we provide a perspective regarding the development and integration of modern agrolandscapes based on the ratio of primary productivity (P) to maintenance costs (R) at the agro–urban landscape scale. This perspective is intended to provide long-term sustainability and increased biodiversity. We discuss

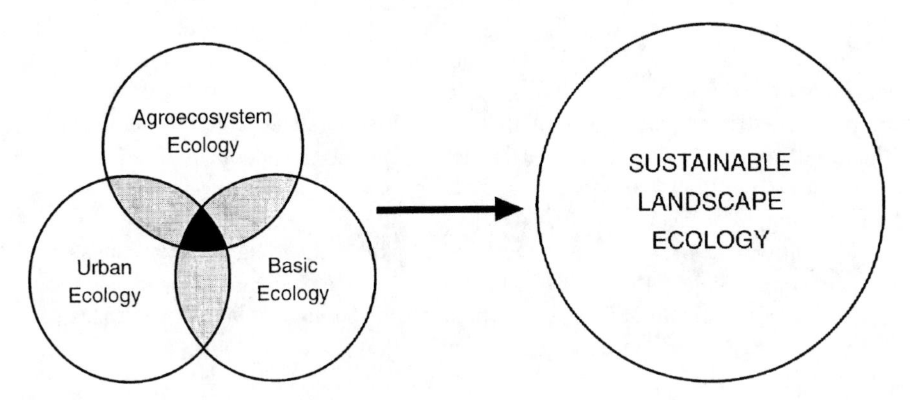

Figure 1 Diagram depicting the new integrative field of study termed *sustainable landscape ecology.*

the importance of providing linkages between agricultural systems and urban systems and note the importance of developing land-use policies necessary to manage for sustainability and biodiversity based on a total landscape approach.

CULTURAL AND HISTORICAL PERSPECTIVES OF THE PRESENT AGROLANDSCAPE

The Roman writer Cicero termed what is currently considered the cultural landscape "a second nature" (*alteram naturam*). This cultural landscape, or second nature, comprised all the elements introduced into the physical world by humankind to make it more habitable. Hunt (1992) interprets Cicero's phrase, a second nature, as implying a first, or primal nature before humans invaded, altered, or augmented the unmediated world.

Various ideologies resulting from this second nature, especially how nature should be managed or controlled, have contributed to the present fragmented landscape. The evolutionary significance of the mature (model) system, including how natural selection has resulted in the evolution of efficient mechanisms for insect pest control, nutrient recycling, and mutualistic behavior, is often poorly understood. A hallmark of these mature and sustainable ecological systems is also maximum biological diversity (Moffat, 1996; Tilman et al., 1996; Tilman, 1997). Environmental literacy must increase if societies are to develop sustainable agriculture and sustainable agrolandscapes (Barrett, 1992; Orr, 1992). For example, natural processes and concepts such as pulsing, carrying capacity, natural pest control, nutrient cycling, positive and negative feedback (cybernetics), and net primary productivity must be understood by ecologically literate societies in order to provide a quality environment for future generations. There exists an urgent need to understand these processes and concepts better, and to manage agroecosystems at the agrolandscape level (Barrett, 1992). It is now imperative to couple the heterotrophic urban environment with the autotrophic agricultural environment if societies are to establish or manage sustainable landscapes on a meaningful regional or global scale.

Environmental literacy also includes the aesthetic languages of diverse cultures and histories that determine what a people traditionally considers essential and nonessential resources within cultures. Shifting economic, social, political, or artistic perspectives, for example, affect the definition of what is considered a resource and what is perceived as a nonresource. The encoded messages endemic to these cultures influence human thought in the determination of what is of value in the life of a human being.

The cultural landscape is an integral part of the holistic agro–urban landscape perspective. Nassauer (1995), for example, recognized the need to investigate the relationship between cultural landscape patterns and ecological landscape processes. The aesthetics that are intrinsic to various cultures have influenced the present agrolandscapes. Acknowledging these relationships, including their present and future influences, will increase dialogue among biological, physical, and social scientists; among resource managers, landscape engineers, and urban planners; and among scholars investigating the role of sustainability at the landscape and global levels (Huntley et al., 1991; Lubchenco et al., 1991). The resulting interfaces among fields of study will lead to a deeper understanding of why and of how landscape elements (patches, corridors, and the agromatrices or urban matrices) are related to present regional/global patterns of belief systems, an understanding necessary to conserve biological diversity.

THE CREATION OF AN "OXBOW" URBAN AREA

Figure 2 depicts an urban environment, including the relationship of the inner urban landscape to the outer agricultural landscape. Although much has been written regarding the pattern and shaping of the landscape from prehistory to present day (see review by Jellicoe and Jellicoe, 1987, for details), there exists the need to address and quantify the concept of landscape sustainability from an energetic (solar energy and energy subsidy) perspective. One objective of this chapter is to increase trans-disciplinary dialogue concerning this need. Although we recognize that markets have become increasingly global in structure and function (Brady, 1990), it appears that management practices, for example, integrated pest management and information processing, will be conducted on a regional basis (Elliott and Cole, 1989).

Traditionally, towns and cities were integrated in a sustainable manner (Figure 3). The town served as the marketplace for farmers to sell their goods and products (Mumford, 1961; Hough, 1995); goods and services radiated from the city to support the agricultural landscape, including providing cultural, educational, and social benefits (Le Corbusier, 1987). During the early part of this century in the agricultural Midwest, crop diversity was high (Barrett et al., 1990), as was species and habitat diversity (Barrett and Peles, 1994). The shift from a biologically diversified and, perhaps, a sustainable landscape to monoculture or diculture crops (especially corn and soybean) in the rural landscape was accompanied by the development of sub-urban areas that reduced not only the amount of arable land, but also the diversity of wildlife habitats and cultural linkages between the inner city and the agricultural landscape. This created what we term an *oxbow city,* analogous to the creation of

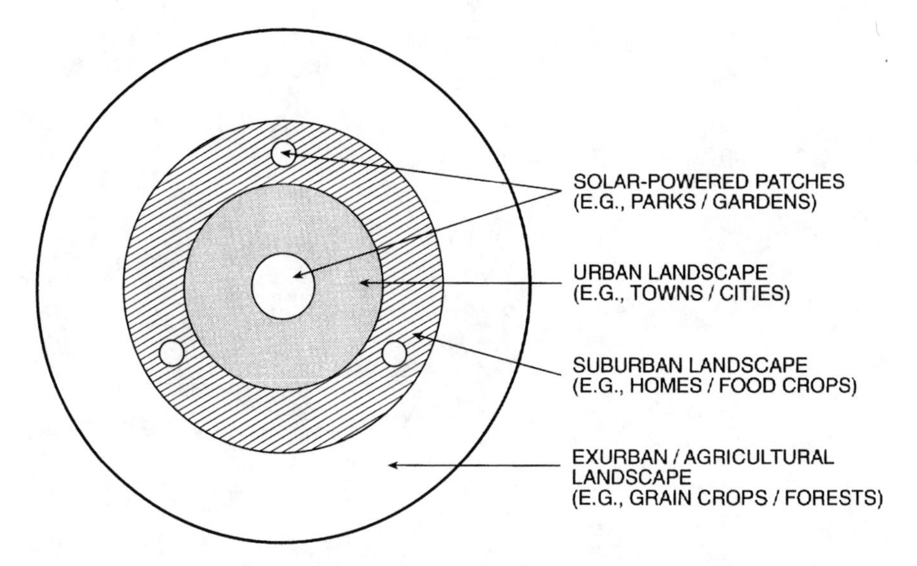

Figure 2 Diagram depicting urban, suburban, and exurban/agricultural systems. Solar-powered (autotrophic) patches are shown within urban and suburban (heterotrophic) systems.

an oxbow lake when it becomes separated (physically and functionally) from a flowing meandering stream once the stream changes its course. This isolated city develops different functional processes (i.e., provides different services), resulting in changes in niche and biodiversity (i.e., the inner city creates different occupations and provides habitats for different species of flora and fauna). The integrity of the city frequently becomes less closely related to the total watershed from which it evolved. This developmental process is depicted in Figure 4.

LINKAGES BETWEEN AGRICULTURAL AND URBAN COMPONENTS OF THE LANDSCAPE

Odum (1997) classified ecosystems based on the proportions of solar and fossil fuel energy used to drive the system. Most natural ecological systems are driven entirely by solar energy. Subsidized systems depend, to varying degrees, on the input of subsidies such as fossil fuel energy, fertilizers, and/or pesticides. Agroecosystems, for example, are driven by both solar energy and subsidies; urban systems depend mainly on enormous inputs of fossil fuel subsidies (Odum, 1989).

These ecosystems may also be classified based on the ratio of energy produced by primary productivity (P) to energy used for respiration or system maintenance (R). Natural and agricultural ecosystems, especially during ecosystem growth and development, represent autotrophic systems where $P/R > 1$. In contrast, urban areas have increasingly become heterotrophic ($P/R < 1$). We define sustainable systems as those systems or landscapes where long-term P/R ratios equal 1. During the growth and development of autotrophic systems (i.e., during ecological succession),

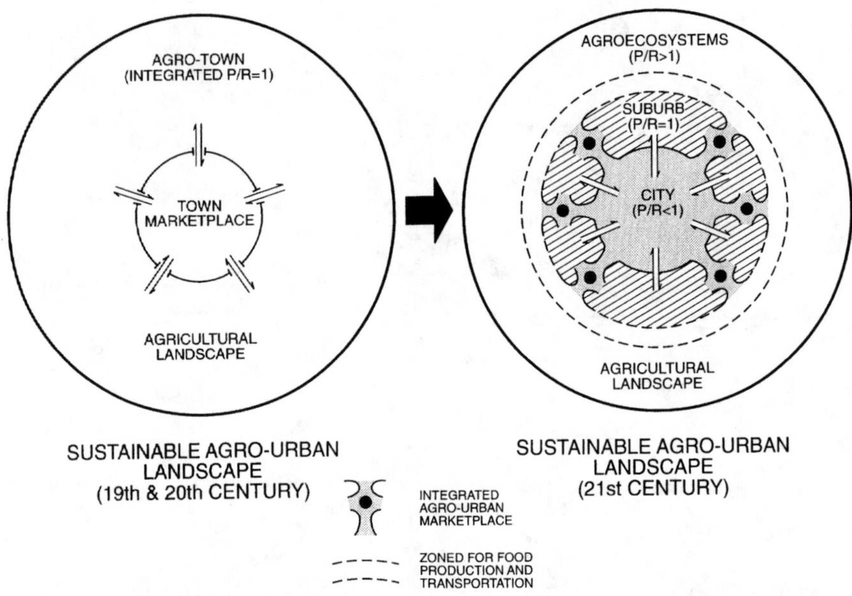

Figure 3 The development of an agro–urban sustainable (P/R = 1) landscape. The town marketplace historically was closely linked to the agricultural landscape. Sustainability in the modern agro–urban landscape increasingly must be based on the management of suburban areas (ecotones) as natural linkages between urban and agricultural systems.

P/R decreases as biological (organic) materials accumulate (Figure 5). This results in a balance between productivity and respiration (P/R = 1) in the climax stage of succession (Odum, 1969). An increase in physical (inorganic) materials in urban systems coincides with a decreasing P/R ratio (i.e., a significant increase in maintenance costs). Thus, the result of urban succession is a city where energy demands greatly exceed productivity. During the past century, the large numbers of people living in urban and suburban areas have led to increased need for food produced in rural areas (Steinhart and Steinhart, 1974; Odum, 1989). This demographic and cultural transition has led to increased reduction of P/R ratios in urban areas, as well as increased subsidization of agriculture (including economic subsidies) to maximize crop yield (National Research Council, 1989; 1996).

More recently, suburban expansion has led to increased pressure on rural land used for agriculture (Lockeretz, 1988). A result of urban sprawl has been an increase in the proportion of the agrolandscape occupied by heterotrophic systems. This has serious implications regarding the conservation of biodiversity from a sustainable agrolandscape perspective (Rookwood, 1995). In addition, urban expansion into agricultural land has important consequences for aesthetic, social, and economic values (Lockeretz, 1988), including the need to understand more fully how human

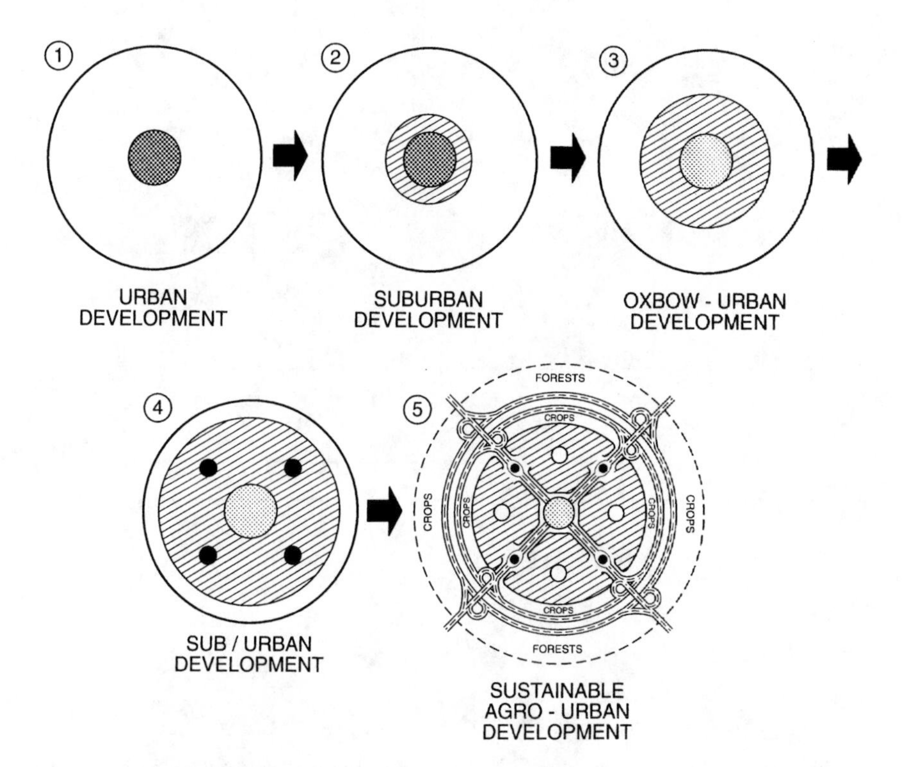

Figure 4 Diagram depicting the development of urban areas into oxbow cities (1 to 3) and then, we hope, into a modern, sustainable agro–urban landscape. We argue that the modern landscape must be increasingly based on the ecological/economic management of suburban areas as natural linkages between urban and agricultural systems.

values will likely define future landscape boundaries and resources, especially those values that relate to ecosystem/landscape sustainability.

The present challenge for agrolandscape management is to minimize the infringement of urbanization on agricultural land, to restore biological diversity (genetic niche, species, and landscape) at greater temporal and spatial scales, to establish linkages (ecological and economic) between urban and rural (heterotrophic and autotrophic) patch elements, and to achieve sustainable productivity ($P/R = 1$) at agro–urban (regional) scales. Goals for achieving sustainable agrolandscape management should focus on (1) achieving stability regarding P/R ratios among heterotrophic and autotrophic systems at these scales; (2) creating both natural corridor and human transport linkages between rural and urban systems; (3) protecting the integrity of ecosystem/watershed processes, such as nutrient recycling and primary productivity; and (4) establishing management policies for optimal land use within transition suburban areas that ecologically and economically form an interface between urban and agricultural landscape systems. As previously noted, sustainable agriculture is based on the coupling of agricultural ecosystems with natural ecosystems (Barrett et al., 1990). Here, we stress the need to integrate natural, agricultural,

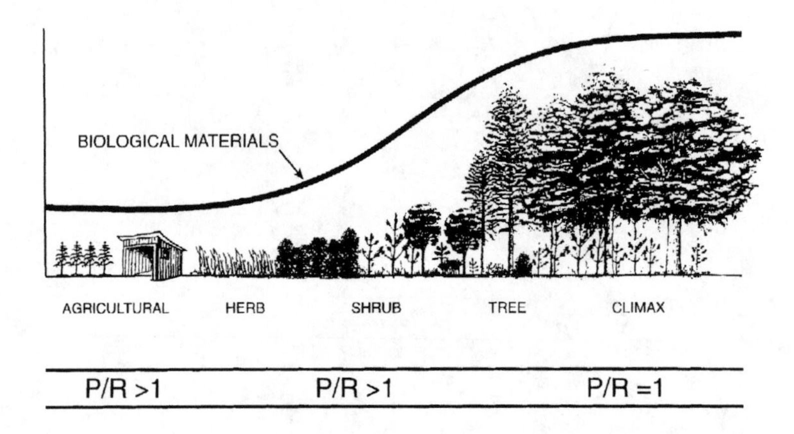

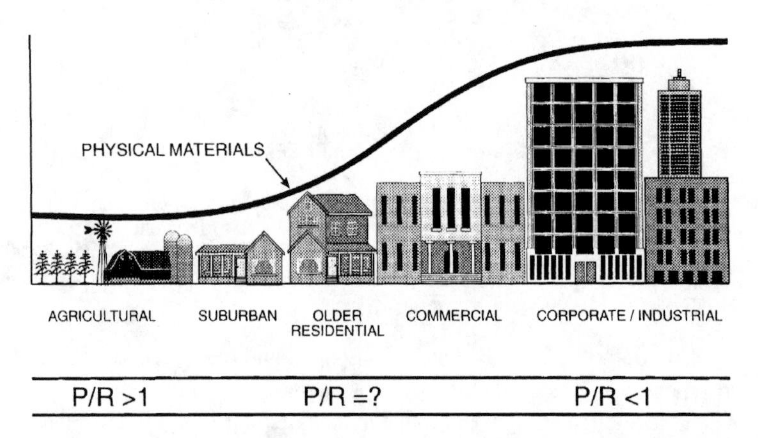

Figure 5 Changes in *P/R* ratios during ecological (autotrophic) and urban (heterotrophic) succession. Although *P/R* decreases as the amount of biological material increases during ecological succession, the result is a mature (climax), sustainable community. In contrast, the accumulation of physical material during urban succession frequently results in a fragile, nonsustainable community.

and urban components if societies are to design and implement the concept of sustainability at the agrolandscape scale (see Figure 1). This approach should simultaneously conserve and enhance biotic diversity at greater temporal and spatial scales.

LINKAGES BETWEEN BIODIVERSITY AND SUSTAINABILITY

Management of agrolandscapes for sustainability both influence and is influenced by biodiversity (Paoletti, 1995). Landscape planning is a process through which the conservation and management of biodiversity can be pursued (Rookwood, 1995).

Turner et al. (1995) stressed that there exists a three-way interaction of biodiversity, ecosystem processes, and landscape dynamics at greater scales. Sustainable agricultural practices leading to increased crop and genetic diversity have resulted in increased agroecosystem stability (Cleveland, 1993). For example, increasing crop diversity benefits agriculture by reducing insect pests (Altieri et al., 1983). Other sustainable agricultural practices, such as conservation tillage, are known to increase habitat diversity, wildlife diversity, and numbers of beneficial insect species (Barrett, 1992; McLaughlin and Mineau, 1995).

Although the importance of a landscape approach for management of biodiversity is well recognized (Franklin, 1993), little is known regarding how biodiversity affects landscape pattern and dynamics (Turner et al., 1995). Turner et al. (1995) have suggested that there exists a three-way interaction among biodiversity, ecosystem processes, and landscape dynamics. In addition, it is well documented that there exists a reciprocal relationship between sustainability and biodiversity (Paoletti, 1995). Thus, as biodiversity and sustainability are increased by management practices at the landscape level (e.g., agrolandscape management), the resulting increase in biodiversity will likely have important benefits concerning the conservation and efficiency of these processes and dynamics (Culotta, 1996; Tilman et al., 1996).

It is important to recognize that, by definition, the agrolandscape approach requires the consideration of biological diversity in the management of agroecosystems (Paoletti et al., 1992). Paoletti et al. (1992) and Paoletti (1995) note that sustainable strategies in food production in agriculture improve the existing biodiversity. These strategies include proper management of natural vegetation, better use and recycling of organic residues, introduction of integrated farming systems, reduced tillage, intercropping, crop rotation, biological pest control, and increased number of biota involved in human food webs. McLaughlin and Mineau (1995) point out, however, that agricultural activities such as tillage, drainage, rotation, grazing, and extensive usage of pesticides and fertilizers have significant implications for wild species of flora and fauna. Therefore, reduced or (no-till) farming, in contrast to conventional tillage, benefits biological diversity in terms of maintaining wild or native species populations.

Increased biodiversity at the landscape level (in the form of increased habitat or agroecosystem diversity) will play a key role in protecting diversity and in providing a linkage between urban and rural areas in our sustainable landscape approach. For example, that an optimum balance between solar-powered and subsidized systems in suburban areas is critical to this linkage. The success of obtaining this optimum balance will almost certainly be enhanced by increasing the diversity of habitat types across the total landscape. Greenways or natural corridors will also enhance the linkage and conserve biological diversity between urban and rural areas (Little, 1990). Management of agroecosystems and agrolandscapes for sustainability will lead to increased habitat and genetic diversity, which, in turn, will lead to increased agroecosystem stability (Altieri et al., 1983; Cleveland, 1993). Likewise, increased biodiversity within individual systems should also increase the diversity and stability of these systems within the total landscape or watershed.

Agrolandscapes should also be managed to increase species diversity within landscape patches and to increase and/or to conserve genetic material among landscape patches (Barrett and Bohlen, 1991). In recent years, there has been increased emphasis on the connectivity and integration of the agricultural landscape with the urban landscape (see Lockeretz, 1988, including a special issue of *Landscape and Urban Planning,* Volume 16, for details). However, most studies at the watershed or agrolandscape levels have failed to encompass or integrate the urban environment into the agrolandscape concept. Since the approach to sustainable agriculture is based on natural ecosystems serving as the model system for efficient agricultural management (Jackson and Piper, 1989; Barrett, 1990), urban systems should also be designed and based on natural ecological systems serving as model systems to ensure maximum ecological and economic efficiency. Thus, sustainability and biodiversity share an important interrelationship that is more fully understood when questions are addressed at the agro–urban landscape scale, and when research designs and management strategies are based on ecological theory.

Importantly, there is growing recognition that cities need to be managed based on the concept of sustainability (Stren et al., 1992). This approach is based on an ecological understanding of how natural ecological systems are organized, and most important, how they function. As with sustainable agriculture, an urban perspective based on sustainability means working with, rather than against, nature. Recently, there has been increased effort in urban areas to maximize the efficiency of energy use, to increase the rate of recycling of goods and materials, and to reduce pollutants entering the system. In addition, "green city" movements have placed emphasis on preservation of natural areas, and on the establishment of vegetable gardens (i.e., solar-powered patches, Figure 2) in urban areas (Stren et al., 1992). Continued efforts to increase solar-driven productivity, while simultaneously decreasing maintenance costs, in urban areas will greatly enhance the sustainability of the total agro–urban landscape. Equally important is the need to plan for (or to zone) future suburban areas, encompassing a productivity/maintenance ratio equal to 1, if we are to achieve regional landscape sustainability (Figure 4). This strategy must also make every effort to increase biodiversity at the genetic, species, and landscape levels.

Lockeretz (1988) stressed the importance of protecting and creating natural linkages between urban and rural areas. Urban greenways (Little, 1990) provide noteworthy examples of these natural linkages. Although greenways take many different forms, they are primarily natural areas set aside for their ecological, recreational, or aesthetic value within urban areas. Greenways also provide natural corridors for movement of wildlife species and transfer of genetic materials between urban and rural areas.

Suburban areas also represent a vital linkage (transition or ecotone) between urban and rural systems. Therefore, it is important to manage these areas as a "transitional zone" between urban population centers and rural farmland (Figure 4) with an optimum balance existing between land area devoted to natural (solar-powered) systems and those managed as subsidized systems. Management plans aimed at establishing a new approach to suburban development should strive to attain an integration of autotrophic and heterotrophic systems. The success of these management plans will greatly depend on an increased ecological literacy of people living within and in consort with future land-use policies (Orr, 1992; Barrett et al., 1997).

In recent years, numerous forms of land-use policies have been instituted to protect agricultural land at urban-fringe areas (Luzar, 1988) and to promote sustainable agriculture research and education (Hess, 1991). Although these programs have been moderately successful (Harsch, 1991; National Research Council, 1996), there still is much to learn regarding the most effective combination of strategies for a particular region. There is urgent need for a new integrative "greenway" strategy. For example, the words *ecology* and *economics* are both derived from the Greek root *oikos*, meaning "household" (*logos* meaning "study of" and *nomics* meaning "management of"). Thus, ecology is the study of the household or the study of natural systems, whereas economics is the management of the household or the management of natural/socioeconomic systems. We must now integrate *greenways, green* meaning primary production or *green* meaning "greenbacks" or dollars, with *ways*, meaning progress in a specific direction, as a new integrative ecologic/economic strategy.

Determining optimum land use is a difficult task (Werner, 1993). For example, the most effective land-use strategy for urban–rural fringe areas depends on agricultural, cultural, demographic, and socioeconomic characteristics of an area (Lockeretz, 1988). The development of appropriate management strategies at the agrolandscape level will require the cooperation of diverse fields of study (Barrett, 1992). This cooperation will enhance biotic diversity at the landscape level, will provide numerous societal benefits, and will protect both natural capital (goods and services provided by nature) and economic capital (goods and services provided by socioeconomic systems) as a long-term strategy (Costanza et al., 1997). The longer the wait, the greater are the long-term costs regarding the loss of biodiversity, the continued loss of soil and nutrient resources, the degradation of water quality, and an increased cost to restore habitat quality.

TOWARD SUSTAINABILITY OF AGROLANDSCAPES

There exists the need to integrate and interface natural, agricultural, urban, and suburban systems if humans are to manage agrolandscapes in a truly sustainable manner. How might we approach this problem? Agricultural practices should increasingly be devoted to the preservation of natural areas (e.g., forests) and to the planting of a diversity of food and grain crops near the outer perimeter of our large cities (i.e., should view agricultural planning in conjunction with, rather than against, the planning of perimeter highways and other modes of transportation). Traditionally, urban and rural landscapes were interrelated in a mutual, sustainable infrastructure. Urban areas were frequently directly linked to the watershed or drainage basin because of transportation needs (rivers or lakes) or geological nodes (mineral deposits) of significance. Urban "sprawl" into the agricultural landscape tended to decouple this connectivity and sustainability. Rather than serving as an ecological/economic barrier, the suburban patch should provide a critical link to restore habitat patch connectivity (Fahrig and Merriam, 1985) and to promote societal sustainability (Barrett, 1989).

Urban areas contain human and natural resources that once again must become integrated with the agrolandscape to the extent that the city can be sustained by both economic capital (gross national product, employment, property values) and natural capital (primary productivity, biotic diversity, solar energy, and sludge usage as natural fertilizers). We refer to this perspective as "dual capitalism." Economic and ecological capital must become integrated; otherwise the resulting oxbow urban areas will increasingly become drained of aesthetic and natural resources. Ecosystem processes will increasingly become disconnected and/or diminished from the vital flow of energy, information, or biotic diversity that once connected the landscape system as an integrated whole.

The integrity of the architecture, the efficiency of the natural processes, and the quality of materials typically found within city boundaries (Kaplan and Kaplan, 1982) also have been reduced or degraded during this fragmentation process. Ecologists frequently note "that the ecosystem is greater than the sum of its parts." Perhaps the landscape level of organization is now no longer greater than the sum of its parts. Isolated urban patches exhibit greater entropy (poverty and crime) and require greater subsidies (financial and human) to maintain their existence (Jake and Wilson, 1992). Isolation and fragmentation impede the attainment of ecological literacy (Orr, 1992) necessary to understand how natural regulatory mechanisms enhance the reestablishment of diversity (ethnic, biological, and ecological) which, in turn, is necessary to maintain agrolandscape functional processes. It has become increasingly critical to understand these relationships and regulatory mechanisms more fully. To do so, greater emphasis should be placed on suburban patches in an effort to integrate and reconnect urban and agricultural processes. Suburban "ecotones" provide the opportunity to help transform the total landscape unit. Suburban areas are not marginal or isolated places, but increasingly central locations in the contemporary world (Baumgartner, 1988). Table 1 contrasts urban, agricultural, natural, and suburban systems and suggests that suburban areas be designed and managed based on our understanding of how natural ecosystems are structured and how they function.

Figure 4 depicts how this suburban area might be designed to interface with a transportation network, including the establishment of an enriched food crop, forestry, and recreational diversity that function as an economic/ecological transition area between the urban and rural landscapes. Landscape corridors, both natural (streams and trails) and human built (highways and mass transportation), will further enhance these linkages (Little, 1990).

Landscape corridors (linkages) manifest various configurations depending on natural phenomena, cultural preferences, or historical development (Hough, 1990). Corridors may connect a sequence of congruent patches. For example, corridors of an ecological mosaic have been defined as disturbance corridors (e.g., power lines), planted corridors (e.g., shelterbelts), regenerated corridors (e.g., fence-row successional vegetation), remnant corridors (e.g., strips of native or climax vegetation) and resource corridors (e.g., riparian areas) (see Barrett and Bohlen, 1991, for details regarding corridor types). These corridors also may reflect or parallel land-use

Table 1 Contrasts among a Traditional Heterotrophic Urban System, an Autotrophic Conventional Agricultural System, a Natural Mature Ecological System, and a Well-Planned Integrated Suburban System

Attribute	Urban	Agricultural	Natural	Suburban
Primary productivity	$P/R < 1$	$P/R > 1$	$P/R = 1$	$P/R > 1$
Nutrient source	Commercial	Commercial sludge	Nutrients recycled	Wastewater effluents, sludge, and manures
Goal-driven	r-selected[a]	r-selected[a] (yield)	K-selected[a]	K-selected[a]
Pest control	Pesticides	Pesticides and Integrated Pest Management (IPM)	Natural	Integrated Pest Management (IPM)
Input energy	Fossil fuels	Solar and fossil fuels	Solar	Solar
Biodiversity	Low	Low	High	High
Sustainability	Short-term	Short-term	Long-term	Long-term

[a] r-selected: favors lifestyle in which resources are diverted to high growth rate and fecundity rather than persistence. K-selected: favors lifestyle in which high growth rate and fecundity are reduced to direct resources to persistence under carrying capacity conditions.

policies, a mapping of sacred spaces, or the inventory/acquisition of natural resources. Corridors often reflect, or parallel, political, social, economic, and artistic preferences. Thus, the management of sustainable agrolandscapes and their restored connectivity will require a better integration of aesthetic and ecological principles and theory. This integration process is one of society's greatest challenges for the 21st century.

CONCLUDING REMARKS

A telescopic/microscopic view of the landscape provides a mosaic of patches (isolated and connected) that form a complex network of elements constituting our 20th-century landscape. The fragmentation of science and the proliferation of management policies indicative of the 20th century must be replaced by integrated approaches during the 21st century if societies are to achieve the goal of sustainable societies and to protect biotic diversity at regional and global scales. Lubchenco et al. (1991) outlined a research initiative necessary to establish a sustainable biosphere initiative (SBI) at greater scales. Programs or initiatives, such as the SBI, should focus on the landscape level (including human settlements) with special attention directed to protecting and restoring the integrity of ecosystem and landscape processes and to conserving biotic diversity. Only with long-term land-use planning can humans hope to protect and restore sustainable landscapes for future generations.

An agro–urban landscape perspective is important for ecologically self-sufficient and economically cost-effective management of agricultural systems. Management decisions based on this perspective should be aimed at achieving sustainability of the total landscape. We define sustainability in energetic terms in which primary productivity of the total landscape is in balance with community maintenance (i.e., $P/R = 1$). To achieve such a landscape sustainability, the rural autotrophic landscape ($P > R$) must balance the heterotrophic urban–suburban landscape ($P < R$). Biotic and cultural diversity must be maintained within these systems to protect and/or conserve regulatory feedback mechanisms (Barrett et al., 1997).

Because it is a difficult task to determine optimal land use, the development of appropriate management strategies at the agro–urban level will require the cooperation of diverse fields of study. The political, social, artistic, and economic components must become integrated with ecological theory in order to optimize land-use planning and, consequently, determine the landscape mosaic for the 21st century (Barrett and Peles, 1994). The integration of the ecological, cultural, and historical spheres of knowledge should generate new theories and methodologies necessary for the appropriate management of these agrolandscapes. The coupling of theory with practice must become an integral part of this integrative planning and management strategy.

The human process of orientation by another can have a significant influence on what human beings perceive as a resource and a nonresource and, consequently, on how humans view the total landscape. What people consider a resource changes with purpose or intent while negotiating their environment (Hunt, 1992). This perception, in turn, changes the status of other resources. For example, humans increasingly must combine ecological "capital" with economic "capital" (dual capitalism) in order to manage these systems and landscapes best for future generations.

The integration of the agricultural, urban, and natural landscapes, landscapes that transcend cultural (i.e., intragenerational) and historical (i.e., intergenerational) boundaries, must be more fully understood in order to optimize biological and landscape diversity. Linking an increasing number of diverse cultural and historical perspectives with the future design of the total landscape, through dialogue, changes the pattern of our understanding of landscapes. As Nassauer (1995) notes, changing perceptions change the landscape. The role of biodiversity in total landscapes, viewed and researched based on this perspective, can best be managed within this continuing dialogue.

Future generations will require this type of mutualistic behavior and transdisciplinary cooperation to manage increasingly complex regional and global landscapes. Maintaining biodiversity is paramount to the sustainability (health) of our planet and is by necessity transgenerational.

ACKNOWLEDGMENT

We thank Eugene P. Odum and an anonymous reviewer for providing comments and suggestions regarding this chapter.

REFERENCES

Altieri, M. A., Letourneau, D. K., and Davis, J. R., 1983. Developing sustainable agroecosystems, *BioScience*, 33:45–49.

Barrett, G. W., 1984. Applied ecology: an integrative paradigm for the 1980's, *Environ. Conserv.,* 11:319–322.

Barrett, G. W., 1989. Viewpoint: a sustainable society, *BioScience*, 39:754.

Barrett, G. W., 1990. Nature's model, *Earthwatch*, 9:24–25.

Barrett, G. W., 1992. Landscape ecology: designing sustainable agricultural landscapes, in *Integrating Sustainable Agriculture, Ecology, and Environmental Policy,* R. K. Olson, Ed., Haworth Press, New York, 83–103.

Barrett, G. W. and Bohlen, P. J., 1991. Landscape ecology, in *Landscape Linkages and Biodiversity,* W. E. Hudson, Ed., Island Press, Washington, D.C., 149–161.

Barrett, G. W. and Peles, J. D., 1994. Optimizing habitat fragmentation: an agrolandscape perspective, *Landscape Urban Plann.,* 28:99–105.

Barrett, G. W., Rodenhouse, N., and Bohlen, P. J., 1990. Role of sustainable agriculture in rural landscapes, in *Sustainable Agricultural Systems,* C. A. Edwards, R. Lai, P. Madden, R. H. Miller, and G. House, Eds., Soil and Water Conservation Society, Ankeny, IA, 624–636.

Barrett, G. W., Peles, J. D., and Odum, E. P., 1997. Transcending processes and the levels-of-organization concept, *BioScience*, (in press).

Baumgartner, M. P., 1988. *The Moral Order of a Suburb,* Oxford University Press, New York.

Brady, N. C., 1990. Making agriculture a sustainable industry, in *Sustainable Agricultural Systems,* C. A. Edwards, R. Lai, P. Madden, R. H. Miller, and G. House, Eds., Soil and Water Conservation Society, Ankeny, IA, 20–32.

Callicott, J. B., 1989. *In Defense of the Land Ethic: Essays in Environmental Philosophy,* SUNY Press, Albany, NY.

Cleveland, D. A., 1993. Is diversity more than the spice of life? Diversity, stability, and sustainable agriculture, *Culture Agric. Bull.,* 45:2–7.

Costanza, R., d'Arge, R., de Groot, R., Farber, S., Grasso, M., Hannon, B., Limburg, K., Naeem, S., O'Neill, R. V., Paurelo, J., Raskin, R. G., Sutton, P., and van den Belt, M., 1997. The value of the world's ecosystem services and natural capital, *Nature*, 387:253–260.

Culotta, E., 1996. Exploring biodiversity's benefits, *Science*, 273:1045–1046.

Daniels, T. L. and Reed, D. E., 1988. Agricultural zoning in a metropolitan county: an evaluation of the Black Hawk County, Iowa, program, *Landscape Urban Plann.,* 16:302–310.

Elliott, E. T. and Cole, C. V., 1989. A perspective on agroecosystem science, *Ecology,* 70:1597–1602.

Fahrig, L. and Merriam, G., 1985. Habitat patch connectivity and population survival, *Ecology,* 66:1762–1768.

Franklin, J. F., 1993. Preserving biodiversity: species, ecosystems, or landscapes?, *Ecol. Appl.,* 3:202–205.

Goldsmith, E., 1996. *The Way: An Ecological World-View,* Themis Books, Dartington, Totnes, Devon.

Goodland, R., 1995. The concept of environmental sustainability, *Annu. Rev. Ecol. Syst.,* 26:1–24.

Harsch, J. H., 1991. Assessing the progress of sustainable agriculture research, in *Sustainable Agriculture Research and Education in the Field: A Proceedings,* National Academy Press, Washington, D.C., 387–392.

Hess, C. E., 1991. The U.S. Department of Agriculture commitment to sustainable agriculture, in *Sustainable Agriculture Research and Education in the Field: A Proceedings,* National Academy Press, Washington, D.C., 13–21.

Hough, M., 1990. *Out of Place: Restoring Identity to the Regional Landscape,* Yale University Press, New Haven, CT.

Hough, M., 1995. *Cities and Natural Process,* Routledge Press, New York.

Hunt, J. D., 1992. *Gardens and the Picturesque: Studies in the History of Landscape Architecture,* MIT Press, Cambridge, MA.

Huntley, B. J., Ezcurra, E., Fuentes, E. R., Fujii, K., Grubb, P. J., Haber, W., Harger, J. R. E., Holland, M. M., Levin, J. A., Lubchenco, J., Mooney, H. A., Neronov, V., Noble, I., Pulliam, H. R., Ramakrishman, P. S., Risser, P. G., Sala, O., Sarukhan, J., and Sornbroeck, W. O., 1991. A sustainable biosphere: the global imperative, *Ecol. Int.,* 20:1–14.

Jackson, W. and Piper, D., 1989. The necessary marriage between ecology and agriculture, *Ecology,* 70:1591–1593.

Jake, J. A. and Wilson, D., 1992. *Derelict Landscapes: The Wasting of America's Built Environment,* Rowman and Littlefield, Savage, MD.

Jellicoe, G. and Jellicoe, S., 1987. *The Landscape of Man: Shaping the Environment From Prehistory to the Present Day,* Thames and Hudson, New York.

Kaplan, S. and Kaplan, R., 1982. *Humanscape: Environments for People,* Ulrich's Books, Ann Arbor, MI.

Le Corbusier, 1987. *The City of Tomorrow and Its Planning,* Dover Publications, New York.

Little, C. E., 1990. *Greenways for America,* Johns Hopkins University Press, Baltimore, MD.

Lockeretz, W., 1988. Editor's introduction to special issue: sustaining agriculture near cities, *Landscape Urban Plann.,* 16:291–292.

Lowrance, R., Hendrix, P. F., and Odum, E. P., 1986. A hierarchical approach to sustainable agriculture, *Am. J. Alternative Agric.,* 1:169–173.

Lubchenco, J., Olson, A. M., Brubaker, L. B., Carpenter, S. R., Holland, M. M., Hubbell, S. P., Levin, S. A., MacMahon, J. A., Matson, P. A., Melillo, J. M., Mooney, H. A., Peterson, C. H., Pulliam, H. R., Real, L. A., Regal, P. J., and Risser, P. G., 1991. The sustainable biosphere initiative: an ecological research agenda, *Ecology,* 72:371–412.

Luzar, E. J., 1988. Strategies for retaining land in agriculture: an analysis of Virginia's agricultural district policy, *Landscape Urban Plann.,* 16:319–332.

McLaughlin, A. and Mineau, P., 1995. The impact of agricultural practices on biodiversity, *Agric., Ecosyst. Environ.,* 55:201–212.

Moffat, A. S., 1996. Biodiversity is a boon to ecosystems, not species, *Science,* 271:1497.

Mumford, L., 1961. *The City in History: Its Origins, Its Transformations, and Its Prospects,* Harcourt Brace, San Diego, CA.

Nassauer, J. I., 1995. Culture and changing landscape structure, *Landscape Ecol.,* 10:229–237.

National Research Council (NRC), 1989. *Alternative Agriculture,* National Academy Press, Washington, D.C.

National Research Council (NRC), 1996. *Use of Reclaimed Water in Sludge in Food Crop Production,* National Academy Press, Washington, D.C.

Odum, E. P., 1969. The strategy of ecosystem development, *Science,* 164: 262–270.

Odum, E. P., 1989. Input management of production systems, *Science,* 243:177–182.

Odum, E. P., 1993. *Ecology and Our Endangered Life-Support Systems,* Sinauer Associates, Inc., Sunderland, MA.

Odum, E. P., 1997. *Ecology: A Bridge between Science and Society,* Sinauer Associates, Inc., Sunderland, MA.

Orr, D. W., 1992. *Ecological Literacy: Education and the Transition to a Postmodern World,* State University of New York Press, Ithaca, NY.

Paoletti, M. G., 1995. Biodiversity, traditional landscapes and agroecosystem management, *Landscape Urban Plann.,* 31:117–128.

Paoletti, M. G., Pimentel, D., Stinner, B. R., and Stinner, D., 1992. Agroecosystem biodiversity: matching production and conservation biology, *Agric. Ecosyst. Environ.,* 40:3–23.

Risser, P. G., Karr, J. R., and Forman, R. T. T., 1984. Landscape Ecology: Directions and Approaches, Illinois Natural History Survey Special Publication No. 2, Champaign, IL.

Rookwood, P., 1995. Landscape planning for biodiversity, *Landscape Urban Plann.,* 31:379–385.

Steinhart, J. S. and Steinhart, C. E., 1974. Energy use in the U.S. food system, *Science*, 184:307–316.

Stren, R., White, R., and Whitney, J., 1992. *Sustainable Cities: Urbanization and the Environment in International Perspective,* Westview Press, Boulder, CO.

Tilman, D., 1997. Community invasibility, recruitment limitation, and grassland biodiversity, *Ecology*, 78:81–92.

Tilman, D., Wedin, D., and Knops, J., 1996. Productivity and sustainability influenced by biodiversity in grassland ecosystems, *Nature*, 379:718–720.

Turner, M. G., Gardner, R. H., and O'Neill, R. V., 1995. Ecological dynamics at broad scales: ecosystems and landscapes, *BioScience*, Suppl.:29–35.

Werner, R., 1993. Ecologically and economically efficient and sustainable use of agricultural landscapes, *Landscape Urban Plann.,* 27:237–248.

CHAPTER 12

Local Management of Biodiversity in Traditional Agroecosystems

Robert E. Rhoades and Virginia D. Nazarea

CONTENTS

INTRODUCTION

The development goal of increased food and fiber production to match the needs of growing populations and their rising expectations may often conflict with the *in situ* conservation goal of preserving plant genetic diversity (Williams, 1986; Alcorn, 1991). As broadly adapted high-yielding varieties — coupled with input packages

1-56670-290-9/99/$0.00+$.50

of irrigation, fertilizers, and pesticides — find their way into traditionally diverse, marginal agroecosystems, the number and diversity of local landraces as well as associated local knowledge may erode. This process is global, although its dynamic and intensity may vary from place to place and across time (Ford-Lloyd and Jackson, 1986; Oldfield and Alcorn, 1991; Brush, 1992; Dove, 1996).

One of the premises of sustainable agriculture is that this trade-off between increasing productivity and loss of biodiversity is not inevitable (National Research Council, 1992; Thrupp, 1997). Precisely how the two goals are simultaneously achieved, however, is not an insignificant research problem or an easily answered policy issue (see Williams, 1986). Present demographic and economic global trends require more food per unit area and unit time, but the necessary yield increases will not be forthcoming unless sufficient biodiversity is continuously available to plant and animal breeders. Our primary thesis is that one useful but neglected strategy to achieve sustainable food production lies in supporting traditional *in situ* biodiversity management. We argue that many local populations have historically managed biodiversity, that the associated knowledge is valuable and irreplaceable, and that both management practices and knowledge should be enhanced through policy and technology initiatives.

Specifically, we address our thesis by exploring three major themes related to indigenous management of germplasm and the potentialities for the localized creation, maintenance, and enhancement of biodiversity. First, we place biodiversity management by traditional agroecosystems in global historical context, especially as it relates to the major food crops. Second, based on our own research experiences we outline some principles of *in situ* biodiversity maintenance within traditional, marginalized agroecosystems and contrast these to the scientific, *ex situ* approaches of formal, external input-dependent and market-linked approaches. Third, we examine the social, economic, and political dimensions of marginal communities managing *in situ* agrobiodiversity. Finally, we conclude with some observations on future research and action needs.

HISTORICAL PERSPECTIVE ON AGROBIODIVERSITY

Landrace-based genetic materials available for plant breeding or biotechnology programs have already been purposely manipulated by traditional cultivators over centuries and even millennia (Ucko and Dimbleby, 1969; Struever, 1971; Altieri and Merrick, 1987). Although archaeological debates continue over precisely where and how nomadic hunters and gatherers finally began the conscious planting of seeds, roots, or tubers and thereby ushered in the "Agricultural Revolution," the outcome of early farmers' efforts cannot be denied (Hobhouse, 1985; McCorriston and Hole, 1991). The historic tendency for preindustrial agricultural communities has been to foster and increase landrace diversity, rather than decrease it (Harlan, 1995). As long as 8 to 12 thousand years ago, "primitive" farmers had already successfully experimented with invading wild "weedy" species in their settlement clearances and domesticated the first crops (Harlan, 1975). Not only did prehistoric cultivators give humanity the major food crops and animals which nourish us today, they simultaneously created their own specialized knowledge systems about the food, fiber, and medicinal values of thousands of plant and animal species (Schery, 1972; Fowler and Mooney, 1990). While modern science has been appreciative of and concerned about the supply of the genetic raw material provided by farmer curators, much less interest has been shown in the local knowledge or management strategies which underpinned *in situ* landrace development in the first place (Nazarea-Sandoval, 1990).

De Candolle (1885) and later Vavilov (1926; 1949) were the first to observe that the density of interspecific and intraspecific variation of crop species was found in "centers of domestication" which tended to be in the ecologically complex mountainous regions or areas of marked dry–wet seasons in Africa, Asia, and Latin America (Rhoades and Thompson, 1975). Due to a variety of causes, major ancient civilizations — such as the Andean, Mesopotamian, Mesoamerican, Indus, and Chinese — evolved near these centers in close association with diverse plants and animals. In complex ecological settings under conditions of human population expansion, the coevolution of human culture and plant populations led to a level of people–plant interdependency so high that some modern crops — such as maize — cannot even reproduce themselves without purposeful human intervention (Iltis, 1987). The historical and ethnographic records are rich with data on how cultural knowledge intertwines with the biological to the degree they cannot be separated and still maintain dynamic evolutionary-ecological systems (Nazarea, 1998a). This detailed knowledge not only focused on production but also storage, processing, cooking, and utilization qualities needed for the survival and rejuvenation of crops and humans (Nazarea-Sandoval, 1992). As a result, domesticated crops can be understood as culturally created and conceived human artifacts — valued for multiple qualities such as utility, taste, color, shape, and symbolism (Zimmerer, 1991). Indeed, foods or other materials derived from crops or animals are not just calories for the human body but are integral parts of daily social and cultural lives (Brush, 1992).

The original diversification of crops in the centers of domestication was further enhanced after the Age of Discovery when plants were transferred by explorers between the Old and New Worlds (Hobhouse, 1985). In their new homes, migrating plants were further manipulated by curious cultivators and horticulturists who tested the exotic materials, selected those that did well, and then integrated them into their local farming and gardening systems. However, problems in the transplanted plants soon became evident. Since only a small amount of the genetic variability found in the agroecosystems of domestication made it to the new environments, resistance to disease and pests was often lacking and collapse under the onslaught of disease or pests was devastatingly frequent (Rhoades, 1991). The most famous documented case is that of the widespread potato crop failure in mid-19th-century Europe due to late blight (*Phytophtera infestans*), a fungus likely introduced from the Americas. Most of Europe had come to depend on a few varieties; in Ireland there was total dependence on a single variety. Combined with political exploitation by the British government, the unfortunate timing of the crop failure led to the deaths of millions of Irish (Woodham-Smith, 1962). A few years later, the grape crop on mainland Europe succumbed to a minute, aphidlike insect (*Phylloxera vitifoliae*) accidentally introduced from wild American grape stock (*Vitis labrusca*) (Olmo, 1977). These two incidents spurred a tremendous interest on the part of European scientists to understand not only the nature of disease (plant pathology was born of these efforts), but also the relationship among the centers of genetic origin, natural range of variation, and disease resistance. Although neither Mendelian genetics nor the theory of disease was understood by the 1870s, European farmers appreciated that "renewed" seed stock from the regions where the crops originated brought bloom back to their crops. In the case of postfamine potatoes, a single small Andean tuber direct from South America fetched its weight in gold, thereby creating a potato seed craze as intense as the tulip craze in Holland in earlier times (McKay, 1961). Likewise, European and American plant scientists came to appreciate the link between the well-being of their farmers' crops and the genetic diversity in the homelands of the crops themselves.

Although the importance of genetic material from the centers of diversity and domestication remains highly appreciated by geneticists and crop scientists, there is less awareness of the curator role of extant farmers or pastoral communities and their knowledge which makes it possible for this valuable diversity to be maintained *in situ* and passed on to the global human family. Historically, most governments have seen marginal tribal and peasant communities as practicing a backward, primitive agriculture ripe for "modernization" through information and technology (Rogers, 1969). As industrial developments in Europe, North America, Japan, and the cities of the Third World attracted wage labor from the countryside, planners and agricultural scientists sought ways to provide cheap and abundant food for the growing urban areas. This cheap food policy, which has intensified in the post–World War II era, meant that the potentially productive agroecological zones (flat, fertile, and hydrologically favorable) were to become targets of planned agricultural change to make them more productive through genetic uniformity and mechanization of the agroecosystem for the purpose of achieving higher yields. One outgrowth of this simplification of the agricultural landscape was the renowned "Green Revolution," which combined scientific plant breeding with input packages for favorable environments (Plucknett et al., 1987). The dramatic increases in world food supply witnessed in the 1960s and 1970s are directly traceable to these crop improvement programs which focused on increasing the productivity of plants though breeding for high response to inputs such as fertilizer (Mellor and Paulino, 1986).

The role ascribed by scientists to local cultivators and their communities during the Green Revolution was that of recipients of "technological" packages of improved seeds, fertilizers, and other inputs, as well as infrastructure development. "Transforming traditional agriculture," as Nobel Peace Prize–winner Theodore Schultz (1964) called the effort, was promoted as the motor for global growth and the most efficient exit from agricultural stagnation and famine. Farmers were seen as individual rational decision makers who only needed to be provided the necessary inputs and knowledge by governments and scientists to get the job accomplished. Hence, breeders made selections and crosses from advanced breeding lines derived from landraces. These lines were tested on experiment stations or controlled farm conditions, and, after a dozen or more years, these materials were released to farmers through certified seed programs, extension efforts, and other mechanisms (Duvick, 1983). Rather than breed for local conditions, breeders aimed for broad adaptability of high-yielding fertilizer-responsive varieties in irrigated, fertile zones. Feedback from farmers in on-farm trials rarely provided information on the suitability of selection for specific locations. The seeds were delivered to farmers largely through the patron–client extension model which focused on the individual farm enterprise, not the community or social groupings of farmers (Duvick, 1986).

The "success" of the Green Revolution was double-edged. Signficant increases in food production were achieved in a short time, leading to the alleviation of food shortages and famine in critical areas (Mellor and Paulino, 1986; Plucknett et al., 1987). However, with this success (along with other forces such as urbanization, out-migration, grazing) and as farmers responded to markets, growth, and development programs by adopting a few high yielding varieties, many landraces were abandoned. Concern over genetic erosion by national and international agencies has led to the creation of a global network of *ex situ* gene banks and living collections where landraces and wild materials are kept in short- and long-term seed storage (Plucknett et al., 1987). Fewer resources, however, have been given to support *in situ* conservation by native communities, and even less attention has been aimed at preserving the knowledge of local peoples regarding plants, a critical legacy just as vulnerable to erosion (Nazarea, 1998a).

This historical ecological–evolutionary trajectory of traditional *in situ* management of landraces underscores the following points:

1. The often controversial proposal to maintain the dynamic evolutionary management of landraces within traditional landscapes is based on the historical reality of marginal farmers as folk curators;
2. Despite the tendency of modern agricultural science to separate the genetic "resources" from the local knowledge base, both are essential components of *in situ* maintenance of diversity and, by extension, a well-supplied *ex situ* system; and
3. Despite the loss of diversity in the "favored" environments, rich gene pools still exist in many farming communities which survive along the margins of the world economic order. These marginal rural populations are often seen — even by conservationists — as a threat to biodiversity in protected areas and surrounding buffer zones. Our approach is to see them as part of the solution.

STRATEGIES FOR *IN SITU* AGROBIODIVERSITY CONSERVATION BY INDIGENOUS COMMUNITIES

Over the years, we have spent a great deal of time working with subsistence farmers in Asia, Latin America, and the American South (Nazarea-Sandoval and Rhoades, 1994). We have studied these "native curators" intensively as anthropologists, worked with farmers as members of interdisciplinary teams at International Rice Research Institute (IRRI) and International Potato Center (CIP), and — more recently — as anthropologists attempting to document and revive landrace use in the southern U.S. through support of traditional means of use and exchange of heirloom varieties. Whether Andean farmers, Filipino rice cultivators and sweet potato growers, or Appalachian gardeners, they share common characteristics. Universally, regions of rich biodiversity exist along the margins of their economic and political worlds. Landrace cultivators are typically found in more remote mountains, islands, rain forests, or desert agroecosystems which are momentarily insulated from the dominant forces of the outside world economy (Dasmann, 1991). Communities — and households within communities — with a propensity to maintain diverse systems tend to be disenfranchised from the dominant order surrounding them. Even the individuals who tend to be key native curators are marginal within their own households. Thus, *marginality* at various scale levels is a key common designator of landrace *in situ* curation.

Agroecologists, along with ethnoecologists who focus on the cognitive underpinnings of human–biological system interactions, have pioneered studies which show that farmers pursue various strategies in using biodiversity as a way to meet their basic physical, social, and spiritual needs (Hecht, 1987; Oldfield and Alcorn, 1991; Nazarea-Sandoval, 1995). This body of research points to fundamental differences between informal and formal models of genetic resource/biodiversity management (Altieri, 1987; Altieri and Merrick, 1988). First, traditional producers/communities use a different set of selection and evaluation criteria for germplasm management than modern breeding or commercial seed programs. Second, their methods of experimentation and testing are fundamentally different, although there are some points of common interest. Third, the strategies which preserve biodiversity are often embedded in community action which channels and encourages individual households to act in such a way as to foster biodiversity.

Multidimensional Criteria for Selection and Maintenance of Landraces

Scientists find the tremendous diversity of landraces in marginal agroecosystems useful and valuable in developing new and better varieties. To the practicing subsistence farmer, however, it is strange — probably inconceivable — that one would be so foolish as to risk this diversity with the narrow selection of just a few varieties and species. In maintaining a wide range of varieties and species, traditional farmers use multidimensional decision-making criteria which holistically involve ecology, the complete food system from seed handling to consumption, and cultural aspects such as culinary qualities, ritual, and cosmology. This complex decision-making process may often be poorly understood by the formal scientific sector which tends to be largely market oriented.

Farmers opt for an adaptive strategy of using biodiversity in such a way that it spreads production risk and labor scheduling across the landscape. In the Cusco Valley of Peru, for example, we found farmers who plant up to 50 different varieties as well as several species of potatoes at different time intervals in 20 to 30 scattered fields characterized by different altitude, soil types, and orientations to the sun. This principle of diversity to spread risk is simply an Andean version of "don't put all your eggs in one basket." This dispersion pattern reduces the risk that one disease outbreak or an unpredicted frost will devastate an entire crop. Simultaneously, by using different varieties a continuous flow of production through time and space can be realized so that different markets, household needs, or labor supplies can be accommodated. Interspecific and intraspecific variation is also used for agronomic control of weeds and pests, microclimatic variation through shading, as well as a buffer against climatic and pest damage. Andean potato farmers' strategies are based on a long-term, detailed knowledge of specific plant–environment interaction. Any variety is tested against several seasons of variable frosts and rainfall as well as performance in different soils.

In the Philippines, market forces are as salient as ecological factors in farmers' decision-making frameworks, and the cognizance of instability and unpredictability of both leads to constant experimentation, information and germplasm exchange, and hedging. A sweet potato farmer in Bukidnon resists the pull toward monoculture because of his perception that environmental flux and economic trends are beyond his control. According to him,

> I ask for different planting materials from our neighbors but I don't mix them up. I plant at least five different varieties of sweet potatoes at any one time to experiment from which ones I get the most benefit. At different seasons, we should plant different varieties because we never know which ones would be most productive (Nazarea, 1998b).

Some rice farmers, integrated as they are to the market system and credit infrastructure, still plant their favored varieties in the middle of clumps or at the borders of agriculture extension and credit-backed varieties, thus managing to have their credit, and eat, too.

In localized agroecosystems, household production units are also direct consumption units; thus, they have a vested interest in carefully linking production and consumption in a way not found in commercial systems where different activities are typically carried out by separate groups. In subsistence systems, the household unit manages all stages of the food system, including seed selection, production, storage, processing, and marketing. Even when there is a need for interhousehold exchange of genetic material, the linkages are generally along kin-based and community networks. There are no "formal" seed certification systems and the people who select cultivars are the same ones who grow, process, store, eat, and exchange/sell them. When the consumption unit and the production unit are coterminus, a more-refined and more-detailed set of criteria is used compared with when these two functions are separated.

In the Andes, an interdisciplinary research team from the CIP discovered some 39 criteria that farmers consider in their evaluation of varieties (Prain et al., 1992). This led to the conclusion that farmers do not seek an ideal variety. Instead, farmers seek to manage an ideal range of varieties that address their food system requirements related to cash and subsistence needs (Prain et al., 1992). These requirements were highly local and specific to household needs. In one of the research sites, for example, farmers would grow "improved" varieties for subsistence while in another village farmers cultivated folk varieties for the marketplace (Bidegaray, 1988). These unexpected uses were tied to certain local realities which only the farmers fully appreciated. In one case, there was a shortage of land and wage opportunities so they used their land to produce high-yielding varieties for food, while in the other case, a nearby market provided higher prices for the valued native varieties (Brush, 1992; Prain et al., 1992).

Another aspect of diversity maintenance involves postproduction activities (storage, seed selection, processing, and cooking). Women, who are often in charge of these nonfield activities, handle materials in such as way as to increase aspects of diversity further. Shapes and colors proliferate in landrace material since these are used as perceptual signals for sorting and identification. Most published research on potato selection makes reference to the significance of differences in color, shape, texture, and taste. Selection for "storability" or "culinary quality" occurs in the hands of women who are acknowledged by the men to have superior knowledge of the crop. Andean farmers are connoisseurs of potatoes which they evaluate with a wide range of cooking descriptors as well as taste labels such as "flouriness," "stickiness," "wateriness," and so on. Native potatoes are universally recognized as superior to improved varieties in terms of culinary quality.

Among Philippine sweet potato farmers, characteristics such as cooking quality, aesthetic appeal, storability, and propensity of mixing well with other cultivars are valued as much as yield or disease resistance by households surviving in the marginal zones. Morphological, gastronomic, life habit attributes, familiarity gradients, and functional criteria were used in distinguishing and prioritizing among varieties, and were far from being mutually exclusive. Interestingly, local criteria for evaluation of sweet potato varieties tend to be fuzzy or to trail off into gray areas as to which properties or traits are positive or preferred and which ones are negative or not preferred. For example, people would say they prefer sweet varieties but bland ones are good to eat with fish and are a good substitute for rice during lean times, or that newer varieties are desirable because they produce bigger roots but older varieties produce tastier though smaller roots. The result of this "fuzziness" is that it is impossible to construct a hierarchy of sweetpotato varieties from the most desirable to the least, and, as a consequence, people retain different varieties in their farms and home gardens.

Another dimension of genetic resource diversity in traditional societies often overlooked by scientists and planners from more "utilitarian" urban-dominated societies is the interconnectedness between plants and cosmology, that part of culture which deals with perception, ritual, religion, and worldview. Given the intimacy of daily contact between cultivators and their biological environment, especially plants and animals, a cultural interplay is not uncommon during which the domesticates are assigned significant symbolic roles in the lives of the people themselves (Zimmerer, 1991). Therefore, plants are more than just food. Plants are also ascribed gender, spiritual qualities, mystical powers, and important religious roles in the lives of the people (*Down to Earth,* 1994). People of many cultures believe they originated from certain sacred plants (e.g., Mayan creation story and maize). In the case of southwestern Native American groups, the diversity of maize types (and colors) reflects group relationship, ethnic origins, cardinal directions, and a reverence for diversity (Ford, 1984; Sekaquaptewa and Black, 1986).

Evidence from many cultures around the world points to a playfulness and appreciation of landrace diversity as expressed not only through color and shape, but also reflected in complex folk taxonomies and cultural identity related to landraces. In the Andes, certain potato varieties are valued more for their symbolic role in gift exchange and honoring guests at ritualized meals than for any agronomic and economic values. Brush (1977) reports that the most highly prized varieties are often the most delicate and least productive (see also Carter and Mamani, 1982). One study from Bolivia pointed to the importance of potato diversity to the cultural identity of the Aymara (Johnsson, 1986). In some parts of the Andes, the most prestigious meal one can serve is made up of native cultivars, especially of potato. Although such beliefs are frequently disregarded by scientists as superfluous, the ethnographic record shows that such beliefs play a major shaping role in creating variability among cultures (Zimmerer, 1991).

Try as we might, as scientists, to coax the fan of strategies into a logical, universal framework, none seems to provide greater exploratory power than the "framework" of expediency — of hedging, making do, and muddling through. By this, we mean the development and maintenance of plant genetic diversity in local agroecosystems based on day-to-day pragmatic concerns and the natural inertia that preserves diversity due to the existence of a multiplicity of local demands and preferences but cannot be fully satisfied by any one "ideal" or "best" variety. The decision-making process, in other words, is characterized by conflicting demands, complementation, and compromise, resulting in behavioral outcomes that augur well for the maintenance of a wide variety of cultivars.

Comparison of Scientific/Formal Approaches to Biodiversity Maintenance

Contrasting the approaches of traditional farmers and scientists in methods of varietal selection can clarify the reasons plant-breeding programs often fail to reach farmers with new genetic materials (Berg et al., 1991). Since traditional farmers deal with holistic systems and multiple selection criteria they do not normally think in terms of formal dichotomies like "improved" vs. "local" varieties. Farmers select varieties that perform well in certain areas (e.g., agronomic, yield, marketability, culinary) important to the context of their localized food system. Although farmers do not use the agronomists' multiple replications side by side, the folk selection process is far from haphazard. Like breeders, traditional farmers have a systematic way of seeking and integrating materials into their living, working informal gene banks. Farmers are fanatic seekers of new varieties, and they will eagerly seek materials wherever they can be found (e.g., formal seed programs, neighbors, markets). Once a new variety is obtained, it is generally planted on a small scale in a kitchen garden or in a single row along the margins of a regular field. If the variety proves itself, farmers amplify their production as the amount of seed allows. The variety is observed and evaluated for multiple qualities relevant to the local food system (see Table 1). All the while, they continue to maintain their own "germplasm" bank which is constantly being replenished and experimentally culled (Rhoades, 1989a).

Many farmers are avid experimenters by nature (Richards, 1985; Rhoades and Bebbington, 1995). The "atmosphere of experimentation" which characterized the neolithic farmer since the earlier stages of cultivation is one of the foundations upon which agriculture advances (Braidwood, 1967), and farmers are as creatively involved in this ongoing process as are scientists. A key difference, however, between

Table 1 Breeders' and Farmers' Cultivar Selection Methods

Breeders	Farmers
1. Genetically uniform cultivars (pure lines, clones, hybrids)	Genetically diverse cultivars
2. Test under ideal conditions	Real-world conditions
3. Yield and disease/climate tolerance	Multiple criteria; fuzziness
4. Widely adapted; agroecological target zones (flat, irrigated, fertile, homogenous, inputs)	Niche-specific (rain-fed, poor soils, inaccessible, local inputs)
5. Formal structures; highly centralized; top-down	Informal; kin/community based; gendered
6. Negative attitude toward G × E	Positive attitude toward G × E

formal and informal cultivar selection is that breeders tend to narrow the genetic alternatives in search of yield and disease or climatic resistance while marginal, subsistence farmers tend to broaden their choices by seeking more diverse varieties to fit their overall needs (Soleri and Smith, 1998; Nazarea-Sandoval and Rhoades, 1994). Indigenous cultivators do not design, perceive, or manage plots or zones in isolation of surrounding areas. To the contrary, they manage for diversity along continuous boundaries by pursuing opportunities creatively to mix genetic resources and inputs to meet their household and community needs. Farmers use diverse criteria in selection and adoption decision making which does not necessarily end up with the intentional elimination of "less desirable" options. What is desirable or not desirable to local farmers may be a matter of taste, a matter of timing, and sometimes a matter of mood. In other words, they use fuzzy multiple criteria; if not, the diverse cultivars would likely have disappeared long ago (Nazarea, 1996).

One of the reasons that small farmers in marginal environments have benefited little from the yield and disease resistance achieved by formal breeding programs is precisely because of the real-world interaction between genotype and environment (G × E). Breeding programs typically assume agricultural scientists know better than farmers the characteristics of a successful cultivar (Witcombe et al., 1996). Breeders select under favorable growing conditions, and, if adoption does not occur, the cause is frequently assumed to be ineffective extension or insufficient seed production (Ceccarelli, 1995). Breeding for broad adaptation to agroecological zones requires large-scale centralized seed production and distribution which in turn further promotes genotypes that might be inferior to the landraces they are replacing under stressful conditions. This formal approach contrasts with that of marginal farmers who have traditionally relied on a strategy based on both intraspecific diversity (crop mixes and landraces on the same farm) and where seed is produced either on the farm or obtained from neighbors through community-based informal seed networks.

In bridging the gap between breeding programs and farmers in marginal areas, breeders have begun to think innovatively about marginal farmers, experimental designs, field plot techniques, and landraces (Maurya et al., 1988; Galt, 1989; Ceccarelli, 1995). As a result, participatory breeding programs have begun to emerge in which farmers are encouraged through support and partnership with scientists to exchange knowledge and test, under farmer experimental conditions and designs, cultivars early in the breeding-selection process (Prain et al., 1992; Joshi and Witcombe, 1996). These participatory programs have already generated varieties that match farmers' needs and increase production simultaneously (Maurya et al., 1988; Sperling et al., 1993; de Boef et al., 1993; Sperling and Scheidegger, 1995; Witcombe et al., 1996).

SOCIAL, ECONOMIC, AND POLITICAL DIMENSIONS OF *IN SITU* CONSERVATION

The Social Context of Community-Based Biodiversity Management

While unusual innovativeness in biodiversity preservation finds manifestation in experimental, individualistic farmers, ultimately the social context of local communities shapes *in situ* biodiversity maintenance. In communities that have not yet been fully incorporated into commercial markets and still manage high levels of landrace biodiversity, future protection of such genetic resources requires values compatible with locally defined social and economic goals. Whether biodiversity is decreased or enhanced in the future may depend on the degree of self-determination of local communities to attain these locally defined goals. In many traditional, closed corporate communities (i.e., membership determined by birth), intergenerational equity or "bequeath value" is as important as it is in developed countries where heads of farm enterprises expect their families to continue to operate the business well into the future. Many indigenous communities with a firm sense of place are aware of the value and role of land and diverse crop inventories to their cultural survival, and communally strive to guard these resources. Andean communities, for instance, carefully regulate, through annual village assemblies, the rotation of land parcels, the use of communal pastures, and the complex of species and varieties planted. In other cases, such as in the South American Amazon, Amerindian groups purposely increase biological diversity in many locations through shifting cultivation which involves systematically transplanting crops throughout the forest (Castilleja et al., 1995). The landscape in this setting is in fact a cultural creation by the populations who have over the centuries altered the natural landscape through clearing, burning, planting, and other mechanisms of diversifying resource clusters for their use. A similar effect is obtained around the Awa Ethnic Forest Reserve in Ecuador where the Awa Indians have planted "forest belts" of gardens and fields as a territorial signal to logging companies and other outsiders (Castilleja et al., 1995).

Indigenous peoples value their community resources and typically practice collective decision making at a much higher intensity than found in open, Western societies. To maintain and exploit communal resources requires group values, dedication, and willingness to follow village leadership faithfully (Rastogi and Pant, 1996). In most cases, regulation is enforced and punishment meted out by the community itself. Leadership often rotates among households so that all will have a degree of responsibility through time. Tribes of Arunachal Pradesh in India, for example, have a self-governing system wherein the tribal council is responsible for societal decision making (Rastogi and Pant, 1996). Their councils (variously known as *buliangs* of the Apatanis, the *kegangs* of the Adis, the *nyels* of the Nishings) are informal in nature and led by elders who command a great deal of respect. Any conflicts, such as over boundaries or in sharing of resources, are handled in the village courts through mediation. The main threat to these indigenous local institutions has been the national and state governments which have declared control of much of the ancestral lands by overpowering and disbanding local institutions (Fisher, 1995). In Nepal, for instance, when the government seized all community forest lands in the 1950s and superimposed outside government agencies to manage the forest, widespread deforestation and land exploitation resulted (Ives and Messerli, 1989). Even local people who formerly protected their lands participated in the exploitation. Only in the 1990s, after a great deal of destruction, has the government tried to reverse itself and return the forests back to local community-based management (Griffin et al., 1988; Gilmour and Nurse, 1991).

The power of communities in the social creation and maintenance of biodiversity can be clearly seen through their role in seed or germplasm exchange of cultivated crops through carefully organized fairs or market days. For centuries, people from the native communities have congregated in spatially rotating markets on established days during the year for the purpose of exchanging goods from different regions or valleys. At Andean local markets and fairs, women with distinctive village apparel sit at local markets behind sacks and baskets full of brightly colored beans, maize, or potatoes. There are weekly fairs, which are tied to normal marketing of household goods, and regional fairs held only once a year. The latter correspond to a religious holiday and bring together communities from different zones and villages. These festivals are held at the end of harvest so that a maximal diversity of crops, technologies, and knowledge is available for exchange prior to the commencement of the next season. These regularly scheduled fairs, and the ritualistic manner in which they are organized, in fact function for the systematic exchange of genetic materials — a true manifestation of "conservation through use." Similar markets are found throughout the tropical world, such as certain mid-hill markets in the Nepal Himalaya which are renowned for the diversity of crops and the excellence of landrace planting materials which arrive from distant valleys on a predictable schedule (Rhoades, 1985).

The social locus of biodiversity management within households resides mainly with women in households of marginal individuals. In our 1992 UPWARD project in Bukidnon, Philippines, we compared two models for *in situ* conservation of traditional root crops. One system was dominated by the traditional political hierarchy, mainly comprising men, while the other was managed by an informal network of migrant women called the "industrious mothers" (Nazarea, 1996). Through time, the importance of the more informal, egalitarian approach of the women for biodiversity enhancement as opposed to the authoritarian formally organized male structure became apparent. The male organization could not sustain interest in the maintenance and enhancement of crop germplasm. In fact, genetic material in the male-dominated garden would have perished had a female relative not rescued it by transplanting surviving plants to her garden. The women's garden, however, flourished amid an atmosphere of lighthearted communal spirit from the beginning. All members of the group, including children and men, joined in the process. A tracing of the exchanges and information flows show that the women were connected in layered vertical and interconnected horizontal relationships along which germplasm and associated information were continually and reciprocally exchanged (see also Prain and Piniero, 1994). One might surmise that the informal female network replicates how germplasm has been created by traditional communities for centuries. The germplasm is enhanced and conserved by different social "pathways" such as between blood relations or ceremonial kinship networks or exchange between market associates. Strong reliance on interpersonal and familial ties is important and a great deal of germplasm changes hands between neighbors. Any new variety that is acquired is soon spread, and redundancy considerably buffers the system from genetic erosion. A well-entrenched cultural ethic of sharing, coupled with an interest in seeking and soliciting materials to secure household needs, helps create and maintain diversity.

Global Change and Plant Genetic Resources

The vast majority of marginalized indigenous communities in the world are still unaware that during the past few years they have been thrust center stage in the political debate over access, control, and ownership of the genetic resources found within their traditional agroecosystems (Mooney, 1979; Fowler and Mooney, 1990; Shand, 1991; Brush, 1993). Most are unaware of the concerted efforts of trade liberalization and political policies aimed at their political and economical integration into the global "village" (Hall, 1991). However, this situation is rapidly changing. Due to a combination of rising awareness of the possibility of economic value of locally controlled genetic materials as well as increasing legal and political debate over the same resource on a global scale, marginalized farming communities are no longer content to be partners voluntarily assisting plant collectors or ethnobotanists in finding and identifying useful genetic materials in their territories. This newly found awareness for traditional peoples further complicates an already entangled global dynamic in which many stakeholders — private companies, national and international breeding programs, activists, and academics — must now contend with the role of indigenous communities that goes beyond their previous uncompensated provisioning and protection of the germplasm in their traditional agroecosystems (Rhoades and Nazarea, 1996). Indigenous communities can have crucial roles in biodiversity protection if their ancestral territories and knowledge are legally recognized, if they are provided effective control over the resources — including their own knowledge — within their environments, and if they are not hampered by external extractive policies which interfere with traditional life in a destructive manner (Cunningham, 1991; Martin et al., 1996).

The modern thrust in development to empower local communities has been generated, ironically, by a process called globalization. These changes are also related to the new regional and supranational monetary and financial arrangements (e.g., World Trade Organisation, North American Free Trade Agreement, etc.) which break down trade barriers, open new markets, and bring democracy to formerly highly authoritarian areas (Hall, 1991). One result of the global liberalization movement is that, increasingly, local communities, even in formerly highly centralized countries like China, are given more and more freedom to manage their local affairs.

These new arrangements signal possible new alliances between conservationists and the international movement for local rights (e.g., human rights, community empowerment, self-determination). This often means that global environmental organizations endorse local community rights while the community adopts the slogans of international environmental protection (even to the point of writing grant proposals for funding).

In recent years, the policy recommendations and funding initiative following the publication of *Our Common Future* have stimulated agricultural and development agencies to show more interest in developing marginal areas of the developing world (World Commission on Environment and Development, 1987). In many of these hard-to-reach regions, a Green Revolution has not occurred, much to the frustration of central governments, international agricultural research centers, and input suppliers. Since such regions are not well integrated into national economies, they also tend to be areas where political resistance, civil unrest, and narcotic trafficking or illicit trade are common. They are also rich in many natural resources such as economically significant flora (especially medicinals and timber), minerals, and water. Despite their "invisibility," the sheer numbers of such marginalized people are not insignificant. Perhaps over 1 billion people live in the more marginal diversified ecosystems that are somewhat insulated from "roadside" development — mountain, uplands, arid and semiarid deserts, and tropical lowlands — where most of the remaining world biodiversity is found.

Much of the development thinking about these marginal areas has been guided by a "blame the victim" mentality; that is, too many unplanned children lead to more poverty which in turn creates more land degradation which leads the vicious spiral inevitably downward (Rhoades and Harwood, 1992). One widely promoted solution to such land degradation, including loss of biodiversity, is to transform or modernize such zones either through the transfer of technology or by enhancing income through creation of enterprises which would allow the populace to market their local resources on the outside. The farming system research and extension movement was an effort in this direction, as is the more-recent participatory sustainable agriculture and natural resource management model (Rhoades, 1989b). Despite an emphasis on the value of indigenous systems in these well-intended approaches, the prevailing belief among most agricultural scientists is that improved agricultural systems should replace those traditional systems which are not capable of producing sufficient food and income. Therefore, a great deal still hinges on the development of external inputs and technologies (especially high-yielding varieties), commercial markets, and road building.

The neoclassical economic approach to transforming agricultural communities outlined in the previous paragraph has been questioned from the perspective of agroecology and ethnoecology. Dove (1996), in discussing the granting of intellectual property rights to farmers, even goes so far as to argue that well-meaning projects which aim to help marginal producers maintain diversity actually serve to undermine the diversity to be preserved. By pulling the marginal areas into the mainstream, one destroys the context which fostered diversity in the first place, that is, marginality. While other agro- and ethnoecologists may adopt a more moderate position than Dove's, many question the neo-Malthusian assumptions about population–poverty–degradation by arguing instead that traditional communities possess creative and dynamic strategies for purposeful resistance to outside influences (Nazarea-Sandoval, 1995; 1998a). Indigenous peoples are seen as creative actors who are becoming increasingly knowledgeable about global politics and economics. This, in turn, has given them new external alliances which allow them to add value to the resources they control. In other words, traditional communities are no longer seen as "passive victims" in the global/political biodiversity game (Colchester, 1994). In fact, many local communities are either closing their doors to or negotiating with outside "bioprospectors" (Shyamsundar and Lanier, 1994). Recently, we witnessed in Ecuador the federation of Amazonian native peoples halting all foreign collecting in response to an attempt by a U.S. researcher to secure legal rights to a medicinal plant given by a shaman. Native peoples are examining options by which they can gain economic leverage with both their knowledge and the genetic resources found within their territories. This story is only now unfolding in many previously isolated communities around the world (Martin et al., 1996).

PATHWAYS TOWARD THE FUTURE

In the preceding sections we have outlined many of the problems and potentialities of local *in situ* management and conservation of plant genetic resources. In proposing positive future steps for effective conservation, we utilize a dichotomy articulated by Janice Alcorn as a discussant at a symposium on "Local-global (Dis)articulation of Plant Genetic Resources" (Rhoades and Nazarea, 1996). Alcorn noted that in the rush to address critical issues of genetic resources, farmers' rights and intellectual property rights, poverty causal links, and other concerns, much more emphasis has been placed on Conservation with a "big C" vs. conservation with a "little c." That is to say that more energy and debate are being directed toward the macrolevel (e.g., international forums, world organization policies, national laws) and much less is devoted to thinking about or working at the critical juncture, at the microlevel. Alcorn concludes that we need more case studies, documented experience, and community models for working with farmer curators at the grassroots level. Research on such case studies and models is now emerging at the global level through a number of international programs (Martin et al., 1996).

Biodiversity will not be preserved *in situ* unless local communities see it in their best interest to do so (Norgaard, 1988). However, this in itself is complicated. On the one hand, marginality and poverty can drive communities to act in ways which are not necessarily in accordance with the global agenda for biodiversity conservation (Agenda 21 of the United Nations Conference on Environment and Development). Many objectives of Agenda 21–inspired projects (nature reserves, endangered species, clean air and water, natural resource management) may not be congruent with the interests and priorities of local farmers (Reardon and Vosti, 1995). On the other hand, subsidies, price supports, and other developed country approaches (e.g., such as applied in soil conservation programs in the U.S.) will probably not work. Solutions such as intellectual property rights, patents, and trade secrets will only benefit local communities if the nation-state allows income from these legal arrangements to reach local populations (Vogel, 1994).

While conservation of genetic variability has received national and global attention, funding continues to be inadequate to assure protection of *ex situ* collections and many nations give low priority to genetic resource efforts. However, even less recognition and funding support is given to preservation of the cultural knowledge which underlies the genetic material (Nabhan et al., 1991). Interest in this activity is relegated to a few token social scientists in international agencies or advocacy groups such as local and international non-governmental organizations.

To address knowledge loss, Nazarea (1995; 1998a) developed an approach called "memory banking," based on the principle that effective germplasm conservation should involve cultural dimensions along with the biological. She points to parallels between memory banking and gene banking. While germplasm encodes genetic information that has evolved as a response to selection pressures, cultural information in the minds of local farmers who have had considerable experience in growing these crops is likewise a repository of coded, time-tested adaptations to the environment. Moreover, genetic information embodied in folk varieties is threatened with erosion because of pressure toward more intensive, monocultural production which favors the adoption of newer, higher-yielding crop varieties in the same way that cultural knowledge and practices associated with traditional varieties are in imminent danger of being swamped by modern technologies (see also de Boef et al., 1993). This loss of germplasm, and associated knowledge, has been referred to in a report of a panel of the Board on Science and Technology for International Development (National Research Council, 1992) as the "real tragedy of the commons; that is, within a few decades traditional management systems in effect for thousands of years become obsolete, replaced by systems of relentless exploitation of rural people and rural countries."

One of the most important issues for biodiversity conservation is how to give farmers incentives to maintain diversity in light of these external pressures (Brush, 1993; Vogel, 1994). Although many marginal regions remain somewhat impervious to the introduction of improved varieties as a result of inaccessible land and isolation from markets, this same marginality leaves farmer curators normally politically powerless and struggling with poverty. In such settings, the answer to *in situ* biodiversity preservation does not entirely lie with projects which focus on commercial farms as private enterprise, but rather with the strength and interest of the traditional community. In this chapter, we have already discussed such mechanisms as informal curation, seed fairs, seed exchange gardens, and memory banking, and access and control by local people of their native patrimony. If only a small percentage of development funds could be earmarked to support individuals and the community's sense of pride in diversity, a significant amount of genetic variation might continue to be nurtured *in situ* (Shulman, 1986).

Formulating effective guidelines or action plans for locally based conservation is more easily written on paper than actually carried out in the field. We are fully aware of the complications, trade-offs, and unanticipated effects, not to mention the power of homogenizing forces of the global markets swirling around the remaining pockets of agrobiodiversity on this earth. However, we cannot accept the conclusions of the cynics who argue that diversity, once lost, cannot be restored or if the market penetrates at all then the very comparative advantage of the marginal area for diversity creation is gone. Brush (1993), Cleveland et al. (1994), and Nazarea (1996), among others, show that diversity can indeed be reintroduced to regions and there are many individual and project cases (e.g., Seed Savers Exchange, Native Seed Search, Southern Seed Legacy) of both restoration and preservation.

In response to the apparent widespread loss of *in situ* genetic resources of heirlooms (vegetables, fruit trees, and ornamentals) in the rapidly changing deep south of the U.S., we established in 1996 a program called the Southern Seed Legacy. What began as a limited project to test the memory-banking approach mentioned above has taken on dimensions of a social movement itself. Farmers and gardeners of diverse ethnic origins who were born in the earlier part of this century and whose land is being enclosed by urban development still maintain diverse stocks of local varieties of beans, melons, peas, okra, pecans, ornamentals, and dozens of other disappearing varieties. Very little of this material is maintained within the region and accessible to the local people in the U.S. Department of Agriculture germplasm system. Likewise, other seed-saving organizations (for profit or not for profit) often collect or solicit the southern seeds but maintain or market them from locations in the northern U.S. or on the West Coast. Although the Southern Seed Legacy is a young project, the enthusiasm for exchange and maintenance of the old varieties is widespread and intense among both the old-timers and a new generation of small-scale growers. A key to the success of the project so far has been the enthusiasm on the part of farmer–gardener curators for documenting and passing along the crucial link between cultural knowledge and preservation of the genetic material itself.

Regardless of whether *in situ* conservation is an indigenous event or one stimulated by an outside-funded program, one principle must be preserved. That is, the degree of independence, even irreverence, necessary for the persistence of diversity is inversely proportional to the degree of integration into the market system and the degree of capture by the political vortex. *In situ* conservation is a promising channel of genetic conservation that is entirely compatible with *ex situ* gene conservation. It already exists in the form of home gardens, polycultures, and traditional agroecosystems. The next step is to make this conservation strategy more systematic, more sustainable, less risky, and to link it to imperatives beyond the local and even the regional scale. Finally, if the system of access and rewards can be restructured, we feel that local populations can use conventions that exist in the global marketplace to their benefit — and for the benefit of humankind — instead of being subjected to these conventions (Moran, 1998). Our argument is that marginal populations know how to preserve biodiversity, have been doing it for centuries, and do not need a great deal of outside direct assistance to get the job done. However, they cannot continue to preserve the essential diversity solely for the sake of conservation if there is no compensation for their labors or access and control over their ancestral resources. These are the factors which need to be restructured. If not, we will "kill the goose that laid the golden egg" and not even realize what we have done until it is much too late.

REFERENCES

Alcorn, J., 1991. Epilogue, in *Biodiversity, Culture, Conservation, and Ecodevelopment*, M. L. Oldfield and J. B. Alcorn, Eds., Westview Press, Boulder, CO, 317–349.

Altieri, M. A., 1987. *Agroecology: The Scientific Basis of Alternative Agriculture*, Westview Press, Boulder, CO.

Altieri, M. A., and Merrick, L. C., 1987. Peasant agriculture and the conservation of crops and wild plant resources, *Conserv. Biol.,* 1:49–58.

Altieri, M. A. and Merrick, L. C., 1988. Agroecology and *in situ* conservation of native crop diversity in the Third World, in *Biodiversity*, E. O. Wilson, Ed., National Academy Press, Washington, D.C., 361–370.

Berg, T., Bjornstad, A., Fowler, C., and Skroppa, T., 1991. *Technology Options and the Gene Struggle*, Development and Environment No. 8, NORAGRIC Occasional Papers Series C.

Bidegaray, P., 1988. Potato Adoption and Diffusion in Peruvian Andean Communities, Master's thesis, Department of Anthropology, University of Kentucky, Lexington.

Braidwood, R. J., 1967. *Prehistoric Men*, Scott, Foresman and Co., Glenview, IL.

Brush, S., 1977. *Mountain, Field, and Family: The Economy of an Andean Valley*, University of Pennsylvania Press, Philadelphia.

Brush, S. B., 1992. Ethnoecology, biodiversity, and modernization in Andean potato agriculture, *J. Ethnobiol.,* 12:161–185.

Brush, S. B., 1993. Indigenous knowledge of biological resources and intellectual property rights: the role of anthropology, *Am. Anthropol.,* 95:653–686.

Carter, W. and Mamani, M., 1982. *Irpa Chico: Individuo y Communidad en la Cultura Andina*, Liberia-Editorial Joventud, La Paz, Bolivia.

Castilleja, G., Poole, P. J., and Geisler, C. C., 1995. *The Social Challenge of Biodiversity Conservation*, Working Paper No. 1, The World Bank, Washington, D.C.

Ceccarelli, S., Grando, S., and Booth, R. H., 1995. Are small farmers too small or are international breeding programmes too large? Unpublished manuscript, ICARDA, Allepo, Syria.

Cleveland, D., Soleri, D., and Smith, S. E., 1994. Do folk crop varieties have a role in sustainable agriculture? *Bioscience,* 44: 740–751.

Colchester, M., 1994. Salvaging Nature: Indigenous Peoples, Protected Areas and Biodiversity Conservation, Discussion Paper DPSS, UNRISD, Geneva and WWF, Gland.

Cunningham, A. B., 1991. Indigenous knowledge and biodiversity: global commons or regional heritage? *Cult. Survival Q.,* 15: 4–8.

Dasmann, R. F., 1991. The importance of cultural and biological diversity, in *Biodiversity, Culture, Conservation, and Ecodevelopment*, M. L. Oldfield and J. B. Alcorn, Eds., Westview Press, Boulder, CO, 7–15.

de Boef, W., Amanor, K., Wellard, K., and Bebbington, A., 1993. *Cultivating Knowledge: Genetic Diversity, Farmer Experimentation and Crop Research*, Intermediate Technology Publications, London.

De Candolle, A., 1885. *Origin of Cultivated Plants*, D. Appleton, New York.

Dove, M., 1996. Center, periphery, and biodiversity: a paradox of governance and a developmental challenge, in *Valuing Local Knowledge: Indigenous People and Intellectual Property Rights,* S. Brush and D. Stabinsky, Ed., Island Press, Washington, D.C., 41–67.

Duvick, D. N., 1983. Improved conventional strategies and methods for selection and utilization of germplasm, in *Chemistry and World Food Supplies: The New Frontiers*, L. W. Shermit, Ed., Pergamon Press, Oxford, 577–584.

Duvick, D. N., 1986. Plant breeding: past achievements and expectations for the future, *Econ. Bot.,* 40:289–297.

Fisher, R. J., 1995. *Collaborative Management of Forests for Conservation and Development*, Issues in Forest Conservation series, IUCN — The World Conservation Union and Worldwide Fund for Nature, SADAG, Valserine, France.

Ford, R. I., 1984. The evolution of corn revisited, *Q. Rev. Archaeol.*, 4:12–13, 16.

Ford-Lloyd, B. and Jackson, M., 1986. *Plant Genetic Resources: An Introduction to Their Conservation and Use*, Edward Arnold, London.

Fowler, C. and Mooney, P., 1990. *Shattering: Food, Politics, and the Loss of Genetic Diversity*, The University of Arizona Press, Tucson.

Galt, A., 1989. Joining FSR to Commodity Programme Breeding Efforts Earlier: Increasing Plant Breeding Efficiency in Nepal, ODI Agricultural Administration Unit, Network Paper 8.

Gilmour, D. A. and Nurse, M. C., 1991. Farmer initiatives in increasing tree cover in central Nepal, *Mountain Res. Dev.*, 11:329–337.

Griffin, D. M., Shepherd, K. R., and Mahat, T. B. S., 1988. Human impact on some forests of the middle hills of Nepal — Part 5. Comparisons, concepts, and some policy implications, *Mountain Res. Dev.*, 8:43–52.

Hall, S., 1991. The local and the global: globalization and ethnicity, in *Culture, Globalization, and the World System: Contemporary Conditions for the Representation of Identity*, A. King, Ed., Macmillan, London, 19–39.

Harlan, J., 1975. *Crops and Man*, American Society of Agronomy, Madison, WI.

Harlan, J., 1995. *The Living Fields: Our Agricultural Heritage*, Cambridge University Press, New York.

Hecht, S. B., 1987. The evolution of agroecological thought, in *Agroecology: The Scientific Basis of Alternative Agriculture*, M. Altieri, Ed., Westview Press, Boulder, CO, 1–20.

Hobhouse, H., 1985. *Seeds of Change*, Harper & Row, New York.

Iltis, H. H., 1987. Maize evolution and agricultural origins, in *Systematics and Evolution of the Gramineae*, T. Soderstrom, K. Hilu, M. Barkworth, and C. Campbell, Eds., Smithsonian Institution Press, Washington, D.C., 195–213.

Ives, J. D. and Messerli, M., 1989. *The Himalayan Dilemma: Reconciling Development and Conservation*, Routledge, London.

Johnsson, M., 1986. *Food and Culture among the Bolivian Aymara: Symbolic Expressions of Social Relations*, Uppsala Studies in Cultural Anthropology 7, Almquist and Wiksell International, Stockholm, Sweden.

Joshi, A. and Witcombe, J. R., 1996. Farmer participatory crop improvement. II. Participatory varietal selection, a case study in India, *Exp. Agric.*, 32:469–485.

Martin, G. J., Hoare, A. L., and Posey, D. A., Eds., 1996. Protecting rights: legal and ethical implications of biodiversity, *People and Plants Handbook: Sources for Applying Ethnobotany to Conservation and Community Development*, UNESCO, Paris.

Maurya, D. M., Bottral, A., and Farrington, J., 1988. Improved livelihoods, genetic diversity and farmer participation: a strategy for rice-breeding in rainfed areas of India, *Exp. Agric.*, 24:311–320.

McCorriston, J. and Hole, F., 1991. The ecology of seasonal stress and the origins of agriculture in the Near East, *Am. Anthropol.*, 93:46–69.

McKay, R., 1961. *An Anthology of the Potato*, Allen Figgis and Co., Ltd., Dublin.

Mellor, J. W. and Paulino, L., 1986. Food production needs in a consumption perspective, in *Global Aspects of Food Production*, M. S. Swaminathan and S. K. Sinhu, Eds., Tycooly International, Oxford-Riverton, NJ, 1–24.

Mooney, P., 1979. *Seeds of the Earth: A Private or Public Resource?*, Inter Pares for the Canadian Council for International Co-operation and the International Coalition for Development Action, Ottawa.

Moran, K., 1998. Toward compensation: returning benefits from drug discoveries to native communities, in *Ethnoecology: Situated Knowledge/Located Lives,* V. Nazarea, Ed., University of Arizona Press, Tucson, in press.

Nabhan, G. P., House, D., Suzan, H., Hodgson, A. W., Hernandez, L. S., and Malda, G., 1991. Conservation and use of rare plants by traditional cultures of the U.S./Mexico borderlands, in *Biodiversity, Culture, Conservation, and Ecodevelopment,* M. L. Oldfield and J. B. Alcorn, Eds., Westview Press, Boulder, CO, 127–146.

National Research Council, 1992. *Conserving Biodiversity: A Research Agenda for Development Agencies*, Academy Press, Washington, D.C.

Nazarea, V., 1995. *Local Knowledge and Agricultural Decision Making in the Philippines: Class, Gender, and Resistance*, Cornell University Press, Ithaca, NY.

Nazarea, V., 1996. Fields of memories as everyday resistance, *Cult. Survival Q.,* 20:61–66.

Nazarea, V., 1998a. *Memory Banking: The Cultural Dimensions of Biodiversity*, University of Arizona Press, Tucson. (In Press).

Nazarea, V., 1998b. *Ethnoecology: Situated Knowledge/Located Lives*, University of Arizona Press, Tucson. (In Press).

Nazarea-Sandoval, V., 1990. Memory banking of indigenous technologies of local farmers associated with traditional crop varieties: a focus on sweet potato, in *Proceedings of the Inaugural Workshop on the Users Perspective with Agricultural Research and Development*, UPWARD/International Potato Center, Los Banos.

Nazarea-Sandoval, V., 1992. Ethnoagronomy and ethnogastronomy: on indigenous typology and use of biological resources, *Agric. Human Values,* 8:121–131.

Nazarea-Sandoval, V., 1995. *Local Knowledge and Agricultural Decision Making in the Philippines*, Cornell University Press, Ithaca, NY.

Nazarea-Sandoval, V. D. and Rhoades, R. E., 1994. Rice, reason, and resistance: a comparative study of farmers' vs. scientists' perception and strategies, in *Rice Blast Disease*, R. S. Zeigler, S. A. Leong, and P. S. Teng, Eds., CAB International, Wallingford, England, 559–575.

Norgaard, R. B., 1988. The rise of the global exchange economy and the loss of biological diversity, in *Biodiversity*, E. O. Wilson, Ed., National Academy Press, Washington, D.C., 206–211.

Oldfield, M. and Alcorn, J., 1991. Conservation of traditional agroecosystems, in *Biodiversity, Culture, Conservation, and Ecodevelopment,* M. L. Oldfield and J. B. Alcorn, Eds., Westview Press, Boulder, CO, 37–58.

Olmo, H. P., 1977. Introduction of disease and insect resistance, *Calif. Agric.,* 31:24–25.

Plucknett, D. L., Smith, N. J. H., Williams, J. T., and Anishetty, N. M., 1987. *Gene Banks and the World's Food*, Princeton University Press, Princeton, NJ.

Prain, G. and Piniero, M., 1994. Community curatorship of plant genetic resources in southern Philippines: preliminary findings, in *Local Knowledge, Global Science and Plant Genetic Resources: Towards a Partnership, Proceedings of the International Workshop on Genetic Resources,* G. Prain and C. P. Bagalanon, Eds., UPWARD, Laguna, Philippines.

Prain, G., Uribe, F., and Scheidegger, U., 1992. "The friendly potato": farmer selection of potato varieties for multiple uses, in *Diversity, Farmer Knowledge, and Sustainability*, J. L. Moock and R. E. Rhoades, Eds., Cornell University Press, London, 52–68.

Rastogi, A. and Pant, R., 1996. Case study of eastern Himalaya, in *Changing Perspectives of Biodiversity Status in The Himalaya*, British Council Division, British High Commission, New Delhi, 93–110.

Reardon, T. and Vosti, S. A., 1995 Links between rural poverty and the environment in developing countries: asset categories and investment poverty, *World Dev.,* 23:1495–1506.

Rhoades, R. E., 1985. Traditional potato production and farmer selection of varieties in eastern Nepal, in *Potatoes in Food Systems Research Series*, Report No. 2, International Potato Center, Lima, Peru.

Rhoades, R. E., 1989a. The role of farmers in the creation of agricultural technology, in *Farmer First: Farmer Innovation and Agricultural Research*, R. Chambers, A. Pacey, and L. A. Thrupp, Eds., Intermediate Technology Publications, London.

Rhoades, R. E., 1989b. *Evolution of Agricultural Research and Development since 1950: Toward an Integrated Framework*, Gatekeeper Series No. SA12, Sustainable Agriculture Programme, International Institute for Enviornment and Development, Calvert's Press, London.

Rhoades, R. E., 1991. The world's food supply at risk, *Natl. Geogr.,* 179:74–107.

Rhoades, R. E. and Bebbington, A., 1995. Farmers who experiment: an untapped resource for agricultural research and development, in *The Cultural Dimension of Development: Indigenous Knowledge Systems*, D. M. Warren, L. J. Slikkerveer, and D. Brokensha, Eds., Intermediate Technology Publications Ltd., London, 296–307.

Rhoades, R. E. and Harwood, D., 1992. A framework for sustainable agricultural development: synthesis of workshop discussions, in *Proceedings of the Regional Workshop on Sustainable Agricultural Development in Asia and the Pacific Region*, D. F. Bryant, Ed., Asian Development Bank and Winrock International, Manila, Philippines, 107–120.

Rhoades, R. E. and Nazarea, V., 1996. Local-global disarticulations in plant genetic resources conservation, *Diversity,* 12:5–6.

Rhoades, R. E. and Thompson, S., 1975. Adaptive strategies in Alpine environments, *Am. Ethnol.,* 2:535–551.

Richards, P., 1985. *Indigenous Agricultural Revolution: Ecology and Food Production in West Africa*, Westview Press, Boulder, CO.

Rogers, E. M., 1969. *Modernization among Peasants: The Impact of Communication*, Holt, Rinehart, and Winston, New York.

Schery, R. W., 1972. *Plants for Man*, Prentice-Hall, Englewood Cliffs, NJ.

Schultz, T., 1964. *Transforming Traditional Agriculture*, Yale University Press, New Haven, CT.

Sekaqueptewa, E. and Black, M., 1986. Corn and linguistics: two accounts, in *Hopi: Songs of the Fourth World*, Ferrero Films, San Francisco, CA, 11–17.

Shand, H., 1991. There is a conflict between intellectual property rights and the rights of farmers in developing countries, *J. Agric. Environ.,* 1:131–142.

Shulman, S., 1986. Seeds of controversy, *BioScience,* 36:647–651.

Shyamsundar, P. and Lanier, G. K., 1994. Biodiversity prospecting: an effective conservation tool?, *Trop. Biodiversity,* 2:441–446.

Soleri, D. and Smith, S., 1998. Conserving folk crop varieties: different agricultures, different goals, in *Ethnoecology: Situated Knowledge/Located Lives*, V. Nazarea, Ed., University of Arizona Press, Tucson. (In Press.)

Sperling, L. and Scheidegger, U., 1995. *Participatory Selection of Beans in Rwanda: Results, Methods and Institutional Issues*, IIED Sustainable Agriculture Programme Gatekeeper Series No. 51, Calvert's Press, London.

Sperling, L., Loevinsohn, M. E., and Ntabomvra, B., 1993. Rethinking the farmer's role in plant breeding: local bean experts and on-station selection in Rwanda, *Exp. Agric.,* 29:509–519.

Struever, S., Ed., 1971. *Prehistoric Agriculture*, The Natural History Press, New York.

The spirit of the sanctuary, *Down to Earth,* 2:21–36, 1994.

Thrupp, L. A., 1997. Linking biodiversity and agriculture: challenges and opportunities for sustainable food security, *WRI: Issues Ideas,* March:9–19

Ucko, P. J. and Dimbleby, G. W., Eds., 1969. *The Domestication and Exploitation of Plants and Animals*, Aldine, Chicago.

Vavilov, N. I., 1926. Studies on the origins of cultivated plants, *Bull. Appl. Bot. Plant Breeding*, 16:1–245.

Vavilov, N. I., 1949. *The Origin, Variation, Immunity and Breeding of Cultivated Plants*, Chronica Botanica Company, Waltham, MA.

Vogel, J., 1994. *Genes for Sale: Privatization as a Conservation Policy*, Oxford University Press, New York.

Williams, J. T., 1986. Germplasm resources, in *Global Aspects of Food Production*, M. S. Swaminathan and S. K. Sinhu, Eds., Tycooly International, Oxford-Riverton, NJ, 117–128.

Witcombe, J. R., Joshi, A., Joshi, K. D., and Sthapit, B. R., 1996. Farmer participatory crop improvement. I. Varietal selection and breeding methods and their impact on biodiversity, *Exp. Agric.*, 32:445–460.

Woodham-Smith, C., 1962. *The Great Hunger*, New English Library Ltd., London.

World Commission on Environment and Development, 1987. *Our Common Future*, Oxford University Press, Oxford.

Zimmerer, K. S., 1991. Managing diversity in potato and maize fields of the Peruvian Andes, *J. Ethnobiol.*, 11:23–49.

CHAPTER 13

Valuing Genetic Diversity: Crop Plants and Agroecosystems

Douglas Gollin and Melinda Smale

CONTENTS

1-56670-290-9/99/$0.00+$.50

INTRODUCTION

Researchers have long been aware of the importance of genetic diversity in agroecosystems. At one level, diversity can provide ecosystem services to agriculture: for example, beneficial insects and soil organisms can contribute to crop health and reduce the need for agricultural chemicals.[1] At another level, plant breeders and other scientists rely on genetic resources for crop improvement and other technological advances. Traditional plant breeding and newer biotechnologies make it possible to incorporate desirable traits into crop varieties; this implies that crop landraces, wild relatives, and other species have value as sources of desirable genetic traits.

To many agricultural scientists and other biologists, it seems self-evident that genetic diversity has economic value and that genetically diverse ecosystems are valuable resources. Many biologists take the view that, since all life depends on genetic resources, their value must be infinite. The protection of all biodiversity (and certainly of diversity in crop plants) should be a fundamental priority, in this view. For example, Frankel et al. (1995) write that, "[T]he species serving humanity and the communities safeguarding life and its diversity are of immense value. The highest priority the human species can confer must go to their conservation. ..." Other authors voice similar views. Ehrenfeld (1988) argues specifically that economic criteria are inadequate and inapplicable to thinking of the broader "value" of biological diversity:

> Value is an intrinsic part of diversity; it does not depend on the properties of the species in question, the uses to which particular species may or may not be put, or their alleged role in the balance of global ecosystems. For biological diversity, value **is**. Nothing more and nothing less. No cottage industry of expert evaluators is needed to assess this kind of value. (p. 214; author's emphasis)

Ehrenfeld goes on to argue that economists are fundamentally unable to account for many of the most important aspects of value. One reason is the lack of adequate biological knowledge about the functions of genes, species, and ecological commu-

[1] See, for example, Tilman's studies of the contributions that biodiversity makes to the level and stability of biomass production per unit area (e.g., Tilman, 1997).

nities. A second reason that Ehrenfeld cites is the difficulty of putting values on such intangible benefits as the satisfaction that people derive from the continued existence of pristine environments. A third reason he gives is that the utilitarian principle underlying economic valuations (the insistence on measuring net benefits to humans) is inherently inadequate as a way to value the natural world. "The very existence of diversity is its own warrant for survival," Ehrenfeld argues. "As in law, long-established existence confers a powerful right to a continued existence."

Arguments such as Ehrenfeld's have power, dignity, and deep ethical appeal. Without taking issue with such viewpoints, this chapter will add to the "cottage industry" of discussing the economic value of genetic resources. We focus specifically on the case of agroecosystems. Perhaps the justification for an economic perspective is more readily apparent in the case of agroecosystems. Since agriculture inherently involves the management of ecosystems for the benefit of humans, there can be little reason to dispute the utilitarian premise of economic analysis. By the same token, we have relatively little hesitation valuing genetic resources for agroecosystems in terms of their current and potential use values.

Ehrenfeld's first point, however, has substantial relevance for our work. Economists and agricultural scientists do not yet have much information about the extent or potential value of genetic diversity for agroecosystems. There are few data yet available about the degree of genetic variation within the *Oryza* genus for rice, for example, or within the *Triticum* genus for wheat. Researchers are currently working to map the rice genome, but it will be some time before the genetic bases for resistance to major diseases and pests will be well understood. For more-complex traits, such as yield potential, this understanding will require even longer.

In the meantime, important decisions must be made about the allocation of resources for the protection and conservation of genetic diversity for agroecosystems. How much is it worth to collect the remaining varieties of wheat from farms in eastern Turkey? Should we build additional gene banks for *ex situ* storage of barley varieties? Does the rationale for conserving wheat and rice varieties extend to spelt? Does it extend to eggplant? To parsley? Is there a reason to encourage the cultivation of "genetically diverse" plots of crop varieties or to encourage the conservation of traditional varieties in farmers' fields? Economists tend to approach such questions with a particular concern for costs and trade-offs. Unlike some environmentalists, economists generally believe that the costs of conserving genetic resources should be viewed seriously, and that the benefits should be quantified to the greatest possible extent (see, for example, Brown, 1990).[2] Generally speaking, economists have been more skeptical than biologists concerning the need to protect all forms of genetic diversity. As Brown (1990) notes, "[I]f we can't save all species, we need a ranking based on one or more criteria, from which we select the highest ranked for conservation." This view — which would undoubtedly be considered heretical by most biologists — is readily accepted by many economists.

[2] See also Evenson, R. E., Genetic resources: assessing economic value, manuscript, Yale University, Department of Economics, New Haven, CT, 1993; also Wright, B. D., Agricultural genetic resource policy: towards a research agenda, paper prepared for presentation at the Technical Consultation on Economic and Policy Research for Genetic Resource Conservation and Use, International Food Policy Research Institute, Washington, D.C., June 21–22, 1995.

To an economist, it is of central importance to consider the value of genetic resources and genetically diverse agroecosystems (and even to assign dollar values, where possible). Without such information, we are forced to make decisions on an arbitrary basis. Even crude economic methods can offer valuable insights and can help in setting priorities.

This chapter reports on recent conceptual and empirical work concerning the valuation of genetic resources for agriculture, with particular attention to the role of diversity within agroecosystems. The next section reviews current concerns over diversity and summarizes economic theory and concepts pertaining to the valuation of genetic resources. The third section reviews some of the empirical studies that have sought to value genetic diversity for agriculture. The fourth section assesses the implications of these empirical studies and other research for programs of conservation of agricultural biodiversity. The last section offers some assessments and concluding remarks.

VALUING GENETIC DIVERSITY FOR AGROECOSYSTEMS: CONCERNS, CONCEPTS, AND CAUTIONS

Economists have long recognized the importance of biologically diverse farming systems and have helped to draw attention to the multiple objectives of farm households. A substantial literature in household economics and farming systems analysis has analyzed farm-level decision making. Much of this literature has studied the portfolio choice of farm households: in other words, the ways in which different farm and on-farm activities are chosen to achieve multiple objectives, including high economic returns and reductions in risk. Implicit in this literature is an interest in farmers' efforts to diversify production of agricultural crops through mixes of species and varieties. In recent years, a number of studies have sought to assign a value to genetic diversity in agriculture by showing how diversity is a private good; in other words, how individual farmers benefit from incorporating genetic diversity into their farming systems.[3] A related question is the social benefit that is derived from diversity (for example, reductions in aggregate production variability that may be obtained if different farmers choose different varieties, so that an aggregate level of diversity is maintained).

But genetic diversity has values that extend beyond the static function of diversifying production and consumption choices. Increasingly, economists are focusing their attention on the value of genetic diversity as an input into the production of new crop technologies. Genetic resources are used to breed new crop varieties and to improve agricultural technologies (e.g., pest control through *Bacillus thuringiensis*). Landraces and other crop varieties can be important future sources of disease and pest resistance, and hence can be viewed as forms of insurance.[4] For all these reasons, the total economic value of genetic diversity extends beyond the short-run private value that farmers attach to it.

[3] For example, farmers might benefit from growing varieties with different taste or consumption characteristics, or from choosing varieties that are adapted to different microclimates or production locations.
[4] Perrings (1995) articulates the view that biodiversity conservation should be viewed as a form of insurance payment.

The following sections describe some of the concerns and directions of recent economic research. We begin by considering the issues that have focused public attention on genetic diversity in agriculture. We then ask how the concepts and methods of resource economics can help to assess the value of genetic resources in agroecosystems. Finally, we discuss the limitations of these methods.

Popular Concerns: Perception and Misperception

Much of the concern over genetic diversity for agroecosystems involves three widespread (and related) perceptions. The first is the notion that modern agriculture has caused *genetic erosion,* a term which encompasses the loss of traditional varieties. The second is the idea that modern crop varieties and agroecosystems are increasingly uniform, rendering crops extremely vulnerable to pests, diseases, and other pathogens. The third is the view that genetic resources are scarce in agriculture, i.e., that there is a problem with the adequacy of genetic resources for agriculture. This last view is most commonly expressed in terms suggesting that modern agriculture is based on a precariously narrow stock of genetic diversity.

In all three cases, popular concerns are based largely on anecdotes. With regard to the first concern, no causal relationship between the Green Revolution and genetic erosion can be established for bread wheat, given the difficulties in measuring genetic erosion on such a large scale and of demonstrating causality. The patterns of genetic variation in farmers' wheat fields have undoubtedly changed over the past 200 years with increasing cultivation of varieties released by plant-breeding programs, but the implications of these changes for the scarcity of useful genetic resources are unclear. As expressed succinctly by Wood and Lenné (1997), the assumption that the spread of modern varieties has been mainly responsible for an overall loss of traditional varieties "goes beyond our knowledge of the facts of genetic erosion." Historical sources also demonstrate that most of the areas in which the Green Revolution has had its greatest impact are high potential areas (not ancient centers of crop diversity) which were targeted by local plant breeders and widely planted to the products of their efforts since at least the early years of this century (Gill, 1978; Pray, 1983; van der Eng, 1994).

With regard to the second concern, the evidence assembled in Smale (1997) and Smale and McBride (1996) suggests that, in the major wheat-growing areas of the developed and developing world, concentration among leading cultivars has tended to decline as agricultural research and seed systems have matured. Semidwarf varieties of bread wheat are in general more resistant to major pests and diseases, such as the wheat rusts, than either traditional varieties or the taller varieties previously released by breeding programs. They now incorporate broader and more-durable types of resistance (Rajaram et al., 1996).

Finally, the stock of genetic diversity for agriculture remains relatively untapped. There are many varieties (indeed many species) that have not yet been exploited. Given the size of the genomes of agricultural crops and given the increasing feasibility of incorporating genes from wild relatives or unrelated species through biotechnology and other breeding techniques, it does not seem useful to think of genetic combinations as determinate in number. In a recent study, Rasmussen and Phillips

(1997) show that genetic gains were made in barley based on a very narrow genealogical base. They question the generally held belief that the variation on which selection is based in elite gene pools is provided only by original ancestors, and hypothesize that the contributions of newly generated variation and gene interactions are underestimated. On all three counts, then, current concerns appear misplaced.

Conceptualizing the Sources of Economic Value

To point out the possible inaccuracies of popular concerns about the "loss" of genetic material for agriculture is not to argue against the value of crop genetic resources. From an economic perspective, genetic diversity has multiple sources of value. A number of surveys have discussed these sources of value. (See, for example, Brown, 1990; Pearce and Cervigni, 1994; Pearce and Moran, 1994; or Perrings et al., 1995.[5])

Perrings et al. (1995) suggest that it is useful to consider the total economic value of genetic diversity as consisting of "use values" and "non-use values," with perhaps a broader valuation including noneconomic or "nonanthropocentric instrumental value." Use values can be further disaggregated. One form of use value is the direct benefits of genetic diversity to consumers and producers, such as the increased satisfaction that comes from having a dozen apple varieties from which to choose, or the increased productivity that arises through genetic crop improvements. A second category of use value is the indirect benefits that people receive from ecosystem functioning. Examples include the contribution of earthworms to soil tilth, or of wetlands to flood control (e.g., Brown, 1990). A third type of use value would include the future "option value" and "quasi-option value" of retaining for the future the possibility of using a resource or of acquiring information about that resource.

Use values are distinguished from non-use values (sometimes called "existence values"), which reflect the pleasure that people derive from the sheer existence of genetic diversity (without any regard for the usefulness of diversity). Thus, people may derive some value simply from knowing that elephants exist or that rain forests are being conserved. In considering genetic resources for agriculture, however, such existence values are seldom of much importance. Relatively few people derive satisfaction from the sheer existence of 80,000 rice varieties, but the use values are substantial. By the same token, for other biological resources, existence values may greatly outweigh use values. Few people derive much direct productive value from the Bengal tiger, but many people value its continued existence.[6]

In most cases, then, it is use values that are relevant for agricultural genetic diversity. More specifically, we will be interested in genetic diversity as it contributes to expanded consumer choice and satisfaction and as it directly or indirectly enhances

[5] See Swanson, T., The values of global biodiversity: the case of PGRFA, paper prepared for presentation at the Technical Consultation on Economic and Policy Research for Genetic Resource Conservation and Use, International Food Policy Research Institute, Washington, D.C., June 21–22, 1995.

[6] It is worth noting, however, that economists have made great use of models in which consumers are thought to have a "preference for variety," such that they prefer to consume many different varieties of a single product. Thus, for example, many people do appreciate (and are willing to pay for) diversity of coffee varieties, apples, wine, etc.

producer profitability and security. These are private benefits that are obtained by consumers and producers, and economic theory suggests that markets should do an adequate job of meeting the demands for diversity that originate from these sources. For example, consumers will be prepared to pay extra for exotic varieties of fruit or (indirectly) for wheat varieties with special milling and baking characteristics. Farmers will decide efficiently which varieties to cultivate and how to allocate land and resources so as to achieve multiple production objectives.

But markets do not always work efficiently to provide genetic diversity. Genetic diversity may generate benefits that cannot be captured by individual actors. In such cases, markets may not provide diversity at adequate levels. Such cases are discussed below.

The Concept of Public Goods

For most goods, market prices serve as measures of value. Prices reflect the amount that people are willing to pay to purchase goods and services and the amount that others are willing to accept in compensation for producing those commodities.

For some categories of goods, however, market prices are poor measures of value. When consumers can benefit from a good without having to pay for it, the market price will tend to understate the value of the good. Consider, for example, an open-air concert in a public park. Many people can benefit from such a concert without paying, and those who do pay (out of some sense of civic virtue perhaps) will tend to contribute less than they might pay to hear the same concert in a concert hall. Economists characterize goods such as this concert as "public goods." Economic theory predicts that freely operating markets will place too low a price on public goods and will provide them at inefficiently low levels.[7]

Genetic diversity is a classic example of a public good. Although individuals may benefit themselves from maintaining diverse farming systems, with multiple species and varieties, their actions may also benefit others. For example, one farmer may benefit from having her neighbor cultivating a diversified array of varieties, reducing the attractiveness of their adjacent lands to certain kinds of pests or pathogens.[8] But there is generally no mechanism through which diversity can be "purchased" to ensure that it is provided in sufficient quantities.

Similarly, genetic resources themselves can be seen as a public good. Farmers cultivate many traditional landraces of rice, wheat, and other crops. These landraces represent resources that can be used by the whole world in the creation of new varieties with desirable properties. But at present, there are no incentives beyond

[7] The concept of public goods is discussed in any introductory economics textbook. The problem of people sharing the benefits of public goods without paying for them is called the "free rider program."

[8] For example, suppose that there is a particular wheat variety that performs better than all other varieties in a particular region (say, Minnesota). If every farmer in Minnesota grows this high-performance wheat variety, it may increase the chance that a variety-specific pathogen will emerge. Each farmer then benefits if some fraction of the farm population chooses to grow other varieties. But no farmer has an individual incentive to grow a lower-performance variety. Moreover, there is no vehicle through which some farmers can compensate others for growing the lower-performance varieties, even though everyone would expect to benefit from such an arrangement. This is a common problem with public goods: the absence of a market for a public good (genetic diversity, in this case) implies that the good will be provided at inefficiently low levels.

their own private profit for farmers to cultivate these landraces. Although the global community benefits from the conservation of genetic resources, there is at present no market or other mechanism that would enable farmers to receive a share of the global benefits. Without such an incentive, farmers will grow landraces only when these varieties provide sufficiently high private benefits. The result may be inefficiently low rates of conservation of landraces.

A Caution: The Myth of Enormous Value

Markets may undervalue genetic resources and genetic diversity, but an equally serious problem is that many people overvalue genetic resources. It is commonplace for biologists and agricultural scientists to argue that genetic diversity has enormous value. But such overvaluation of genetic diversity poses as much of a problem as undervaluation by the market. If genetic diversity is indeed of enormous value, then it becomes extremely difficult to establish priorities or to make trade-offs regarding diversity. Without some estimates of relative value, we have no way to set priorities for conservation or conversation. Some may argue that setting priorities is, in itself, an unethical act of "picking winners" biologically. But as Swaney and Olson (1992) write, "We are valuing biodiversity. We can choose to continue to undervalue [biodiversity], or we can change our valuations, but we cannot choose to not value it."

Several arguments are advanced for the infinite value of genetic diversity. It is worth looking briefly at these in turn. It is not possible in the space of a chapter to characterize all of the arguments adequately, nor is it possible to refute them convincingly. Instead, what follows is an attempt to explain the reasons economists are skeptical about the alleged enormous value of genetic resources.

All Human Life Depends on Genetic Diversity

Some people argue that genetic diversity for agroecosystems is priceless, essentially that it has infinite value. Without the genetic diversity of our crops, it is claimed, humans would be unable to survive as a species; our agroecosystems depend on the continued availability of a range of genetic resources. Although this is true, the high total value of genetic resources should not be confused with the marginal or incremental value of adding or subtracting one more species or gene. We are in no danger of losing all the genetic diversity available to people, so the issue is best understood as: How costly is it to lose some of the species or varieties now known to us (or, equivalently, to protect some of the species or varieties now in danger)? The cost is surely not infinite, but it is also not negligible.

Some Species and Varieties Have High Value to Humans

Lovejoy (1997) cites a number of instances in which genetic resources have proved extremely valuable, from the directly useful genes found in penicillin and perennial corn to the water filtration services provided by oysters in the Chesapeake. There is no doubt that these examples are valid, but it is misleading to argue on the basis of these examples that all genetic diversity is equally valuable (or even that the average value is high). Many species have little potential for direct contribution to human welfare, and many are sufficiently similar to other species that humans might, realistically, suffer little from their loss. The valuable attributes of a particular species might, upon search, be found to occur elsewhere or in other forms. Thus, it is difficult to extrapolate at all from the "success stories" of biological resources that currently have high values.

Extinction Is Irreversible and, Hence, Infinitely Costly

Biologists usually argue that the extinction of a species imposes losses on humans. Two distinct effects are noted. First, an extinct species is "lost" for future use, in the sense that its genetic materials cannot be put to utilitarian purposes. If a species is extinct, we can never know whether or not it might have offered a cure for cancer or — more prosaically — a gene that could be used in crop improvement. Second, the loss of any species can perturb the delicate ecological balance of a natural system. This in turn can cause damaging effects for humans.

In both cases, however, the costs of extinction can be overestimated if we do not recognize the opportunities for people to find substitutes. As Simpson et al. (1996) have pointed out, people can often find alternative sources of desirable genes. There are arguably very few cases where a particular useful trait can be found only in a single species or variety. More commonly, the compound occurs in several species or varieties; or perhaps similar compounds are found in (related or unrelated) species occupying similar ecological niches; or people can develop synthetic compounds with the same attributes as the natural material; and so forth. The scope for humans to substitute and adapt to extinction is remarkable. From the woolly mammoth to the passenger pigeon, humans have survived the loss of economically important species without irreparable material losses.[9]

Agricultural Genetic Diversity Protects Against Disastrous Disease and Pest Outbreaks

Within the agricultural sciences, a commonly cited justification for conserving genetic diversity is the need to guard against potential outbreaks of diseases or pests. Diversity gives farmers and scientists the resources with which to respond to emerging disease and pest problems. A related issue is the danger of genetic uniformity in crops. Where cultivated varieties of a crop are closely related, it is suggested, new pests and diseases can spread rapidly and with enormous destructive potential. As evidence, several historical episodes are cited: the Irish potato famine, the Southern corn leaf blight epidemic in the U.S., and a handful of other well-documented cases (e.g., National Research Council, 1972; Ryan, 1992).

[9] The passenger pigeon is often cited as an example of an economically important species that was forced into extinction by the failure of markets to reflect its increasing scarcity. Although it is certainly true that market forces did not operate to encourage the conservation of passenger pigeons, it is equally true that little long-term material harm to humans has resulted. This is not to deny that our lives are, in some sense, poorer because passenger pigeons no longer fill the skies of North America.

Although genetic diversity undoubtedly is valuable for "maintenance breeding," it is probably not the case that the associated value of genetic resources is particularly large. Typically, multiple sources of resistance can be tapped, and relatively small arrays of genetic diversity would suffice to provide adequate resistance to many diseases. Gollin et al.[10] investigate this topic quantitatively, looking at the actual distributions of resistance to diseases and pests in bread wheat. They conclude that bread wheat landraces will be used for maintenance breeding on relatively rare occasions, and they calculate as a consequence that these varieties have significant but modest value.

Wright points out further that plant disease epidemics are rare and that their costs are often overstated.[11] Although the Irish potato famine was associated with a blight, recent histories downplay the importance of blight as a cause of the famine.[12] More recently, the Southern corn leaf blight epidemic barely caused a ripple. Although this was due in part to the ready availability of resistant cytoplasm in plant-breeding programs, it also reflects the inherent resilience of modern agriculture. Trade, storage, crop diversification, and consumer substitution all serve to mitigate the impacts of crop failures. Even in developing countries with imperfect markets, farmers and consumers can rely on a variety of *ex post* consumption-smoothing techniques to make up for the income losses associated with crop failures. (See, for example, Rosenzweig, 1988; Rosenzweig and Stark, 1989; Udry, 1990; Alderman and Paxson, 1992; Rosenzweig and Wolpin, 1993; Townsend, 1995.) As Sen (1981) has argued in his seminal study of famines, crop failure does not correspond to famine. Famine instead depends on a variety of other institutional and market failures — often involving war, violence, or deliberate exploitation.

Genetic Diversity for Agroecosystems: Distinctive Features

The analysis presented in the preceding sections suggests that we should be somewhat skeptical of exaggerated claims for the value of genetic diversity. But in recent years, a sober literature has begun to seek theoretical and empirical measures of the value of diversity. The dollar figures are not very precise, but they provide a focus for discussion. The most common approach has been to value genetic resources based on their "rareness" or their potential contribution to the development of new or improved products. Among the more widely known efforts, Weitzman (1992; 1993) focuses on the measurement of diversity and on an application to the conservation of crane species. Simpson et al. (1996) consider the likelihood that a new species discovered in the rain forest will have value for pharmaceutical research.

[10] See Gollin, D., Smale, M., and Skovmand, B., The empirical economics of *ex situ* conservation: a search theoretic approach for the case of wheat, paper presented at the international conference: Building the Basis for Economic Analysis of Genetic Resources in Crop Plants, sponsored by CIMMYT and Stanford University, Palo Alto, CA, August 1997.

[11] Wright, B. D., Agricultural genetic resource policy: towards a research agenda, paper prepared for presentation at the Technical Consultation on Economic and Policy Research for Genetic Resource Conservation and Use, International Food Policy Research Institute, Washington, D.C., June 21–22, 1995.

[12] It is also interesting that few historians of the Irish potato famine consider the potato blight to have been an important factor in the famine. For example, Mokyr (1983) does not include plant disease among the seven "factors of importance" in explaining the Irish famine.

Similarly, Mendelsohn and Balick (1995) attempt to value undiscovered pharmaceuticals in tropical forests. This literature is hampered, however, by the paucity of empirical data, which makes it difficult to move from models to empirically relevant conclusions.

In contrast, the agricultural sector offers unique opportunities for describing the value of genetic diversity. Agriculture has long used genetic resources to achieve increases in productivity. Scientific plant breeding has been well developed in the U.S. and Europe since the start of the 20th century, and organized systems for collecting and conserving new genetic materials date back even further (Plucknett et al., 1987). Government-sponsored systems of plant selection and varietal improvement date back still further in Japan and China. Agriculture thus offers a logical setting in which to examine the value of genetic resources. Moreover, data on plant breeding and crop improvement are readily available for many crops, and much of the research for the most important crops (wheat, rice, and maize) has taken place in the public sector, so that data are accessible to researchers. By contrast, much of the data on the utilization of genetic materials in, say, pharmaceutical research are proprietary.

Studies of genetic diversity for agriculture must recognize some important differences from the broader biodiversity literature. First, genetic diversity in agriculture is in large measure a matter of intraspecies diversity, rather than interspecies diversity. Second, agroecosystems are largely human managed; thus, the extent of diversity is purposefully chosen, not the result of natural accident. Third, the conservation of diversity (and, more generally, the conservation of genetic resources for agriculture) intimately involves human communities. Agricultural diversity cannot be conserved simply by setting aside tracts of uninhabited land; it necessarily involves people. Rain forest biodiversity can be maintained (in principle) by conserving habitat. But agricultural diversity can only be maintained in farmers' fields as long as incentives are appropriate.[13]

These distinctive features of agricultural genetic diversity have been recognized in a growing literature concerned with valuing diversity. The next section of this chapter summarizes some of the approaches and findings of this literature.

APPROACHES TO VALUING GENETIC DIVERSITY FOR AGROECOSYSTEMS

The new economic literature on genetic diversity can be unraveled into four distinct strands. One strand has sought to understand the choices of households as they relate to on-farm diversity. A second strand of literature has focused on the contributions of diversity to farm-level productivity. A third strand of research has tried to measure the private costs of implementing a "socially optimal" level of

[13] For the conservation of wild crop relatives, of course, agricultural diversity does involve the protection of uninhabited lands, as for wheats in the Near East and soybeans in China. Moreover, as W. Collins has pointed out (personal communication, 1997), the protection of uninhabited lands raises problems for agriculture, since protective measures often include prohibitions on plant collection (posing real problems for agricultural scientists who seek to collect wild crop relatives for research purposes).

diversity. Finally, a fourth strand of literature has sought to identify the aggregate value of genetic resources (or global diversity) to agricultural output. The four strands of literature encompass both micro- and macrobehavior and address diversity at the level of the farm, the region, and the globe. We consider these four strands in sequence. The next section will consider the policy implications of these different avenues of research.

Farmer Decision Making and Farm-Level Diversity

Agricultural genetic diversity is ultimately controlled by the decisions of farmers. Farmers choose what crops and animal species to exploit and also select the mix of varieties or breeds that they will use. Understanding farmer behavior is a prerequisite for influencing agricultural diversity through policy, as well as for predicting the impacts on diversity of external forces.

Species Diversity

The issue of species diversity at the farm level has long been viewed as a simple question of production complementarities. Some agricultural commodities can be produced together at lower cost than if they are produced separately. For example, hogs can often be produced most efficiently on farms that also produce corn, soybeans, or other feed ingredients. Likewise, in many traditional Southeast Asian agroecosystems, rice and fish can be produced together relatively efficiently. Moreover, some kinds of species diversity are economically efficient responses to uneven income flows or uneven demand for labor through the year: livestock species may provide off-season cash income; tree crops may offer a productive use for labor outside of peak cropping seasons; and so on. Species diversification offers a shield against production variation, price variation, and other forms of risk. Livestock and tree crops may serve as a store of wealth, particularly in economies where financial institutions are lacking. Species diversification can thus be viewed as an optimal response to farm-level economic incentives.

Varietal Diversity

More of a puzzle has been varietal diversification: the tendency of farmers to cultivate multiple varieties of a single crop. Varietal diversification is widespread. Often, varietal diversification involves both "modern" and "traditional" varieties. A long literature in agricultural economics has focused on farmers' adoption of new crop varieties (and on their continued use of old varieties).[14] Initially, "partial adoption" was seen as a puzzle: why would farmers persist in growing traditional varieties alongside modern varieties? A simple decision-making model in which a profit-maximizing firm chooses its technology from a set of available technologies will give a unique choice, suggesting that a farmer should choose to grow only the

[14]This literature dates back to Griliches's classic article (1957) on the diffusion of hybrid corn technology in the U.S.

single variety with the highest net returns. But other explanations from economic theory have been invoked to explain partial adoption. The most widely accepted explanations involve (1) differentiation of varieties, such that different varieties are in fact seen as different commodities, with partial adoption explained by some of the same factors that explain species diversity; (2) farmers' risk aversion, resulting in portfolio diversification or disaster avoidance; (3) fixity or rationing of seed-related production inputs such as fertilizer, soil type, or credit; and (4) learning behavior or experimentation.

By examining the question in different terms, we can understand varietal "diversity" as the counterpart of "partial adoption." In the adoption literature, adoption is viewed as a response to differences in yields and other characteristics associated with modern and traditional varieties. In this literature, farmers choose the single variety that gives the best set of characteristics. Instead, however, we can view each variety (whether modern or traditional) as a bundle of characteristics or a multidimensional vector of traits (Bellon, 1996). Farmers and their households derive utility from these characteristics. Households seek to obtain desired levels of the characteristics. To attain the desired levels, they may have to grow multiple varieties, so that the allocation of crop area among varieties effectively weights their respective characteristics.

In fully commercialized agroecosystems, market incentives may lead farmers to choose bundles based on few characteristics (such as uniformity, a particular grain quality, grain weight) — resulting in complete specialization among varieties. Varietal diversity tends to be heightened when farmers produce both for their own household consumption and for sale. In such cases, farmers may have multiple uses for their crop and may seek to retain varieties with many different characteristics. The "discarding" of varieties occurs when changes in the production environment (or the tastes and preferences of the farm household) cause farmers to change the weights that they assign to different characteristics. For example, if changes in the value of women's time encourage households to purchase commercial bread instead of baking their own bread, farm families may be less inclined to grow wheat varieties that are well suited to home baking and instead may grow more varieties with commercially desirable milling characteristics. Varietal "loss" can also occur if many characteristics of value to the farmer are bundled into fewer varieties, so that a new variety satisfies many household objectives (Bellon, 1996).

Meng et al.[15] suggest that partial adoption can be used to give a measure of the value that farmers attach to genetic diversity. If farmers are continuing to cultivate traditional varieties, even when modern varieties are available, it indicates that they are willing to sacrifice the higher yield of modern varieties to gain some other characteristics. The amount of yield forgone is, in this sense, a measure of farmers' willingness to pay for retaining their traditional varieties. Meng et al. use data on wheat cultivation from Turkey to consider a number of possible attributes or characteristics that farmers may be "purchasing" when they choose to grow traditional

[15] See Meng, E., Taylor, J. E., and Brush, S., Incentives for on-farm crop genetic diversity: evidence from Turkey, paper presented at the symposium The Economics of Valuation and Conservation of Genetic Resources for Agriculture, May 13–15, University of Rome Tor Vergata, 1996.

varieties: taste and baking properties, yield stability, and location-specific agronomic features. The study concludes that farmers have a nontrivial willingness to pay for the diversity that traditional varieties afford.

Spatial and Temporal Diversity

Farm-level diversity need not necessarily consist of multiple varieties or species being utilized simultaneously on an individual farm. Diversity can equally be achieved across space or over time. For example, if farmers grow one crop at a time but alter varieties from year to year, and if different farmers are at different stages in this cycle at a particular moment in time, then aggregate species or varietal diversity may be high. Likewise, if farmers in different locations specialize in different species or varieties, the aggregate diversity may be high even though no individual farmer has a diversified farming system.

Spatial distributions are often affected by seed industries and delivery systems. Spatial diversity can be represented by numbers of cultivars or percent distributions by area planted to cultivars at a particular point in time. Changes in these counts or distributions express the temporal diversity of varieties, or "diversity in time" (Duvick, 1984). In mature commercial seed systems, temporal diversity, or high rates of varietal turnover among farmers, substitute for spatial diversity (Plucknett and Smith, 1986). Various measures of temporal diversity have been reviewed and applied by Brennan and Byerlee (1991) and determinants of varietal replacement have been treated in an analytical model by Heisey and Brennan (1991).

Diversity, Productivity, and Stability: Hedonic Valuation and Related Approaches

The previous approaches are well suited to analyzing the impact of farm-level incentives on diversity. But we can also ask how diversity contributes to productivity (reduces productivity) at the farm level. Evenson (1996) notes that hedonic valuation techniques may be useful in valuing genetic diversity and genetic resources as producer goods. Hedonic valuation uses statistical techniques to assign value to the characteristics of goods; it is the same approach, in effect, used by appraisers to place a value on a house. The underlying principle is to observe how the value of the final good changes depending on its characteristics. For example, appraisers might observe how the sale price of a house depends on the roofing material; this implicitly assigns a value to different types of roofing.

Similarly, it is possible to look at the productivity of rice in different localities and to associate productivity levels with the characteristics of the breeding stock used by plant breeders in that locality. Gollin and Evenson[16] used this approach to analyze the productivity of alternative categories of rice germplasm in India over the period 1956 to 1983. The study sought to measure the relative contributions of different types of genetic resources to varietal improvement and indirectly to productivity change, using a two-stage estimation process that included clusters of

[16] See Gollin, D. and Evenson, R. E., Genetic resources and rice varietal improvement in India, unpublished manuscript, Yale University, Department of Economics, New Haven, CT, 1991.

genetic resource variables. Over the period 1972 to 1984, they estimated that varietal change in rice contributed more than one third of realized productivity gains, while public research and extension explained much of the remaining growth. Gollin and Evenson found that the contribution of certain types of germplasm to rice productivity gains in India was very high. In particular, the early semidwarfing genes and genetic resources associated with disease and pest resistance showed up as having high value.

In the study by Hartell et al. (1997), the number of generations of plant breeding and number of landraces in the genealogy of varieties were positively associated with mean yield in rain-fed areas. These are indicators of genealogical complexity, which can be viewed as a form of diversity. In irrigated areas, the concentration of wheat area among fewer varieties was positively related to yield, while increasing varietal age depressed yields.

Widawsky (1996) estimated the effects of varietal diversity on rice yields and yield variability among townships in eastern China. He measured varietal diversity using genealogical data and data on planted areas and concluded that diversity reduced rice yield variability and only slightly reduced mean yields for the time period under study.

In a sense, any study investigating the impact of plant breeding on yield is analyzing the effects of genetic resources on productivity, broadly defined. Recent studies of agricultural research impact, for example, have differentiated among varieties based on their ancestry or the source of the germplasm. Bagnara et al.[17] estimated the effects of local germplasm and international germplasm on the adaptability, yield, grain quality, and yield stability of Italian durum wheats. Other examples include Byerlee and Traxler (1995), Pardey et al. (1996), and Brennan and Fox (1995).

Genetic Diversity as a Public Good

As noted above, genetic diversity can be considered a public good, in the sense that aggregate diversity may inhibit the evolution of new disease and pest biotypes and may lead to greater aggregate stability in production and prices. Individuals have no incentive, however, to consider the "socially optimal" pattern of diversity when they make their varietal selections. Instead, they choose the variety or portfolio of varieties that is individually optimal. At a regional or national level, the aggregate of these individual decisions results in a level of diversity that may differ from the level that is socially optimal.

Heisey et al. (1997) considered the case of wheat cultivation in the Pakistani Punjab, where wheat rusts are an important source of yield losses. The rusts are a family of pathogens noted for evolving rapidly in response to selection pressures. In particular, planting of large contiguous areas with cultivars carrying the same genetic base of resistance speeds the evolution of new rust biotypes. In turn, the

[17] See Bagnara, D., Bagnara, G. L., and Santaniello, V., Role and value of international germplasm collections in Italian durum wheat breeding programmes, paper prepared for the CEIS-Tor Vergata Symposium on the Economics of Valuation and Conservation of Genetic Resources for Agriculture, Rome, Italy, Tor Vergata University, 13–15 May, 1996.

emergence of new rust biotypes can cause substantial losses to farmers and high social costs if epidemics occur. From a social standpoint, then, it is desirable to maintain some degree of diversity in the rust resistance genes incorporated in farmers' portfolio of varieties.

In practice, however, farmers do not choose to grow wheat cultivars with the level of rust resistance that would be socially desirable. First, farmers choose to grow high-yielding cultivars whether or not they are known to be susceptible to rust. Second, farmers choose to grow high-yielding cultivars whether or not they have the same basis of genetic resistance as those grown by other farmers. When many farmers choose to grow the same higher-yielding cultivars, or when they grow different higher-yielding cultivars with similar resistance genes, there is a lower level of genetic diversity in farmers' fields than the level that would most effectively protect against the emergence and spread of new strains of rust.

Heisey et al. (1997) compared the portfolio of wheat varieties actually cultivated by Pakistani farmers with an alternative portfolio and area distribution of cultivars that maximized diversity, as measured by genealogical indicators. Switching from the cultivars and areas actually planted to a more genetically diverse portfolio would have generated yield losses worth tens of millions of dollars annually, even without considering the costs of the policy interventions required to achieve it.

This research thus focuses attention on the supposed aggregate benefits from diversity. Would the recommended portfolio actually perform enough better over time to warrant the foregone yields? Does it matter how the recommended portfolio is achieved? That is, does it matter whether every farmer grows each variety in the recommended portfolio, or can farmers specialize entirely in the production of a single variety? Are there other, more-effective ways of dealing with production variability (e.g., through trade or stabilization policies) than with accepting reductions in yield? Are farmers individually better off growing lower-yielding (but more stable) portfolios of varieties, or are they better off using *ex post* methods of income smoothing to deal with production shocks? In other words, is on-farm diversity optimal compared with alternative forms of individual or social insurance?

To address these questions, we clearly need some measurement of the benefits of diversity as a public good. With good measures of these benefits, we could attempt to assess the overall value of diversity. For now, this approach appears to offer an innovative and intriguing framework for measuring the value of genetic diversity as a public good. Continued research along these lines may prove fruitful.

Assessing the Value of *Ex Situ* Collections

Instead of considering the value of genetic diversity, per se, for agroecosystems, we can think of valuing genetic resources themselves. What is the value of a collection of genetic resources, which after all represent diversity in a latent form? To date, the only empirical estimates of the value of a germplasm collection are in a study by Evenson and Gollin (1997) that attempts to value the International Rice Germplasm Collection (IRGC) at the International Rice Research Institute (IRRI) in the Philippines. The general approach of the study is to associate the size of the IRGC with international flows of germplasm and hence with increases in productivity.

Evenson and Gollin find that the size of IRGC influences the extent to which national rice-breeding programs are willing to collaborate with IRRI experiments; when countries think that IRRI is a source of valuable genetic material, they are more likely to participate. Participation in the IRRI international experiments and germplasm–sharing arrangements is in turn associated with increases in the rate at which countries develop and release new varieties of rice. Countries that collaborate heavily with IRRI, and that exchange genetic resources with IRRI and with other national research programs, thus accelerate the process of technological change in rice.

Evenson and Gollin compute a value for additional accessions to IRGC based on this relationship: they estimate that adding 1000 more cataloged accessions to IRGC would generate 5.8 additional released "modern" varieties of rice (globally), which would generate a stream of increased productivity worth $145 million annually (for a net present value of about $325 million).[18]

Several difficulties arise with these calculations. First, from a conceptual standpoint, Evenson and Gollin are estimating only an "instrumental" value to IRGC accessions; the value of a larger collection is simply that it stimulates greater participation in the IRRI international collaborative programs. The study does not attempt to value accessions based on the actual use of IRGC materials in the development of new varieties. Moreover, the study assigns a value to accessions based on the "average value" of a released modern variety, rather than on the marginal value. This undoubtedly tends to overstate the value of IRGC.

Estimating Option Values

Other ways of valuing genetic diversity also warrant attention. As discussed above, genetic diversity has an "option value," which can be thought of as the value that society might place on having the possibility of using genetic diversity at some point in the future.[19] The concept of option value is widely used in financial economics. In financial markets, an option gives the purchaser the right to exercise a particular choice at a future date. For example, an option might assign the purchaser the right to buy a given quantity of wheat at a specified price at a time 3 months from the present. Environmental economists have borrowed this idea as a useful general framework for thinking about environmental goods: people may be willing to pay a certain amount today to guarantee their right to make choices in the future.

There is no question that option values exist and are important: options are widely bought and sold on financial markets. In principle, however, they do not confer values distinct from productive values. Option values exist only insofar as the goods or assets in question will have tangible value in the future. For genetic resources, that value could be presumed to be future value in producing new crop varieties or commercial products.

One interesting study that uses this concept was carried out by Brush et al. (1992). This study provides evidence that Peruvian peasants maintain certain thresholds of on-farm diversity even when the immediate advantages of switching to

[18] This assumes a 10% discount rate; at a 5% rate, the figure is $1.45 billion.

[19] Alternatively, we might consider the value that society would be willing to pay for information about genetic materials at some point in the future.

improved varieties are large. These authors suggest that the cost of maintaining these "emergency" stocks of traditional varieties represents a form of option value. This is the amount that peasants are prepared to forego in order to maintain the option of switching to other varieties at a later date.[20]

A related theme is embodied in the work of Gollin et al. (1997; see note 10), who examine the actual distribution of economically important traits in the entire population of wheat varieties. These distributions shape expectations about the value of adding one more variety of wheat to the international *ex situ* collection. The value of a variety, in this approach, is essentially an option value: it reflects the amount that we expect the variety to be worth in terms of future yield losses averted, independent of its current usefulness.

IMPLICATIONS FOR CONSERVATION OF AGRICULTURAL DIVERSITY

Ex situ and *in situ* strategies for genetic resource conservation are increasingly viewed as complements rather than substitutes.[21] *Ex situ* conservation is geared toward a relatively small number of known plants. *Ex situ* conservation also "fixes" the genetic material of the plant at the time that it enters the germplasm bank, although genetic drift may occur over time. By contrast, *in situ* conservation can be aimed at larger collections of species, some of which may not even be known. Thus, protecting a rain forest is a form of *in situ* conservation. A virtue of *in situ* conservation is that it allows adaptive, evolutionary processes to continue and natural prebreeding processes to occur.[22] Since the risk of extinction due to some natural or anthropogenic process is greater *in situ*, *ex situ* collections serve an insurance purpose.

Prospects for *in situ* conservation are likely to vary by species (Dempsey, 1996) and are hotly debated for cultivated crops, on both biological and economic grounds. For domesticated species or subspecies, *in situ* conservation implies farmer management of a diverse set of crop populations in the systems where the crops evolved (Bellon et al., 1997). Conservation in this context refers more to maintenance of key parameters in evolving systems rather than to conservation. Although the historical role of farmers in shaping the evolution of crops and their diversity has long been

[20] It is not clear, however, why individual farmers would not choose to "free ride," in other words, to rely on others to maintain the (unproductive) stocks of traditional varieties. Given that other farmers are continuing to grow traditional varieties, an individual farmer has an interest in not growing them. Instead, we might expect him or her to grow only modern varieties, secure in the knowledge that the varieties will not disappear from existence. If all farmers behaved this way, of course, the traditional varieties might in fact be threatened.

[21] In common usage, *in situ* conservation involves the conservation of genetic materials in a natural habitat or under cultivation. By contrast, *ex situ* conservation is the conservation of genetic materials in a specialized storage facility (whether in an arboretum or botanical garden, or in a deep freeze, or in a gene bank). In general, *ex situ* collections are intended to conserve genetic material outside of environmental influences, whereas *in situ* collections are intended to conserve genetic material in its environmental context.

[22] Of course, in some cases, evolutionary processes may lead to the loss of genetic material and, hence, may be undesirable from one perspective.

recognized, rigorous investigations of the complex socioeconomic and scientific issues involved in such farmer-managed conservation efforts have only just begun.[23]

Major Economic Questions Related to On-Farm Conservation

Several economic questions of importance are raised by the prospect of farmer-managed conservation of diversity. First, economic analysis of the costs and benefits of conserving the intraspecies diversity of cultivated plants requires a different conceptual approach than that proposed for *in situ* conservation of wild species. For wild species, geographical areas may be isolated and contained in reserves, and human populations paid or compensated for not harvesting, not grazing, or not otherwise disturbing the plant populations.

The evolution of cultivated crops cannot be separated from that of humans, however. Even in a center of origin and diversity, the crop genetic systems managed by farmers are "open," and their diversity relies on infusion of new genetic material through seed exchange and varietal introductions, at least in the case of an outcrossing crop like maize in Mexico (Louette et al., 1997). Direct payments or other forms of compensation to farmers for continuing to grow certain landraces have not yet, in general, been proposed.

Conservation and Economic Development

The relationship of conservation to economic development is unclear. Many have argued that on-farm conservation implies arrested development or, at least, cultural stasis. Identifying conservation methods that would allow crop populations to evolve without arresting economic development or penalizing farmers seems a laudable goal (but it is by no means clear how to achieve it). Questions have been raised about whether or not conservation can coexist with the integration of communities into commercial markets.

Certainly the replacement of landraces with modern varieties is not an inevitable process (Brush et al., 1992); farmers often continue to grow landraces even after adopting modern varieties on part of their land. If farmers' seed selection performs a stabilizing function, then experimentation with introduced varieties, seed replacements, and maintenance of landraces and their characteristics can coexist (Louette and Smale, 1997).

Incentives for *In Situ* Conservation

Even if development is compatible with conservation under some circumstances, what are farmers' incentives to maintain the diversity they grow? From an institutional

[23] At IRRI, the project "Safeguarding and Conservation of the Biodiversity of the Rice Genepool, Component II: On-Farm Conservation"; in Mexico, the McKnight Foundation project "Conservation of Genetic Diversity and Improvement of Crop Production in Mexico: A Farmer-Based Approach" and, at CIMMYT, the project "Maize Diversity Management and Utilization: A Farmer–Scientist Collaborative Approach"; in Turkey, the project "Ecology and Ethnobiology of Wheat Landrace Conservation in Central Turkey"; a longitudinal study undertaken by the Institut National de la Recherche Agronomique (INRA) in France; see other initiatives for Ethiopia and Andean crops described in Maxted et al. (1997).

perspective, Qualset et al. (1997) discuss various forms of direct and indirect incentives for local conservation efforts. The research by Meng et al.[24] is an attempt to identify feasible policy incentives that support the management of genetic diversity by Turkish farmers. In the Philippines, Bellon et al.[25] were able to identify with multidisciplinary methods a cluster of rice varieties that were both genetically diverse and highly regarded by farmers in terms of consumption characteristics, disease resistance, and other characteristics. These varieties were still cultivated in the rainfed system but had been "discarded" in the irrigated system. In this case, when both the private and public benefits of growing the traditional varieties are high, the aim of an on-farm conservation program would be to reduce the opportunity cost of growing them, by, for example, shortening their growing cycle. This might be accomplished through rice breeding.

On-Farm Crop Improvement as an Incentive

In recent years, "participatory" plant breeding has been proposed as a means of providing economic incentives for farmers to continue cultivating genetically desirable crop populations (see Eyzaguirre and Iwanaga, 1996). According to this point of view, certain techniques used by professional plant breeders may assist farmers to become more efficient in meeting their own seed selection criteria. Closer farmer–breeder collaboration could promote yield increases or other improvements in marginal environments where modern varieties have not been adopted for agronomic, social, or economic reasons. Proponents of this approach argue that while professional plant breeders have conventionally sought to develop fewer varieties adapted to a wider range of environments, on-farm improvement can support the maintenance of more-diverse, locally adapted plant populations — while including farmers who had been "left out" of the development process.

Such approaches may only make sense in "marginal environments," or, more specifically, when the private benefit–cost ratio associated with growing traditional varieties as compared with modern varieties is high. This would be true when farmers do not have the choice of adopting an introduced cultivar that produces more benefits (with respect to yield and/or any other combination of characteristics) than existing, local cultivars — either because seed systems (formal or informal) don't deliver them or because their own materials perform better with respect to a range of characteristics. Such a scenario appears to be the case in the Perales et al.[26] study of maize landraces in the Amecameca and Cuautla valleys of Mexico.

[24] See Meng, E., Taylor, J. E., and Brush, S., Incentives for on-farm crop genetic diversity: evidence from Turkey, paper presented at the symposium The Economics of Valuation and Conservation of Genetic Resources for Agriculture, May 13–15, University of Rome Tor Vergata, 1996.

[25] See Bellon, M. R., Pham, J. L., Sebastian, L. S., Francisco, S. R., Erasga, D., Sanchez, P., Calibo, M., Abrigo, G., and Quiloy, S., Farmers' perceptions and variety selection: implications for on-farm conservation of rice, paper presented at the International Workshop on Building the Basis for Economic Analysis of Genetic Resources in Crop Plants, CIMMYT and Stanford University, Palo Alto, CA, August 17–19, 1997.

[26] See Perales, H., Brush, S., and Taylor, E., Agronomic and economic competitiveness of landraces and *in situ* conservation in the Amecameca and Cuautla valleys of Mexico, paper presented at the International Workshop on Building the Basis for Economic Analysis of Genetic Resources in Crop Plants, CIMMYT and Stanford University, Palo Alto, CA, August 17–19, 1997.

Since most of the approaches to on-farm crop improvement are labor-intensive, the opportunity cost of labor is likely to play a decisive role in their attractiveness to farmers. Labor is likely to be a factor in the attractiveness of selection techniques for farmers. Rice et al. (1997) suggest that seed selection practices introduced for maize competed for labor with coffee. Referring to a labor-intensive, on-farm selection program to improve seed quality and resistance to disease in beans, Sperling et al. (1996) reported that the yield advantage of 14% was "not very convincing for those who had to do the extra maintenance." Zimmerer (1991) has argued that acute labor shortages resulting from seasonal migration have undermined the management of multiple crops and varieties in a traditionally complex cropping system in Peru. In an Oaxacan community in Mexico, García-Barrios and García-Barrios (1990) identify "diversity management" as first in a list of three agronomic characteristics of pre-Hispanic maize systems that labor migration seriously threatens.

Research Impact

Even if a feasible innovation that is attractive to farmers can be identified, what will be its research impact? Recommending strategies for on-farm maize improvement or modified seed selection practices raises some very familiar issues related to the adoption and diffusion of any agricultural technique, such as

1. Who adopts the practice and who does not?
2. If we observe farmers only for several seasons, can we conclude that they have adopted the practice?
3. Is the practice adopted on both traditional and modern varieties?
4. Is the practice diffused from farmer to farmer or by other means?
5. Does adoption of the practice affect members of the farm household differentially?
6. Does the adoption of the practice generate any observable economic benefits, and if so, for whom?

The benefits from any innovation depend on the way it diffuses among farmers and the longevity of the innovation. Consider, however, the difficulty of assessing the benefits of something like a new seed selection practice. To discern the benefits of such a practice, we are concerned with the diffusion of new seed for the same variety rather than new seed for a new variety. The conceptual frameworks and analytical models social scientists use to analyze the factors that affect seed flows (within varieties) among farmers are not nearly as well developed as those commonly used to analyze the adoption of varieties. Further, as compared with the adoption of variety, the adoption of a seed selection practice affects the characteristics of the germplasm itself. To develop sensible approaches or models, we need answers to basic questions, such as

1. How does seed flow among farmers, and to what extent is seed saved from generation to generation or exchanged among farmers?
2. If seed is exchanged, what social "infrastructure" affects the direction and magnitude of its flows?

An appropriate analytical model must be based on the identification of the essential social unit of seed conservation — which is not likely to be an individual farmer or an individual household.

The research conducted by Aguirre,[27] Louette et al. (1997), and Rice et al. (1997) in Mexico demonstrates the high frequency of seed loss, seed replacement, and seed introductions, both for seed of the same maize variety and for new varieties, including both modern varieties and landraces. These studies, conducted in different regions of Mexico with different research methods, reveal that only a subset of farmers actually retains seed year after year (in spite of common stereotypes). Some farmers replace the seed for their landraces deliberately, explaining that the seed is "tired" or needs to be "rejuvenated." Such practices have been cited in other literature, for other crops (see Wood and Lenné, 1997). Other farmers mix seed from their own and other sources, in search of "vigor."

These findings are fundamental to understanding the genetic composition of a crop and its diversity, and to understanding the diffusion and impact of techniques to improve crops on-farm. They imply that the impact of on-farm improvement is likely to be diffuse and difficult to observe, predict, and measure. To do so, we will need to understand better the "social infrastructure" that shapes seed and information flows, since in the diffusion of innovations of this type, the seed system is based entirely on farmers and their interactions. The investigations of Ashby et al. (1995), Sperling et al. (1996), and Almekinders et al. (1994) are examples of related research.

Property Rights and Conservation

An alternative approach to promoting *in situ* or on-farm conservation of diversity is to invoke changes in property rights regimes that would allow farmers to claim legal compensation when outsiders make use of landraces, traditional varieties, or other genetic resources that have been developed by farmers. Farmers and indigenous people have, in many respects, created the array of genetic diversity for agriculture that now exists. But at present, international property rights regimes allow private sector firms to claim property rights over plant varieties created through "scientific" breeding methods but make it impractical for traditional farmers to claim comparable rights.

Several previous studies have documented the present system of property rights for plant genetic resources and have surveyed the alternative forms of protection that might be extended to indigenous peoples and to farmers in poor countries. Specifically, Brush (1992), Gollin (1993a), Swanson (1994), and Walden (1995), among others, offer overviews of current property rights regimes and analyses of alternatives; Pray and Knudson (1994) investigate the impact of intellectual property regime shifts on the composition of U.S. wheat production.[28]

[27] Aguirre's findings are reported in Analisis Regional de la Diversidad del Maiz en el Sureste de Guanajuato, draft Ph.D. thesis, Universidad Nacional Autonoma de Mexico, Facultad del Ciencias, Mexico, D.F., 1997.

[28] See also a comprehensive review in Wright, B. D., Intellectual property and farmers' rights, paper presented at the symposium The Economics of Valuation and Conservation of Genetic Resources for Agriculture, May 13–15, University of Rome Tor Vergata, 1996.

A number of changes in property rights regimes have raised the prospect that farmers may soon be able to claim increased compensation for the use of traditional varieties. Under pressure from developing countries, the 22nd session of the United Nations Food and Agriculture Organization (FAO) arrived at an "International Undertaking on Plant Genetic Resources" in 1983. This document declared that there is a "universally accepted principle that plant genetic resources are a heritage of mankind and consequently should be available without restriction." This principle was held to extend even "to newly developed varieties and special genetic stocks (including elite and current breeders' lines and mutants)." As Reid et al. (1993) note, "[N]eedless to say, few developed countries with established seed industries supported the Undertaking." In 1989, an FAO Conference rejected the previous formulation and reached a compromise arrangement. Under the compromise, it was accepted that property rights could legitimately be extended to breeders' lines. At the same time, the concept of "farmers' rights" was recognized as a form of communal rights "arising from the past, present and future contributions of farmers in conserving, improving and making available plant genetic resources, particularly those in the centers of origin/diversity" (cited in FAO, 1996).

Subsequently, the Convention on Biological Diversity, which took force at the end of 1993, recognizes "the sovereign rights of States over their natural resources" (cited in FAO, 1996). The convention also called for unspecified national measures that would require signatories to share "in a fair and equitable way the results of research and development, and the benefits arising from the commercial and other utilization of genetic resources" with the countries providing the genetic resources. The mechanisms by which such sharing would take place are not specified and are taken to be worked out at the discretion of each pair of countries. It is unclear at this point how far the convention actually extends; the U.S., for one, has not yet ratified the convention. But it appears that the convention will work in the direction of strengthening property rights protection for genetic resources (Gollin, 1993b).

The Convention on Biological Diversity specifically identified farmers' rights as an outstanding issue in need of resolution through the FAO Global System.[29] Farmers' rights are defined in an FAO resolution as "rights arising from the past, present and future contribution of farmers in conserving, improving and making available plant genetic resources, particularly those in the centres of origin/diversity. These rights are vested in the International Community, as trustees for present and future generations of farmers, for the purpose of ensuring full benefits of farmers and supporting the continuation of their contributions."[30] The FAO Commission on Genetic Resources for Food and Agriculture is at present active in delineating the scope of these rights, and it is likely that at some point a statement of farmers' rights will be proposed as a protocol for the Convention on Biological Diversity. At present, it appears that if (when) farmers' rights are given full international force, mechanisms will be set up to provide compensation to farmers for the use of genetic materials based on varieties that they have developed. There is currently much active debate

[29] See Esquinas-Alcàzar, J., Farmers' rights, paper presented at the symposium The Economics of Valuation and Conservation of Genetic Resources for Agriculture, May 13–15, University of Rome Tor Vergata, 1996.

[30] Ibid.

about the mechanisms through which such compensation would be paid. Key issues include the actual recipients of the compensation; the assessment of value; the determination of genetic content; and the means by which "sharing of benefits" is actually implemented.

Wright and Gollin[31] suggest, however, that systems of farmers' rights may in fact provide little net compensation to farmers in developing countries. Many farmers in developing countries are net "borrowers" of varieties from other parts of the world and hence would lose rather than gain from a system of international compensation. Moreover, the difficulties of implementing compensation systems are formidable. For the purposes of on-farm conservation, it is unclear how monitoring would proceed or how individual farmers (rather than governments or groups) would collect compensation for the conservation or use of traditional varieties. Thus, on-farm conservation may not be greatly enhanced by changes in international legal systems.

CONCLUSIONS

Genetic diversity has enormous value for agroecosystems, both as a resource that can contribute to productivity change and as a source of "diversity services" that aid farmers and consumers alike. Increasingly, economists have made advances in understanding key issues involving genetic diversity for agroecosystems. These include

Understanding the motivations and incentives that lead farmers to use diverse arrays of species and varieties in their agroecosystems;
Recognizing the "services" that diversity offers within agroecosystems and the contributions of diversity to increased productivity;
Accounting for the positive social benefits that farmers create through maintaining diverse agroecosystems and through the conservation of landraces and traditional varieties;
Understanding and estimating the prospective future value of genetic resources as a source of desirable characteristics for varietal improvement;
Estimating the value of gene banks and germplasm collections; and
Identifying the key obstacles and policy issues relating to *in situ* conservation efforts.

In spite of these advances, many issues of importance remain unanswered. What forms of diversity are most valuable? What are the priorities for research and for conservation? How will changes in biological technology affect the importance of diversity (both at the farm level and for researchers)? How best can individual incentives be altered to reinforce individual incentives for conserving diversity?

[31] See Wright, B. D., Intellectual property and farmers' rights, paper presented at the symposium The Economics of Valuation and Conservation of Genetic Resources for Agriculture, May 13–15, University of Rome Tor Vergata, 1996; and Gollin, D., Conserving genetic resources for agriculture: local farmers, international organizations, and intellectual property rights, paper presented to Globalization and Sustainable Livelihood Systems Workshop, Institute for Social, Economic and Ecological Sustainability, University of Minnesota, April 11–12, 1997.

Questions such as these will undoubtedly remain on the agenda for policy makers in the years and decades ahead. Increasingly, however, economists are discovering the methodologies and data needed to address these questions, and the prospects are bright for future advances in understanding.

ACKNOWLEDGMENT

This chapter has grown out of the authors' collaborations and conversations with a number of colleagues over the years. These colleagues have shaped our thinking and informed our views on genetic resources for agriculture. Among them are Mauricio Bellon, Cheryl Doss, Bob Evenson, Paul Heisey, Dominique Louette, Prabhu Pingali, Brian Wright, and members of the faculty workshop on biodiversity at Williams College. Wanda Collins offered detailed and useful comments on an earlier draft of this chapter.

REFERENCES

Alderman, H. and Paxson, C. H., 1992. Do the poor insure? A synthesis of the literature on risk and consumption in developing countries, in *Economics in a Changing World: Proceedings of the Tenth World Congress of the International Economic Association,* Vol. 10, Edmar L. Bacha, Ed., St. Martin's Press, New York, 48–78.

Almekinders, C. J. M., Louwaars, N. P., and de Bruijn, G. H., 1994. Local seed systems and their importance for an improved seed supply in developing countries, *Euphytica,* 78:207–216.

Ashby, J. A., Garcia, T., del Pilar Guerrero, M., Quirós, C. A., Roa, J. I., and Beltrán, J. A., 1995. Innovation in the organization of participatory plant breeding, in *Participatory Plant Breeding. Proceedings of a Workshop on Participatory Plant Breeding,* P. Eyzaguirre and M. Iwanaga, Eds., IPGRI, Rome, Italy, 77–98.

Bellon, M., 1996. The dynamics of crop infraspecific diversity: a conceptual framework at the farmer level, *Econ. Bot.,* 50:26–39.

Bellon, M., Pham, J. L., and Jackson, M. T., 1997. Genetic conservation: a role for rice farmers, in *Plant Genetic Conservation: The In situ Approach,* N. B. Maxted, V. Ford-Lloyd, and J. G. Hawkes, Eds., Chapman and Hall, London, 263–289.

Brennan, J. P. and Byerlee, D., 1991. The rate of crop varietal replacement on farms: measures and empirical results for wheat, *Plant Varieties Seeds,* 4:99–106.

Brennan, J. P. and Fox, P., 1995. Impacts of CIMMYT Wheats in Australia: Evidence of International Research Spillover, Economics Research Report No. 1/95, NSW Agriculture, Wagga Wagga, New South Wales.

Brown, G., Jr., 1990. Valuation of genetic resources, in *The Conservation and Valuation of Biological Resources,* G. H. Orians, G. M. Brown, Jr., W. E. Kunin, and J. E. Swierzbinski, Eds., University of Washington Press, Seattle, 203–228.

Brush, S. B., 1992. Farmers' rights and genetic conservation in traditional farming systems, *World Dev.,* 20:1617–1630.

Brush, S. B., Taylor, J. E., and Bellon, M. R., 1992. Technology adoption and biological diversity in Andean potato agriculture, *J. Dev. Econ.,* 39:365–387.

Byerlee, D. and Traxler, G., 1995. National and international wheat improvement research in the post-green revolution period: evolution and impacts, *Am. J. Agric. Econ.*, 77:268–278.

Dempsey, G., 1996. *In situ* Conservation of Crops and Their Relatives: A Review of Current Status and Prospects for Wheat and Maize, NRG Paper 96-08, CIMMYT, Mexico, D.F.

Duvick, D. N., 1984. Genetic diversity in major farm crops on the farm and in reserve, *Econ. Bot.,* 38:161–178.

Ehrenfeld, D., 1988. Why put a value on biodiversity?, in *Biodiversity,* E. O. Wilson, Ed., National Academy Press, Washington, D.C., 212–216.

Evenson, R. E., 1996. The economic principles of research resource allocation, in *Rice Research in Asia: Progress and Priorities,* R. E. Evenson, R. W. Herdt, and M. Hossain, Eds., CAB International, Oxford, and International Rice Research Institute, Los Banos, 73–90.

Evenson, R. E. and Gollin, D., 1997. Genetic resources, international organizations, and improvement in rice varieties, *Econ. Dev. Cult. Change,* 45:471–500.

Eyzaguirre, P. and Iwanaga, M., 1996. *Participatory Plant Breeding. Proceedings of a Workshop on Participatory Plant Breeding,* IPGRI, Rome, Italy.

Food and Agriculture Organization (FAO) of the United Nations, 1996. *The State of the World's Plant Genetic Resources for Food and Agriculture,* FAO, Rome.

Frankel, O. H., Brown, A. H. D., and Burdon, J. J., 1995. *The Conservation of Plant Biodiversity,* Cambridge University Press, Cambridge, U.K.

García-Barrios, R. and García-Barrios, L., 1990. Environmental and technological degradation in peasant agriculture: a consequence of development in Mexico, *World Dev.,* 18:1569–1585.

Gill, M. S., 1978. Success in the Indian Punjab, in *Conservation and Agriculture,* J. G. Hawkes, Ed., Duckworth, London.

Gollin, M. A., 1993a. An intellectual property rights framework for biodiversity prospecting, in *Biodiversity Prospecting: Using Genetic Resources for Sustainable Development,* W. V. Reid, Ed., World Resources Institute, Washington, D.C. (in connection with Instituto Nacional de Biodiversidad — Costa Rica, Rainforest Alliance — USA, and African Centre for Technology Studies — Kenya), 159–198.

Gollin, M. A., 1993b. The convention on biological diversity and intellectual property rights, in *Biodiversity Prospecting: Using Genetic Resources for Sustainable Development,* W. V. Reid, Ed., World Resources Institute, Washington, D.C. (in connection with Instituto Nacional de Biodiversidad — Costa Rica, Rainforest Alliance — USA, and African Centre for Technology Studies — Kenya), 289–302.

Griliches, Z., 1957. Hybrid corn: an exploration in the economics of technological change, *Econometrica,* 25:501–522.

Hartell, J., Smale, M., Heisey, P. W., and Senauer, B., 1997. The Contribution of Genetic Resources and Diversity to Wheat Productivity: A Case from the Punjab of Pakistan, CIMMYT Economics Working Paper 97-01, CIMMYT, Mexico, D.F.

Heisey, P. W. and Brennan, J., 1991. An analytical model of farmers' demand for replacement seed, *Am. J. Agric. Econ.,* 73:44–52.

Heisey, P., Smale, M., Byerlee, D., and Souza, E., 1997. Wheat rusts and the costs of genetic diversity in the Punjab of Pakistan, *Am. J. Agric. Econ.,* 79:726–737.

Louette, D. and Smale, M., 1997. Maize seed selection and the maintenance of landrace characteristics in a traditional Mexican community, paper presented at International Workshop on Building the Basis for Economic Analysis of Genetic Resources in Crop Plants, CIMMYT and Stanford University, Palo Alto, CA.

Louette, D., Charrier, A., and Berthaud, J., 1997. *In situ* conservation of maize in Mexico: genetic diversity and maize seed management in a traditional community, *Econ. Bot.,* 51:20–38.

Lovejoy, T. E., 1997. Biodiversity: what is it?, in *Biodiversity II: Understanding and Protecting Our Natural Resources,* M. L. Reaka-Kudla, D. E. Wilson, and E. O. Wilson, Eds., Joseph Henry Press, Washington, D.C., 7–14.

Maxted, N., Ford-Lloyd, B. V., and Hawkes, J. G., 1997. *Plant Genetic Conservation: The In situ Approach,* Chapman and Hall, London.

Mendelsohn, R. and Balick, M. J., 1995. The value of undiscovered pharmaceuticals in tropical forests, *Econ. Bot.,* 49:223–228.

Mokyr, J., 1983. *Why Ireland Starved: A Quantitative and Analytical History of the Irish Economy, 1800–1850,* George Allen & Unwin, London.

National Research Council, 1972. *Genetic Vulnerability of Major Crops,* National Academy of Sciences, Washington, D.C.

Pardey, P. G., Alston, J. M., Christian, J. E., and Fan, S., 1996. *Hidden Harvest: U.S. Benefits from International Research Aid,* International Food Policy Research Institute, Washington, D.C.

Pearce, D. W. and Cervigni, R., 1994. The valuation of the contribution of plant genetic resources, in *The Appropriation of the Benefits of Plant Genetic Resources for Agriculture: An Economic Analysis of the Alternative Mechanisms for Biodiversity Conservation,* T. M. Swanson, Ed., Food and Agriculture Organization of the United Nations, Commission on Plant Genetic Resources, Rome, 40–80.

Pearce, D. W. and Moran, D., 1994. *The Economic Value of Biodiversity,* Earthscan, London.

Perrings, C., 1995. Biodiversity conservation as insurance, in *The Economics and Ecology of Biodiversity Decline,* T. M. Swanson, Ed., Cambridge University Press, Cambridge, U.K., 69–78.

Perrings, C., Barbier, E. B., Brown, G., Dalmazzone, S., Foke, C., Gadgil, M., Hanley, N., Holling, C. S., Lesser, W. H., Mäler, K.-G., Mason, P., Panayatou, T., Turner, R. K., and Wells, M., 1995. The economic value of biodiversity, in *Global Biodiversity Assessment,* V. H. Heywood and R. T. Watson, Eds., Cambridge University Press, Cambridge, U.K., 824–914.

Plucknett, D. L. and Smith, N. J. H., 1986. Sustaining agricultural yields, *BioScience,* 36:40–45.

Plucknett, D. L., Smith, N. J. H., Williams, J. T., and Anishetty, N. M., 1987. *Gene Banks and the World's Food,* Princeton University Press, Princeton, NJ.

Pray, C. E., 1983. Underinvestment and the demand for agricultural research: a case study of the Punjab, *Food Red. Inst. Stud.,* 19:52–59.

Pray, C. E. and Knudson, M., 1994. Impact of intellectual property rights on genetic diversity: the case of U.S. wheat, *Contemp. Econ. Policy,* 12:2–12.

Qualset, C. O., Damania, A. B., Zanatta, A. C., and Brush, S. B., 1997. Locally based crop plant conservation, in *Plant Genetic Conservation: The In situ Approach,* N. B., Maxted, V. Ford-Lloyd, and J. G. Hawkes, Eds., Chapman and Hall, London, 160–175.

Rajaram, S., Singh, R. P., and van Ginkel, M., 1996. Approaches to breed wheat for wide adaptation, rust resistance, and drought, in *Proceedings of the 8th Assembly of the Wheat Breeding Society of Australia,* 29 September–4 October, 1996, R. A. Richards, C. W. Wrigley, H. M. Rawson, G. J. Rebetzke, J. L. Davidson, and R. I. S Brettell, Eds., Australia, 2–30.

Rasmussen, D. C. and Phillips, R. L., 1997. Plant breeding progress and genetic diversity from de novo variation and elevated epistasis, *Crop Sci.,* 37:303–310.

Reid, W. V., Laird, S. A., Gámez, R., Sittenfield, A., Janzen, D. H., Gollin, M. A., and Juma, C., 1993. A new lease on life, in *Biodiversity Prospecting: Using Genetic Resources for Sustainable Development,* W. V. Reid, Ed., World Resources Institute, Washington, D.C. (in connection with Instituto Nacional de Biodiversidad — Costa Rica, Rainforest Alliance — USA, and African Centre for Technology Studies — Kenya), 1–52.

Rice, E., Blanco, J. L., and Smale, M., 1997. Farmers' Use of Improved Seed Selection Practices in Mexican Maize: Evidence and Issues from the Sierra de Santa Marta, Economics Working Paper 97-04, CIMMYT, Mexico, D.F.

Rosenzweig, M. R., 1988. Risk, implicit contracts, and the family in rural areas of low-income countries, *Econ. J.,* 98:1148–1170.

Rosenzweig, M. R. and Stark, O., 1989. Consumption smoothing, migration, and marriage: evidence from rural India, *J. Political Econ.,* 97: 905–926.

Rosenzweig, M. R. and Wolpin, K. I., 1993. Credit market constraints, consumption smoothing, and the accumulation of durable production assets in low-income countries: investments in bullocks in India, *J. Political Econ.,* 101:223–244.

Ryan, J. C., 1992. Conserving biological diversity, in *State of the World 1992: A Worldwatch Institute Report on Progress Toward a Sustainable Society,* W. W. Norton and Company, New York, 9–26.

Sen, A. K., 1981. *Poverty and Famines: An Essay on Entitlement and Deprivation,* Clarendon Press, Oxford.

Simpson, R. D., Sedjo, R. A., and Reid, J. W., 1996. Valuing biodiversity for use in pharmaceutical research, *J. Political Econ.,* 104:163–185.

Smale, M., 1997. The Green Revolution and wheat genetic diversity: some unfounded assumptions, *World Dev.,* 25:1257–1269.

Smale, M. and McBride, T., 1996. Understanding global trends in the use of wheat diversity and international flows of wheat genetic resources, in Part I of *CIMMYT 1995/6 World Wheat Facts and Trends: Understanding Global Trends in the Use of Wheat Diversity and International Flows of Wheat Genetic Resources,* CIMMYT, Mexico, D.F.

Sperling, L., Scheidegger, U., and Buruchara, R., 1996. Designing Seed Systems with Small Farmers: Principles Derived from Bean Research in the Great Lakes Region of Africa, Agricultural Research and Extension Network Paper No. 60, Overseas Development Agency, London, England.

Swaney, J. A. and Olson, P. I., 1992. The economics of biodiversity: lives and lifestyles, *J. Econ. Issues,* 26:1–25.

Swanson, T., 1994. *The International Regulation of Extinction,* New York University Press, New York.

Tilman, D., 1997. Biodiversity and ecosystem functioning, in *Nature's Services: Social Dependence on Natural Ecosystems,* Island Press, Washington, D.C., 93–112.

Townsend, R. M., 1995. Consumption insurance: an evaluation of risk-bearing systems in low-income economies, *J. Econ. Perspect.,* 9(3):83–102.

Udry, C., 1990. Credit markets in northern Nigeria: credit as insurance in a rural economy, *World Bank Econ. Rev.,* 4:251-70.

van der Eng, P., 1994. Development of seed-fertilizer technology in Indonesian rice agriculture, *Agric. History,* 68:20–53.

Walden, I., 1995. Conserving biodiversity: the role of property rights, in I*ntellectual Property Rights and Biodiversity Conservation: An Interdisciplinary Analysis of the Values of Medicinal Plants,* T. Swanson, Ed., Cambridge University Press, Cambridge, U.K., 176–197.

Weitzman, M., 1992. On diversity, *Q. J. Econ.,* 107:363–406.

Weitzman, M., 1993. What to preserve: an application of diversity theory to crane conservation, *Q. J. Econ.,* 108:157–184.

Widawsky, D. A., 1996. Varietal diversity and rice yield variability in Chinese rice production, paper presented at the annual meeting of the American Agricultural Economics Association, San Antonio, Texas.

Wood, D. and Lenné, J., 1997. The conservation of agrobiodiversity on-farm: questioning the emerging paradigm, *Biodiversity Conserv.,* 6:106–120.

Zimmerer, K. S., 1991. Labor shortages and crop diversity in the southern Peruvian Sierra, *Geogr. Rev.,* 8:414–442.

CHAPTER 14

Conserving and Using Crop Plant Biodiversity in Agroecosystems

Wanda W. Collins and Geoffrey C. Hawtin

CONTENTS

INTRODUCTION

Biodiversity is the variability among living organisms from all sources, including terrestrial, marine and other aquatic ecosystems, and the ecological complexes of which they are a part. The interaction of various forms of biodiversity creates and shapes the environment in which we live; it also creates and sustains the agroecosystems on which we depend for food and other basic needs. Diversity within an ecosystem enables that ecosystem to survive and be productive and to produce an

1-56670-290-9/99/$0.00+$.50

enormous range of products and services. Agrobiodiversity is that component of biodiversity that is important to agriculture and agroecosystems. It helps ensure sustainability, stability, and productivity of production systems regardless of the level of complexity of the ecosystem in which it occurs, and, in the final analysis, it contributes to social welfare of the population through its contributions to poverty alleviation and sustainable food security.

Diversity at the agroecosystem level contributes to greater food security, helps increase employment opportunities, and increases local or national self-reliance by allowing a variety of enterprises, based on products and services, to develop on a national, regional, or community scale. A diversity of crop and animal species, at the community, farm, or field level adds to social and economic stability through reducing reliance on a single enterprise. Such diversity can also lead to a more efficient use of natural resources, for example, through providing greater opportunities for nutrient recycling (Carroll et al., 1990). Species diversity can also provide a buffering effect against losses to diseases and pests or adverse weather conditions. Diseases and insects that are major problems in large, single-crop species plantings become less of a problem and cause less damage when additional crop species are added to the system (Alexander and Bramel-Cox, 1991). Even at the field level, diversity generated through planting crop mixtures can reduce losses to pests and diseases. The net result of these types of utilization of crop diversity is resilience and sustainability of agroecosystems.

Within any particular species, genetic diversity is the variation which is most important. It is the variation that enables that species to adapt to new ecosystems and environments through natural and/or human selection. Genetic diversity within a cultivated crop species at the field or farm level helps diminish the risk of losses through diseases or pests, and provides opportunities to exploit different features of the microenvironment through, for example, the presence of diverse growth habits and rooting patterns (Smith and Zobel, 1991). Such factors can contribute both to greater stability and, in many circumstances, greater productivity. Both multilines and variety mixtures have been used effectively for this purpose in grain crops (Matson et al., 1997).

Modern agroecosystems often rely on crop species which are more uniform (i.e., less genetically diverse) than those in traditional agroecosystems. However, the conservation of the existing genetic diversity of species is critical to the modern farmer as well as the traditional farmer. Genetic diversity provides the reservoir of genes for future crop improvement by farmers and professional plant breeders. The ability to continue to rely on uniform, high-yielding crop species in modern agroecosystems depends on the constant new identification and use of genes that are, or have been, found in the genetically diverse crops of the traditional agroecosystems. Similarly, useful genetic diversity can be found in wild relatives of crop species and this diversity must also be conserved and appropriately used for improving crop performance.

Thus, diversity is important to agriculture at all levels and in all agroecosystems. While recognizing the importance to agriculture of diversity at all these levels, this chapter will focus on genetic diversity within crops — the genetic resources that lie at the heart of sustainable agricultural development and provide the basis for the continued evolution and adaptation of crops.

SOURCES OF GENETIC DIVERSITY IN CROPS

Genetic variation within a crop gene pool can be found within and among professionally bred varieties, landraces or farmers' varieties, and nondomesticated relatives. In addition, new genetic variation can be introduced through mutations and the transfer of genes from different gene pools. Commercially released varieties aim to combine genes for high productivity with those required to meet different needs and environments. They contain a wealth of useful genes and gene combinations and normally form the basis for further professional plant-breeding efforts.

Landraces and farmers' varieties tend to be genetically heterogeneous and have proved to be an excellent source of genes for, *inter alia*, adaptive characters and disease and pest resistance. They are still widely grown, especially in marginal environments where they may be more stable, and even more productive, than many modern varieties. Landraces of many minor crop species are also still commonly grown as, in general, they have received relatively little attention from plant breeders and have been less subject to replacement by modern varieties.

Wild, nondomesticated relatives of crops frequently provide useful sources of genes. For example, a wild rice, *Oryza nivara*, was used to introduce resistance to grassy stunt virus in cultivated rice (Khush and Beachell, 1972). In Africa and India, cassava (*Manihot esculenta*) yields increased up to 18 times after genes from wild Brazilian cassava, conferring disease resistance, were incorporated into local varieties (Prescott-Allen and Prescott-Allen, 1982). In the U.S., disease-resistant, wild Asian species of sugarcane (*Saccharum* sp.) helped to save the U.S. sugar cane industry from collapse (Prescott-Allen and Prescott-Allen, 1982). Many other cases that have benefited agriculture in all parts of the world can be cited, as well.

In addition to these sources of genetic diversity, new DNA sequences can be created or introduced into crop species. For example, mutations are a source of new diversity and can be induced by chemical mutagens or ionizing radiation. And with modern genetic engineering techniques, all organisms, at least in theory, can contain potentially useful genes which could be transferred between crops and induced to express themselves. These new genes then become integrated into the plant genome and are passed from generation to generation.

THE NEED TO CONSERVE DIVERSITY

Conservation of genetic resources is essential, both to ensure that professional breeders continue to have access to the genes and gene complexes needed for current and future crop improvement and to enable farmers to continue to select and modify their crops in response to changing environments and circumstances. Plant breeders very effectively used genetic resources, coupled with higher levels of inputs, to meet the demands of food shortages in the 1960s and 1970s. The result was the Green Revolution. However, the lessons learned from that period have contributed to the current emphasis on long-term sustainability of production systems and the protection of the natural resource base, while at the same time maintaining an adequate and healthy food supply. This places new demands on plant breeders to continue to increase productivity in socially and environmentally appropriate ways. Plant breeders will respond to that demand; biodiversity is the means for achieving their goals. Potentially valuable genes and gene combinations, which might at present be unknown or undiscovered, can provide the means to fewer external inputs in crop production systems, lower levels of environmentally toxic pesticides, and internal resilience of agroecosystems. Conserving these needed genetic resources for future use in the face of technical and political obstacles can be an enormous challenge.

The number of higher plant species is estimated to be between 300,000 and 500,000, of which approximately 250,000 have so far been identified (Wilson, 1988; Heywood, 1995). Of these, about 30,000 are edible and an estimated 7000 have been cultivated or collected by humans for food (Wilson, 1992). Despite this, the crops that "feed the world," by providing 95% of dietary energy or protein, number only 30 (McNeely and Wachtel, 1988). Just three of them — rice, wheat, and maize — account for almost 60% of the plant-derived calories in the human diet.

Much of the genetic diversity of major tropical food crops is now believed to be reasonably secure in gene banks, especially of those species having orthodox seeds such as rice, wheat, sorghum, and maize. For some crops, which require conserving bulky vegetative organs or living tissues other than seeds, there are often fewer accessions maintained. Cassava is a crop of world importance, but there are only 23,000 accessions; and yam (*Dioscorea* spp.), an important staple crop in Africa, is represented by only 11,500 accessions. There are still specific gaps in collections even for crops which are well represented. For example, there are over 300,000 accessions of rice in storage, but there is still need for conservation of *O. sativa* from Madagascar, Mozambique, South Asia, and Southeast Asia (FAO, 1996).

The situation is greatly different for most minor crops and those species that are vegetatively propagated or produce seeds that cannot easily be stored for extended periods at subzero temperatures. Crops such as taro (*Colocasia esculenta*), yams, rice beans (*Vigna umbellata*), and breadfruit form part of the staple diet of millions of the world poor, yet relatively little work has been done either to conserve or improve them. In addition, there are many underutilized species of vegetables, fruits, and other species, including nondomesticated plants, which contribute to nutrition and dietary diversification in millions of tropical households. These species often have been largely ignored because they are of little commercial value or because they are at little risk of being lost. The result is that very little is known of the diversity, distribution, and characteristics of such species and so conservation and maintenance efforts are minimal.

Concern about the lack of knowledge and attention to conservation and enhancement was expressed by many countries in the regional and subregional meetings held in preparation for the Food and Agricultural Organization (FAO) International Technical Conference on Plant Genetic Resources, held in Leipzig, Germany in June 1996. For example, representatives of countries in west and central Africa identified a large number of underutilized species which are important to the livelihoods of local populations. These included 7 species of cereals, 8 of legumes, 4 of roots and tubers, 8 of oil crops, 31 of fruits and nuts, 17 vegetables and spices, 4 of beverages, 38 of medicinal plants, and 44 genera of forages.[1]

The situation for minor and underutilized species could be improved in the future, as some gene banks have agreed to accept regional responsibility for long-term *ex situ* storage of some minor crops. The National Bureau of Plant Genetic Resources in India has, for example, accepted responsibility for rice bean, moth bean (*V. aconitifolia*), okra (*Abelmoschus esculentus*), and amaranth (*Amaranthus* spp.), and the Institute of Plant Breeding in the Philippines has accepted responsibility for winged beans (FAO, 1996).

Because of the gaps in collections of both major and minor crops, the added factor of genetic erosion increases the urgency to conserve diversity. The diversity can be immense within the relatively small number of plant species which supply most of the world energy and protein. For example, the International Rice Research Institute gene bank contains about 80,000 accessions of rice. But much of the rice diversity may already have been lost from farmers' fields and may now only exist in gene banks. By 1982, the rice variety IR36 was grown on about 11 million ha in Asia and had replaced many local varieties (Plucknett et al., 1987). Although there are varied reasons for the loss of genetic diversity, there is widespread agreement that one major reason is replacement of local crop varieties by new cultivars. At present, few quantitative data exist to define the extent and rate of genetic erosion of crops and their wild relatives (Ceccarelli et al., 1992). Farmers in traditional systems will routinely and intentionally discard components of local crop varieties as a normal part of their management practices (Wood and Lenné, 1997).

The report State of the World's Plant Genetic Resources (FAO, 1996) also lists a number of other causes of such genetic erosion, including:

- Changes in agricultural systems and the abandonment of traditional crops in favor of new ones;
- Overgrazing and excessive harvesting;
- Deforestation and land clearance, which is cited as being the most frequent cause of genetic erosion in Africa;
- Adverse environmental conditions such as drought and flooding;
- The introduction of new pests and diseases;
- Population pressure and urbanization;
- War and civil strife; and
- Policy legislation (for example, until recently the cultivation of farm landraces was discouraged in Europe).

Regardless of the reasons for disappearance of local crop varieties and their wild relatives, the need to conserve that germplasm must be considered in conservation strategies.

There are additional needs which must be recognized in considering the necessity to conserve genetic resources, such as the high concentration of global collections

[1] See http://www.icppgr.fao.org/srm/srm-syn/caf/E3.html.

for many export crops and commodities concentrated in a small number of countries. For example, Zaire maintains over 80% of the global oil palm accessions. It is critical that more attention be given to managing these collections, to safety duplication, and to establishing new collections, especially in areas of diversity. In addition, some tropical countries need to consolidate national collections to comply with the Convention on Biological Diversity, which emphasizes in-country conservation of indigenous genetic resources, both *in situ* and *ex situ*. Many tropical and subtropical countries report problems in maintaining collections because of the lack of suitable drying facilities and unreliable electricity supplies (FAO, 1996). This situation is particularly serious in coastal regions having 60 to 90% relative humidity. Further constraints to conservation in the majority of developing countries are the lack of adequate human and financial resources and infrastructure.

Overall, the efforts to conserve the most important world agrobiodiversity are impressive and commendable. However, as noted above, there is still much to be done and many questions to be answered about the most effective and efficient conservation strategies. The need exists, as well, to ensure that agrobiodiversity is not just conserved but also fully utilized to serve agroecosystem needs.

APPROACHES TO CONSERVATION

Conservation can be broadly considered in two ways: *ex situ* and *in situ*. *Ex situ* conservation involves removing reproductive plant material from its natural setting for maintenance in seed or tissue banks or plantations. Because of the finite nature of any living plant material, *ex situ* conservation also requires regeneration of the reproductive material at given storage conditions and at species-dependent intervals. *In situ* conservation is accomplished by protecting plant material in the site in which it naturally occurs. For most wild relatives this is in nature preserves or in wild stands. For landraces, or traditional farmer varieties, it occurs in the fields in which farmers grow those varieties (on-farm conservation) or in the communities in which they are grown.

Ex Situ Conservation

Conservation in *ex situ* gene banks ensures that stored material is readily accessible; can be well documented, characterized, and evaluated; and is relatively safe from external threats. When material is stored in this way, plant evolution is effectively frozen at the time of storage.

Of the main *ex situ* methods of conservation, the most common is the storage of dried seeds in gene banks at low temperatures. For recalcitrant seeds, such as those of many tropical perennial species, and vegetatively propagated germplasm, such as *Musa*, cassava, or potatoes, other methods are needed. These include conservation as living collections in field gene banks or *in vitro* either as living plantlets, as plant tissue on appropriate media, often under conditions of slow growth, or by cryopreservation at very low temperatures, generally using liquid nitrogen. Genetic resources are also conserved as frozen or freeze-dried pollen. Increasing use is being made of isolated genomic DNA banks for the storage of germplasm (Adams, 1997) and for most major crop species as cDNA inserts in microbial species such as yeast and bacteria. DNA nucleotide sequence data for functional genes are a relatively new mode of conservation of genetic resources. From these data, polynucleotides can be synthesized in the laboratory.

For breeding programs, the screening of very large numbers of accessions for specific traits can be expensive and time-consuming. Attention has been focused in recent years on the development of core collections as a mechanism to facilitate their use (Hodgkin et al., 1995). Core collections are collections that aim to represent most of the diversity spectrum of the parent collection with a manageable number of accessions, thus improving access to the whole collection. In setting up a core collection, hierarchical approaches may be used, frequently with geographic origin as one of the primary levels of discrimination. Specific adaptive traits (such as maturity groups) can also be used to help stratify collections.

To be most useful to breeders, germplasm collections should be well documented. Accurate passport data, including site descriptions, are useful as a basis for correlating origins with environmental parameters. Characterization data include information on traits that are simply inherited and stably expressed in a wide range of environments, such as major morphological features. These types of data assist in discriminating among accessions and provide information on major adaptive features (e.g., phenological characteristics). Evaluation data involve characters important in crop production, such as yield and its components, resistance to diseases and insect pests, flowering time, and plant height, and are perhaps the most useful overall in the search for special adaptive traits, especially if originating in diverse environments.

With respect to landraces, geographic origin and local knowledge can provide very valuable leads to possible sources of genes. Farmers typically have a good knowledge about the attributes of their varieties — e.g., phenology, reaction to prevalent pests and diseases, and suitability for growing on the different soil types found in the vicinity. Local knowledge is only rarely sought during collecting, and greater efforts are needed to record such information (Guarino and Friis-Hansen, 1995). It has often been argued (IPGRI, 1993) that such information is under as great or greater threat as the germplasm itself and data collection forms for plant genetic resource collectors which provide for notes on ethnobotanical information that should be obtained during collecting missions have been developed (Eyzaguirre, 1995).

With recent advances in computer science, not only are germplasm documentation systems becoming more powerful and user-friendly, but also data exchange and the sharing of information among different systems are becoming easier. One example of how the new technologies are being applied is the information system under development by the International Agricultural Research Centers of the Consultative Group on International Agricultural Research (CGIAR). Collectively, these centers maintain over 500,000 germplasm accessions of most of the major world food and forage crops. The information system is known as SINGER (System-wide Information Network on Genetic Resources) and is available for international access through the Internet.[2]

In Situ Conservation

In situ techniques allow the conservation of greater inter- and intraspecific genetic diversity than is possible in *ex situ* facilities. They also permit continued evolution and adaptation to take place, whether in the wild or on-farm where human selection also plays a critical role. For some species, such as many tropical trees, it is the only feasible method of conservation. Sustaining habitats indefinitely due to hazards such as extreme weather conditions, pests, and diseases is a major concern for *in situ* conservation. Difficulties in mapping, characterizing, evaluating, and accessing genetic resources *in situ* are evident.

As with *ex situ* conservation, the method adopted depends on the nature of the species. Traditional crop cultivars may be conserved on-farm, while undomesticated relatives of food crops may require the setting aside of reserves. Agroforestry species, and other plants which require little maintenance, can be conserved by developing and maintaining sustainable harvesting practices and involving local communities, while forest genetic resources are usually maintained in forest reserves and in areas under specially designed management regimens.

One of the first steps for *in situ* conservation of target species or populations is to determine their status in the area where they exist. It is also necessary to determine the factors known to threaten the survival of the species and its vulnerability at various stages of its life cycle. In the case of species threatened by extinction, the minimum viable population size in the target area needs to be determined. This concept implies that a population in a given habitat cannot persist if the number of organisms is reduced below a certain threshold. The Species Survival Commission Steering Committee of the World Conservation Union (IUCN) has recently developed new categories for threatened species based on population sizes, fragmentation, and population viability analysis (IUCN, 1994). With the growing availability and use of techniques for crossing plants which are distantly related and for transferring genes from non related genera or even kingdoms, the search for useful genes has been broadened. This has resulted in an increase in activities devoted to the collection and maintenance of crop wild relatives (Ingram and Williams, 1987). This, in turn, has led to a greater realization of the value of *in situ* techniques for ensuring the conservation of a large range of potentially useful genes for future use in breeding.

Once considered primarily the domain of environmentalists and conservationists, *in situ* conservation is now also becoming of increasing interest to those concerned with crop improvement (Hodgkin, 1993). However, even though there is this growing interest in the *in situ* conservation of genetic resources, most current *in situ* programs target the preservation of ecosystems (often areas of outstanding natural beauty) or particular species (generally endangered animals or plants) rather than the intraspecific genetic diversity of plant species of potential interest for agriculture.

[2] See http://www.cgiar.org/SINGER.

Options for *in situ* conservation range from nature reserves from which all human intervention is excluded, through national parks in which economic activities with a potential to disturb the natural ecosystems are carefully regulated, to the implementation of special management regimes in areas used primarily for agriculture and forestry. The identification of specific areas in which a deliberate attempt is made to increase and maintain intraspecific diversity of key species is another approach (Krugman, 1984) which is being tried in Turkey and in Mexico. The Man and Biosphere (MAB) program of the United Nations Economic, Social and Cultural Organization is perhaps the largest coordinated global attempt to establish *in situ* reserves, one of the objectives of which is the conservation of natural areas and the genetic materials they contain. Under the MAB program, more than 250 biosphere reserves have been established around the world.

As more attention is paid to *in situ* conservation, more innovative approaches are developed. For example, locally based conservation (Qualset et al., 1997) seeks to conserve biological entities at the farm, community, or regional level. Local issues such as traditional and cultural behavior and knowledge play a large part in the conservation effort and are thus conserved themselves.

Integrated Conservation Strategies

While plant breeders can readily access germplasm maintained in *ex situ* collections, it is far more difficult to do so in the case of material conserved *in situ*. Nevertheless, the amount of inter- and intraspecific diversity that can be conserved *ex situ* is a very small proportion of the total potentially useful variation. And, for technical reasons, some domesticated and many wild species are very difficult to conserve *ex situ*. Thus, to provide a comprehensive conservation program for any particular species, strategies must include both *ex situ* and *in situ* approaches.

The comprehensive conservation of crop gene pools, which often comprise both domesticated and wild forms, may require a combination of different methods, each covering a different part of the gene pool, to enable the total to be conserved in the most cost-effective and efficient way possible. Bretting and Duvick (1997) use the terms *static conservation* and *dynamic conservation* (roughly comparable to *ex situ* and *in situ*, respectively) to denote the purpose of the conservation programs rather than the location. They recommend close collaboration between static conservation, which serves to safeguard genetic resources outside the evolutionary context, and dynamic conservation, which seeks to safeguard genetic resources in nature. In dynamic conservation, the potential for evolution of the resources is conserved as well as the cultural and agroecosystem properties that evolve along with them. The two are not mutually exclusive but are seen to be integral parts of a continuum of conservation.

The choice of appropriate strategies to protect and conserve the full range of diversity in a crop species and its relatives depends on technical factors such as reproductive biology and the nature of storage organs or propagules. It also depends on the availability of human, financial, and institutional resources to sustain a course of action once it is chosen. Such combinations of approaches are often referred to as integrated conservation strategies and are based on the unique complementarity of strengths and weaknesses between different approaches with respect to a single crop (Hawtin, 1994).

APPROACHES TO BREEDING

Approaches to crop improvement generally fall into one or two broadly defined systems:

- Formal systems, in which modern science is brought to bear on crop improvement by institutions such as government plant-breeding stations, university departments, and private breeding companies, with the aim of producing cultivars for wide, often commercial, distribution to farmers.
- Informal systems, in which farmers and local communities, mainly in developing countries, breed and select cultivars primarily to meet their own needs and circumstances.

Both systems coexist in many regions and each depends, to a greater or lesser extent, on the other.

Formal Systems

Formal systems of crop improvement normally aim to produce high-yielding cultivars that are broadly adapted across a wide range of agroecosystems. Special attention is given to breeding cultivars with specific resistances to pests and diseases and tolerances to abiotic stresses. Quality characteristics are generally determined by the preferences of large consumer groups, often in importing countries, or by the demands of processors. New cultivars must meet legal requirements for distinctness, uniformity, and stability to be officially registered. Plant breeding is an expensive process, and there is an ever-growing need to show financial returns on investment in crop improvement from the public as well as private sectors. Under these circumstances, it is perhaps not surprising that the majority of modern cultivars are widely grown or high-value crops and tend to have a relatively narrow genetic base.

In the formal sector, breeding for adaptation will continue to be concerned with improving adaptability in existing environments, extending the areas in which individual crops are grown, or seeking improved stability across a range of environments. New variation will be sought and traditional cultivars and wild relatives will continue to provide the necessary variation so long as their conservation is secured. Improved knowledge of the distribution of diversity and of the effect of specific environmental variables on that distribution will improve our capacity to locate desired characters. Improved understanding of the significance and nature of coadapted gene complexes will enable breeders to use adapted germplasm with much greater efficiency.

In formal systems, breeders are generally concerned with adaptation in one of three ways:

- To develop cultivars that are better adapted to the agroecosystems in which the target crop is currently produced;
- To develop cultivars that are adapted to new agroecosystems (geographic areas or farming systems); or
- To develop cultivars that are competitive over large areas in the hope that they will exhibit greater stability across seasons (by broadening adaptability).

The first of these is a more or less continuing part of most breeding programs and includes breeding objectives such as the incorporation of improved frost resistance in northern areas, better drought tolerance in arid lands, or disease and insect resistance. The second objective has resulted in dramatic increases of area for a number of crops such as sorghum in the U.S. (Maunder, 1992) or canola in Canada.[3] The third has been a major objective of international breeding programs in the 1960s and 1970s and resulted in such cultivars as the IR36 rice with its 75% coverage of the Southeast Asian rice acreage. To achieve such objectives, use may be made of genes having a large effect on specific, identified adaptive features, such as phenological characteristics; photoperiod response (e.g., day length in sorghum); tolerance of extremes of temperature, soil moisture, or soil chemical factors; or resistance to pests and diseases. However, a significant part is played by intensive directional selection for characters under largely additive genetic control, such as flowering time, maturation period, or by selection for broad adaptability as evidenced by low genotype × environment interaction, as in the case of IR36 (Evans, 1993).

Informal Systems

In contrast to formal plant-breeding systems, farmer- and community-level crop improvement efforts are more concerned with adapting cultivars to local conditions and systems. In regions of high ecological diversity, a large number of microenvironments may have to be catered to, and farmer circumstances change over time. In continuing to adapt local cultivars to meet their needs, farmer selection pressure may have the effect of producing a series of small adaptive shifts, primarily within the context of complex gene systems. Their high levels of heterogeneity compared with modern commercial cultivars, comparative stability across seasons, location specificity, and generalized, rather than highly specific, tolerances and resistances often generally characterize farmer varieties or landraces.

Brush (1995) has shown that farmers in widely differing environments continue to grow and use traditional cultivars on at least part of their land, even when modern cultivars are available. Traditional cultivars (landraces) may meet their specific requirements and may be better adapted to the environments in which they are grown. There is growing evidence (Riley, 1996) that throughout the world farmers, particularly in traditional agricultural systems, actively seek to improve their crops through exchange with other farmers and through selection within and among their landraces. There have been reports of farmers making crosses, and gene flow between crops

[3] See http://www.canola-council.org.

and their wild relatives can provide additional sources of variation especially in centers of origin, although the extent to which this occurs has been questioned (Wood and Lenné, 1993). Thus, landraces, or farmer cultivars, are not static, but represent diverse and dynamic gene pools that, under the pressures of both farmer and natural selection, evolve over time.

Smallholder farmers, particularly those in marginal environments, have to cope with great environmental variability. They generally lack the economic and institutional resources to transform their environments to meet the requirements of their crops (Eyzaguirre and Iwanaga, 1995). As a result, they rely on within-landrace variation or the inherent plasticity of their crops to ensure at least a minimum level of production over the seasons. For many, meeting immediate family needs and short- or medium-term survival are primary objectives. Strategies for achieving them frequently depend on the maintenance and management of diversity within and among crops and landraces in their farming systems. They must balance, for example, straw vs. grain yields, high yield vs. the ability to yield under stress conditions, and ease of hand-harvesting vs. optimum plant architecture for mechanical harvesting. Under these circumstances, farmers' strategies for managing their genetic resources have resulted in an enormous diversity among and within their crops and cultivars.

LINKING CONSERVATION AND PLANT BREEDING

To serve the two broad crop improvement systems, the formal and the informal, different approaches are needed in the conservation and supply of the diversity for current and future genetic advance. As a generalization, *ex situ* methods were developed primarily to meet the needs of the formal crop improvement system. The importance of *in situ* conservation of crop wild relatives was recognized, but, until recently, there were very few programs that specifically targeted wild relatives of crops. Within the informal system a more holistic approach has been advocated and activities aimed at supporting the conservation of traditional crop cultivars *in situ* have been initiated. As discussed earlier, there is a great need and opportunity to promote an integration of these approaches.

Conservation per se is rarely a conscious objective of farmers; however, in applying selection pressures for particular traits, traditional farmers are generally aware of the need to maintain high levels of "background" diversity. The ways in which traditional farmers manage their genetic diversity have dual effects of conserving within the gene pool many of the adaptive features of their crops either individually or as coadapted gene complexes, while at the same time allowing that gene pool to evolve in other respects to meet new needs.

Plant breeding in the formal sector depends on four main sources of diversity: wild relatives (and increasingly nonrelated taxa), mutation (often artificially induced), landraces, and modern cultivars. Of these four, the variability in landraces, or farmer varieties, is probably the most underexploited. Isozyme and molecular analyses show that landraces and wild relatives remain the main reservoirs of genetic diversity in crop gene pools (Miller and Tanksley, 1990). In addition, much of that diversity may not be obvious even through traditional genetic manipulations, but may be located and successfully used by employing molecular analyses (Tanksley and McCouch, 1997). Genetic erosion of these reservoirs continues, in some cases at alarming rates and for various reasons, and it is imperative that efforts to conserve them be strengthened. *Ex situ* gene banks alone are not sufficient, and it is in the best interests of both the formal system as well as the traditional farming communities themselves that the dynamic "cauldron" of genetic diversity be maintained *in situ*, and be allowed to continue to evolve.

Formal breeding, for economic reasons, has tended to neglect crops of only local importance and the needs of farmers in diverse marginal environments. However, for farmers in marginal areas, such crops may be the basis of their survival. They have to depend on their own efforts and can rarely expect assistance from the formal system. It is highly unlikely that formal breeding will ever fully meet their diverse needs.

The need to produce more food for the ever-growing world population requires increased productivity from all sectors. Both formal and informal systems of crop improvement have a vital role to play; yet each tends to operate independently of the other. While there is a flow of materials from the informal to the formal system, its genetic diversity remains inadequately exploited by professional breeders. Ways must be found to link these valuable sources of diversity more closely to professional breeding efforts.

Qualset et al. (1997) suggest ways that both farmers and professional breeders can link their plant-breeding efforts in breeding for "locally based conservation and improvement" of crop plant genetic resources. In a joint breeding effort, farmers would continue their tradition of improving their own material within the limits of their resources. Formal breeding programs would contribute by introducing new selection criteria and techniques that would improve performance but not change the basic character of the crop. Single gene traits which could, for example, improve disease resistance could also be introgressed and be made available to farmers in the genetic background of the material they are already using. Sperling et al. (1993) successfully involved farmers in Rwanda in breeding and selecting segregating populations of beans.

Integration of formal and informal breeding efforts are increasing and will stimulate the flow of genetic resources and information in both directions. Additional ways in which the formal system can assist the informal include the following (Hawtin et al., 1996):

- Strengthening links between farmers and gene banks to ensure adequate long-term conservation of landraces, with systems to facilitate the restoration of landraces to communities that have lost them;
- The provision by gene banks of adapted materials collected from one location to farmers in other, environmentally similar locations for local testing and adaptation;
- The provision to local communities of segregating populations and other breeding products derived from their own landraces;
- The introduction into local landraces of specific genes of relevance to local circumstances;
- The provision of a broad range of elite lines, within and from which farmers can select according to their needs; and

- Training farmers in crop improvement techniques that are relevant to their own circumstances.

As already mentioned, more participatory breeding approaches are being developed and tested in which formal sector breeders are working closely with farming communities. Through a greater understanding of the needs and aspirations of such communities, and through working in partnership, scientific expertise can best be brought to bear on the problems faced by those farmers that have been the most neglected up to now. Such approaches are expected not only to contribute to rural community development, but may also help to ensure that the large gene reservoir managed by the farmers continues to exist and to evolve, and remains a resource for all crop improvement efforts in the future.

CONCLUSIONS

The contributions of agrobiodiversity to a variety of agroecosystem functions and characteristics are well documented. The need to conserve genetic resources as the fundamental source of agrobiodiversity is also well documented. Targeted and efficient conservation strategies for the array of plant species that contribute to agrobiodiversity will involve integrated approaches which use the best approach for each species and in each system where diversity is found.

To use existing genetic diversity most effectively in an array of agroecosystems, both formal and informal plant breeding systems are vital. Just as an integrated approach to conservation will result in more efficiency in protecting genetic resources, so will integrated approaches to plant breeding result in efficiency of utilizing genetic resources and better meeting the various needs of agroecosystems with respect to biodiversity.

Research and development activities which serve to link conservation and use more closely will promote the realization of both objectives. The involvement of farmers and farm communities, as holders of valuable genetic resources, in those activities at early stages fosters the dual objectives by preserving and enhancing traits which are often ignored by the formal sector, while at the same time providing access to much-needed productivity increases. The appropriate, locally determined balance between the two will lend stability and sustainability to both crop production and crop genetic resource conservation. By blending the effective use of locally adapted landraces and infusing genes for specifically needed production traits, the valuable, farmer-held genetic material will continue to be conserved, on mutually agreed terms, for potential use by formal sector breeders.

REFERENCES

Adams, R. P., 1997. Conservation of DNA: DNA banking, in *Biotechnology and Plant Genetic Resources: Conservation and Use,* J. A. Callow, B. V. Ford-Lloyd, and H. J. Newbury, Eds., CAB Int'l., Wallingford, Oxon, England, 163–174.

Alexander, H. M. and Bramel-Cox, P. J., 1991. Sustainability of genetic resistance, in Plant Breeding and Sustainable Agriculture: Considerations for Objectives and Methods, Special Publication No. 18, Crop Science Society of America, Inc., Madison, WI, 29.

Bretting, P. K. and Duvick, D. N., 1997. Dynamic conservation of plant genetic resources, *Adv. Agron.,* 61:1–51.

Brush, S., 1995. *In situ* conservation of landraces in centers of crop diversity, *Crop Sci.,* 35:346–354.

Carroll, C. R., Vandermeer, J. H., and Rosset, P. M., 1990, *Agroecology,* McGraw Hill, New York.

Ceccarelli, S., Valkoun, J., Erskine, W., Weigand, S., Miller, R., and van Leur, J. A. G., 1992. Plant genetic resources and plant improvement as tools to develop sustainable agriculture, *Exp. Agric.,* 28:89–98.

Evans, L. T., 1993. *Crop Evolution, Adaptation and Yield,* Cambridge University Press, Cambridge, U.K.

Eyzaguirre, P. B., 1995. IPGRI's revised collecting form: ethnobotanical information in plant genetic resources collecting and documentation, *Plant Gen. Resour. Newslett.,*103:45–46.

Eyzaguirre, P. B. and Iwanaga, M., 1995. Farmers' contribution to maintaining genetic diversity in crops, and its role within the total genetic resources system, in *Participatory Plant Breeding: Proceedings of a Workshop on Participatory Plant Breeding,* P. B Eyzaguirre and M. Iwanaga, Eds., IPGRI, Rome, 9–18.

Food and Agricultural Organization (FAO), 1996. Report on the state of the world's plant genetic resources for food and agriculture, FAO, Rome, Italy.

Guarino, L. and Friis-Hansen, E., 1995. Collecting plant genetic resources and documenting associated indigenous knowledge in the field: a participatory approach, in *Collecting Plant Genetic Diversity: Technical Guidelines,* L. Guarino, V. Ramanatha Rao, and R. Reid, Eds., CAB International, Wallingford, Oxon, U.K., 345–366.

Hawtin, G., 1994. Plant genetic resources, in *Encyclopedia of Agricultural Science,* Vol. 3, 305–314.

Hawtin, G., Iwanaga, M., and Hodgkin, T., 1996. Genetic resources in breeding for adaptation, *Euphytica,* 92:255–266.

Heywood, V. H., Ed., 1995. *Global Biodiversity Assessment,* UNEP, Cambridge University Press, Cambridge, U.K.

Hodgkin, T., 1993. Wild relatives, *Naturopa,* 73:18.

Hodgkin, T., Brown, A. H. D., van Hintum, Th. J. L., and Morales, E. A. V., 1995. Future directions, in *Core Collections of Plant Genetic Resources,* T. Hodgkin, A. H. D. Brown, Th. J. L. van Hintum, and E. A. V. Morales, Eds., John Wiley and Sons, London, 253–259.

Ingram, G. B. and Williams, J. T., 1987. *In situ* conservation of wild relatives of crops, in, *Crop Genetic Resources: Conservation and Evaluation,* J. H. W. Holden and J. T. Williams, Eds., George Allen and Unwin, London, 163–179.

IPGRI, 1993. *Diversity for Development,* The Strategy of the International Plant Genetic Resources Institute, IPGRI, Rome.

IUCN, 1994. *IUCN Red List Categories,* IUCN, Gland.

Khush, G. S. and Beachell, H. M., 1972. Breeding for disease resistance at IRRI, in *Rice Breeding,* IRRI, Manila, 302–322.

Krugman, S. L., 1984. Policies, strategies, and means for genetic conservation in forestry, in *Plant Genetic Resources, a Conservation Imperative,* C. W. Yeatman, D. Kafton, and G. Wilkes, Eds., Westview Press, Boulder, CO, 71–78.

Matson, P. A., Parton, W. J., Power, A. G., and Swift, M. J., 1997. Agricultural intensification and ecosystem properties, *Science*, 277:504–550.

Maunder, A. B., 1992. Identification of useful germplasm for practical plant breeding programs, in *Plant Breeding in the 1990s,* H. T. Stalker and J. P. Murphy, Eds., CAB International, Wallingford, Oxon, 147–169.

McNeely, J. A. and Wachtel, P. S., 1988. *Soul of the Tiger,* Doubleday, New York.

Miller, J. C. and Tanksley, S. D., 1990. RFLP analysis of phylogenetic relationships and genetic variation I: the genus *Lycopersicon*, *Theor. Appl. Genet.,* 80:437–448.

Plucknett, D. L., Smith, N. J. H., Williams, J. T., and Anishetti, N. M., 1987. *Gene Banks and the World's Food,* University Press, Princeton, NJ.

Prescott-Allen, R. and Prescott-Allen, C., 1982. *Genes from the Wild. Using Genetic Resources for Food and Raw Materials,* International Institute for Environment and Development, London.

Qualset, C. O., Damania, A. B., Zanatta, A. C. A., and Brush, S. B., 1997. Locally based crop plant conservation, in *Plant Genetic Conservation: The In Situ Approach,* N. Maxted, B. V. Ford-Lloyd, and J. G. Hawkes, Eds., Chapman & Hall, London, 160–175.

Riley, K., 1996. Decentralized breeding and selection — a tool to link diversity and development, in *Using Diversity: Enhancing and Maintaining Genetic Resources on Farm,* L. Sperling and M. Loevinsohn, Eds., IDRC, New Delhi, India, 140–157.

Smith, M. E. and Zobel, R. W., 1991. Plant genetic interactions in alternative cropping systems: considerations for breeding methods, in Plant Breeding and Sustainable Agriculture: Considerations for Objectives and Methods, Special Publication No. 18, Crop Science Society of America, Inc., Madison, WI, 57.

Sperling, L., Loevinshohn, M. E., and Ntabomvura, B., 1993. Rethinking the farmer's role in plant breeding: local bean experts and on-station selection in Rwanda, *Exp. Agric.,* 29:509–519.

Tanksley, S. D. and McCouch, S. R., 1997. Seed banks and molecular maps: unlocking genetic potential from the wild, *Science*, 277:1063–1066.

Wilson, E. O., Ed., 1988. The current state of biological diversity, in *Biodiversity*, National Academy Press, Washington, D.C., 3–20.

Wilson, E. O., 1992. *The Diversity of Life,* Penguin, London.

Wood, D. and Lenné, J., 1993. Dynamic management of domesticated biodiversity by farming communities, in *UNEP/Norway Expert Conference on Biodiversity,* UNEP, Trondheim, Norway, 1–26.

Wood, D. and Lenné, J. M., 1997. The conservation of agrobiodiversity on-farm: questioning the emerging paradigm, *Biodiversity Conserv.,* 6:109–129.

CHAPTER 15

Implementing the Global Strategy for the Management of Farm Animal Genetic Resources

Keith Hammond and Helen W. Leitch

CONTENTS

INTRODUCTION

Domestic animals meet 30 to 40% of human needs for food and agriculture. Over 1.96 billion people derive some livelihood from farm animals, and, for 12% of the world population, domestic animals are the only assurance of food security. By the year 2030, world food production must increase by more than 75% in order simply to maintain current levels of food availability. Animal production must increase by at least this amount.

Domestic animal genetic resources (AnGR) underpin food security, yet irreplaceable resources are disappearing at an alarming rate: about 30% of the estimated

1-56670-290-9/99/$0.00+$.50

5000 breeds, comprising some 40 species of livestock, are at risk of extinction. About 80% of these indigenous breeds at risk occur only in developing countries and few-to-no conservation programs are in place. The fact that agriculture production depends on so few AnGR magnifies the importance of genetic diversity represented by those remaining — diversity that provides farmers with the raw material to develop livestock to be more productive; more able to resist diseases or better adapted to adverse conditions defined by available feeds, climate, and many other stresses; and more able to respond to the changing needs of modern communities.

AnGR is a key sector of agrobiodiversity. Biodiversity enables the sustainable development of agriculture. By definition, sustainability is production environment specific, and there is and will remain a diverse range of production environments globally. Indigenous breeds are adapted to the low-input to medium-input production systems which account for about three quarters of production in the developing world. However, encouraged by promises of high production, farmers are turning to imported exotic breeds developed for high-input, comparatively benign agroecosystems. This displacement of locally adapted genetic resources often results in reduced productivity and sustainability and higher risk because the exotic breeds cannot tolerate the climatic, disease, and feed constraints and the frequent seasonal swings in these components. Erosion of adapted genetics threatens not only food security and the ability to maintain sustainable agriculture throughout all production environments but also the cultural value of these indigenous resources.

The best livestock development strategy for sustainable food and agriculture production and rural development is to improve or enhance breeds adapted to specific production environments.

EVOLUTION OF THE GLOBAL STRATEGY

The Food and Agricultural Organization (FAO) mandate, which is regularly reviewed by its 174 member governments, covers (1) collating, analyzing, and reporting information; (2) providing technical assistance, with emphasis on the developing world; and (3) providing a forum for intergovernmental discussion and policy development. The member governments have resolved that FAO should lead, coordinate, facilitate and report on the Strategy for the Management of Farm Animal Genetic Resources (hereafter termed the Strategy). The term *management* incorporates the spectrum from surveying and characterizing, through to the development, use, maintenance, and access to these resources. Member governments have accepted a framework for the Strategy, and implementation involving donors and other stakeholders has commenced. The implementation of the Strategy involving all stakeholders is known as the Initiative for Domestic Animal Diversity.

Since the early 1980s, FAO has provided technical assistance to member countries in identifying and assessing their farm animal genetic resources, and in conservation and utilization activities. These activities, however, tended to be breed and country specific and on a scale which bore little relation to the magnitude of the problem. The organization did contribute to the development of practical methodologies for characterization and maintenance of animal genetic resources, and to a growing awareness of the urgency and importance of addressing the problem of biodiversity erosion on an international scale. While interventions ultimately must be at the country level, farming systems and domestic animal breeds, and the threats to their diversity and interest in accessing them, often cross national boundaries. Further, an international approach to the management of this natural capital also offers many opportunities for economies, for understanding and sharing the task, and for the development and application of methodologies.

Major developments, such as the UN Conference on Environment and Development (UNCED) in June 1992, the signing of Agenda 21, and ratification of the Convention on Biological Diversity (CBD) in December 1993, reinforced the need for the Strategy and served to shape the Strategy further. The Commission on Sustainable Development mandated to review implementation of Agenda 21, and the Conference of the Parties (COP) negotiating and implementing the articles of the CBD will also influence the development of the Strategy.

In November 1996, parties to the Rome Declaration on World Food Security recognized the need for sustainable food security and, to achieve this, for urgent action on erosion of biological diversity, pledging to implement the World Food Summit Plan of Action as their response. Also in November, the third meeting of COP considered the Conservation and Sustainable Use of Agricultural Biological Diversity as a major agenda item and decided in relation to AnGR that the Conference:

> 20. *Appreciates* the importance of the country-based Global Strategy for the Management of Farm Animal Genetic Resources under the Food and Agriculture Organization of the United Nations and strongly supports its further development, (UNEP/CBD/COP/3/l.12)

PRINCIPLES OF AnGR AND THE RATIONALE OF THE STRATEGY

Achieving "Food for All" will require sustainable intensification of agriculture in many production environments. The genetic makeup of an animal is the key to how it will respond to different aspects of the total production environment, particularly those aspects related to the uses demanded of the animal, to climate, feed and water, exposure to disease, and to type of husbandry. This tenet must be accepted to achieve and maintain sustainable farming systems, to realize production and productivity increases, to manipulate product quality, and to minimize risk of production losses over time. By definition, sustainability is specific to the production environment. Further, there is and will remain a diverse range of production environments globally.

Animals have been domesticated for thousands of years and have migrated over time as human communities occupied new areas. Once settled in a new environment, little further movement occurred but the local animal populations were changed over time by the selection pressures exerted by the local environment, including the imposition of human needs. The end result is that the gene pool of each farm animal species has been redistributed such that about half the quantitative genetic differences are unique to any one breed while the other half is common to all breed resources. It is these quantitative genetic complexes associated with adaptation and performance that are central to food and agriculture production from animals. Sustainable use of this between- and within-breed genetic diversity must be specific to the production system.

While genetic diversity underpins current production, both in terms of the average level and consistency of production from season to season, diversity also provides the basis for achieving the necessary improvements in production, productivity, and product quality. Activities directed at such sustainable intensification of livestock production should utilize genetic resources which are already adapted to the particular farming system production environments. This is because genetic adaptation takes generations to develop, so that use of resource material which is already adapted offers greater immediate and longer-term potential for achieving real and sustainable gains in system performance. The challenge is to develop practical and affordable means of realizing this goal in responding to the Commitments of the World Food Summit Plan of Action (Rome Declaration).

The maintenance of genetic diversity also forms an essential hedge against future threats to world food security, especially those posed by diseases, climate change, and other modifications to the production environment. The rapid modernization of agriculture, which has been so important in enabling the world to feed its expanding population during this century, has promoted as the solution the development and more or less universal adoption of high-intensity livestock production systems. These systems have been based on a very few species and breeds that have been developed to respond to intensive care, treatment, and feeding. Although less resilient to harsh conditions, these developed breeds are displacing and contributing to the progressive elimination of breeds which have been adapted over time to lower inputs and often harsh environments. Although this widespread displacement is the greatest cause of genetic erosion, there are several factors which place breeds at risk of loss and threaten domestic animal diversity. These factors are listed in Table 1.

Table 1 List of Causes for Risk of Loss or Extinction of Breeds

Reason	Description
Aid	Lack of incentive to develop and use breeds, giving preference to those few developed for use in high-input, high-output relatively benign environments
Product	Undue emphasis placed on a specific product or trait leading to the rapid dissemination of one breed of animal to exclusion and loss of others
Crossbreeding	Indiscriminate crossbreeding which can quickly lead to the loss of original breeds
Storage	Failure of the cryopreservation equipment and inadequate supply of liquid nitrogen to store samples of semen, ova, or embryos; or inadequate maintenance of animal populations for breeds not currently in use
Technology	Introduction of new machinery to replace animal draught and transport resulting in permanent change of farming system
Biotechnology	Poorly interpreted international genetic evaluation; artificial insemination and embryo transfer leading to rapid replacement of indigenous breeds
Violence	Wars and other forms of sociopolitical instability
Disaster	Natural disasters such as floods, drought, or famine

Paradoxically, then, the very success of modern animal-breeding programs is the principal factor redirecting effort from adapted and local breed use and development, eroding the breadth of genetic material available for future breeding work.

A series of expert consultations supported by FAO and the United Nations Environmental Program (UNEP) have identified and described the major issues underlying the rationale for the Strategy,

1. Demand for animal products in most developing countries is increasing more rapidly than for plant products; the main effect of this will be intensification of the range of periurban and mixed farming production systems. This must be sustainable intensification. It will also require that the large pastoral areas, to which local species and breeds are well adapted, are kept in production under sustainable systems.
2. Increasing sophistication of developed country consumers is moving demand away from uniform animal products to more varied products but with consistency of each variant also being increasingly demanded. Animal welfare and human health concerns are also being increasingly highlighted. Consumers are paying premium prices for foods which are grown under lower-intensity systems and which meet current lifestyle nutritional needs. These trends will continue to generate stronger market incentives for the use of more-diverse genetic resources and consequently for the maintenance of some additional farm animal biodiversity.

These observations reinforce the view that the most cost-effective way to maintain animal breeds is the development of production and marketing policies that make it financially attractive for farmers to maintain and improve local breeds. Sound genetic resource utilization policy will be particularly important for the animal species because of the high unit cost of individual animals and the often long generation interval of species.

Preliminary survey results show that most countries possess a number of the 5000 or so remaining breeds of farm livestock, with the majority of these breeds occurring only in developing countries. Most of these animal genetic resources are owned by small farmers, emphasizing the importance of private good to the sound management of these resources. The World Watch List for Domestic Animal Diversity (FAO/UNEP, 1995) lists 3882 breeds for 25 domestic species in over 180 countries. Globally, 30% of breeds are classified as endangered and criticial based on population size (Table 2). FAO defines "endangered" as populations having <1000 breeding females and <20 breeding males, and critical as populations having <100 breeding females and <5 breeding males. Of the breeds listed under these two categories, 36% are managed either through a conservation program or maintained by an institute. Presumably, the risk of loss of these breeds actively managed or maintained is far lower than breed populations outside such management programs. Of the total number of breeds with population data identified globally (2924), 19% are categorized as endangered or critical and lack a breed conservation management program. As such, there is a very high risk of loss of these animal genetic resources. Where pricing systems and allocation of benefits are inadequate, public funding will be required for effective management of these resources.

The development of competitive, sustainable production systems does not warrant the use of all existing breeds for a particular period of time. Consequently,

Table 2 Breeds of Domestic Animals at High Risk[a] by Species

Species	Total No. Breeds on File	No. Breeds with Population Data (of Total)	No. of Breeds Categorized as Critical or Endangered (% of Which Are Maintained)	No. of Breeds at High Risk of Loss (%)[a]
		Mammalian		
Ass	77	24 (31)	9 (0)	9 (38)
Buffalo	72	55 (76)	2 (0)	2 (4)
Cattle[b]	787	582 (74)	135 (41)	80 (14)
Goat	351	267 (76)	44 (16)	37 (14)
Horse	384	277 (72)	120 (20)	96 (35)
Pig	353	265 (75)	69 (25)	52 (20)
Sheep	920	656 (71)	119 (29)	85 (13)
Yak	6	6 (100)	0	0 (0)
Dromedary	50	40 (80)	2 (0)	2 (5)
Bactrian camel	7	7 (100)	1 (0)	1 (14)
Llama	3	3 (100)	0	0
Alpaca	4	4 (100)	0	0
Guanaco	2	2 (100)	0	0
Vicuna	3	3 (100)	0	0
Total	3019	2191 (89)	501 (39)	364 (17)
		Avian		
Chicken	606	512 (85)	334 (61)	131 (26)
Turkey	31	29 (94)	11 (9)	10 (35)
Domestic duck	62	54 (87)	29 (38)	18 (33)
Muscovy duck	14	13 (93)	5 (40)	3 (23)
Domestic goose	59	51 (86)	28 (11)	25 (49)
Guinea fowl	22	17 (77)	4 (0)	4 (24)
Quail	24	23 (96)	16 (100)	0
Pigeon	19	16 (84)	2 (0)	2 (13)
Pheasant	8	7 (88)	0	0
Partridge	11	4 (36)	0	0
Ostrich[c]	7	7 (100)	3 (33)	2 (29)
Total	863	733 (85)	372 (48)	195 (27)
GLOBAL	**3882**	**2924 (75)**	**873 (36)**	**559 (19)**

[a] At risk determined based on breeds with population data having <1000 breeding females or <20 breeding males and for which there is no conservation program in place.
[b] Includes mithan and banteng.
[c] Includes cassowary, emu, and nandu.
Source: Adapted from FAO/UNEP (1995).

utilization does not provide the universal solution for conservation of domestic animal diversity. Nor is it essential to retain all breeds to ensure the ready availability of the breadth of domestic animal diversity. Systematic breed-level characterization of the genetic composition of species would make it possible to develop least-cost strategies for maintaining a maximum of genetic breadth within each species from a limited number of breeds.

Although semen and embryos of only a few of the endangered breeds are held in cryopreservation storage, where the technology has been developed, it appears to offer a viable alternative to live animal conservation when that is not feasible. The nature of cryopreservation of animal cells renders such storage very high risk without sample duplication.

The geographically dispersed nature of the resources and the need to involve a range of stakeholders nationally and internationally strongly suggest that a cost-effective Strategy must also incorporate efficient information-sharing systems. Modern data processing, storage, analysis, and communication technologies offer particularly important new potential for achieving this, and full advantage must be taken of these opportunities in developing the Strategy.

Many policy issues concerning the good management of animal genetic resources remain to be addressed, such as:

1. Ownership and sovereignty;
2. Access rights and sharing of benefits;
3. Facilitating use and development;
4. Roles of culture, gender, and indigenous knowledge;
5. Priorities for conservation of breed-level diversity;
6. Intellectual property rights;
7. Use of biotechnology in characterization, utilization, maintenance, and in research and training;
8. Biosafety concerning the use of modified organisms and alien genetic resources; and
9. Zoosanitary aspects of transport of resources for purposes of research, training, and commercial use.

Because of the transboundary and legal implications which these present, such issues must be dealt with by an intergovernmental mechanism; hence, the framework for the Strategy must provide for an intergovernmental mechanism.

IMPERATIVES FOR ACTION

Based on this rationale for upgrading the management of farm animal genetic resources, FAO has identified the following imperatives for action:

1. **Identify** and **understand** the genetic resources of each domesticated animal species which collectively comprise the global gene pool for food and agriculture;
2. **Develop** and **utilize correctly** the associated diversity to increase production and productivity, to achieve sustainable agricultural systems, and, where required, to meet demands for specific product types;
3. **Monitor**, in particular, those resources which are currently represented by very few animals or which are being displaced by breed replacement strategies;
4. **Preserve** the unique resources which are not currently in demand by farmers;
5. **Train** and **involve** people in all essential facets of management of these resources; and
6. **Communicate** to the world at large the importance and value of the domestic animal genetic resources and of the associated diversity, its current high exposure

Table 3 Working Definitions Proposed for Conservation

Animal Genetic Resources (AnGR) At the breed level, the genetically unique breed populations formed throughout all domestication processes within each animal species used for the production of food and agriculture, together with their immediate wild relatives (here "breed" is accepted as a cultural rather than a technical term, i.e., to emphasize ownership, and also includes strains and research lines).

Domestic Animal Diversity (DAD) The genetic variation or genetic diversity existing among the species, breeds and individuals, for all animal species which have been domesticated and their immediate wild relatives.

Conservation (of DAD) The sum total of all operations involved in the management of animal genetic resources, such that these resources are *best used and developed* to meet *immediate and short-term* requirements for food and agriculture, and to ensure the diversity they harbor remains available to meet possible *longer-term* needs.

Conservation (in general) The management of human use of the biosphere so that it may yield the greatest sustainable benefit to *present* generations while maintaining its potential to meet the needs and aspirations of *future* generations. Thus, conservation is positive, embracing preservation, maintenance, sustainable utilization, restoration, and enhancement of the natural environment (IUCN-UNEP-WWF and FAO-UNESCO, 1980).

***In Situ* Conservation** Primarily the active breeding of animal populations for food production and agriculture, such that diversity is both best utilized in the short term and maintained for the longer term. Operations pertaining to *in situ* conservation include performance-recording schemes and development (breeding) programs. *In situ* conservation also includes ecosystem management and use for the sustainable production of food and agriculture. For wild relatives *in situ* conservation — generally called *in situ* preservation — is the maintenance of live populations of animals in their adaptive environment or as close to it as practically possible.

***Ex Situ* Conservation** In the context of conservation of domestic animal diversity, *ex situ* conservation means storage. It involves the preservation as animals of a sample of a breed in a situation removed from its normal production environment or habitat, and/or the collection and cryopreservation of resources in the form of living semen, ova, embryos, or tissues, which can be used to regenerate animals. While other methods of genetic manipulation, such as the use of various recombinant DNA techniques, may represent useful means of studying or improving breeds, they do not constitute *ex situ* conservation and may not serve conservation objectives. At present, the technical capacity to regenerate whole organisms from isolated DNA does not exist.

Source: Adapted from FAO (1995).

to loss, and the impossibility of its replacement. To facilitate effective communication a clear set of terminology is necessary. A minimum set of terms is given in Table 3.

The Strategy must be country focused both to recognize properly the sovereignty that nations have over their AnGR and because resources are lost or saved at the country level. For success, and in understanding and developing genetic resources, the Strategy must also focus interventions on identifiable production environments.

THE GLOBAL STRATEGY

To assist countries to respond to these imperatives for upgrading the management of animal genetic resources, a framework for a global strategy has been supported and is known as the Strategy for the Management of Farm Animal Genetic Resources.

The framework of the Strategy consists of four fundamental components. The Strategy is designed to be comprehensive to emphasize the balanced approach required to understand, utilize, and maintain AnGR better and more cost-effectively over time. The four fundamental components of the Strategy are

1. A global, country-based structure comprising three elements:
   a. Focal points and networks to assist countries design, implement, and maintain comprehensive national strategies for the management of their AnGR. The need has been clearly demonstrated to distribute the focal points to at least three levels — country, regional and global — although for effective policy and technical development subregional networks are also indicated for some regions.
   b. Stakeholder involvement, which FAO will lead and coordinate. This element provides for a broad range of dimensions of involvement: geographical, AnGR management element, production environment, species, time, and funding level.
   c. A virtual structure to enable collection and use of information, coordination, reporting, and to facilitate the spectrum of management processes. This virtual structure is known as the Domestic Animal Diversity Information System (DAD-IS) (Figure 1). It is being developed as an advanced communications and information system primarily for country use. Its security protocols enable countries to be responsible for validation of data and to determine what and when data are released to the world, emphasizing the country-based concept. DAD-IS also serves as the primary clearinghouse mechanism for this sector of biodiversity as required under the CBD.
2. A technical program of activity aimed at supporting effective management action at the country level, in harmony with the CBD, and comprising a set of six elements:
   a. Characterization, encompassing demographic and environment, phenotypic, and genetic indicators and assessment;
   b. *In situ* utilization and conservation;
   c. *Ex situ* conservation;
   d. Communication and information system development, including the development of DAD-IS and training;
   e. Guidelines development and action planning; and
   f. Collaboration, coordination, and policy instrument development (Figure 2).
3. Cadres of experts to guide development of the Strategy and maximize its cost-effectiveness.
4. An intergovernmental mechanism for direct government involvement, policy development, and support. This is provided by the Commission on Genetic Resources for Food and Agriculture (CGRFA) (Figure 2).

To take full advantage of the Strategy, countries are being invited by FAO to nominate an institution as a National Focal Point (NFP) and to identify a national technical coordinator (NC). The NC serves as the point of contact for the involvement of the country in the Strategy and will assist in organizing the essential in-country networking, facilitating, and coordinating activity. To ensure the country level has access to the necessary level of assistance and to best utilize the limited resources of the Global Focus, the planned coordination structure provides for decentralization to the regional level. Regional Focal Points (RFP) are to be implemented in each major genetic storehouse region of the world.

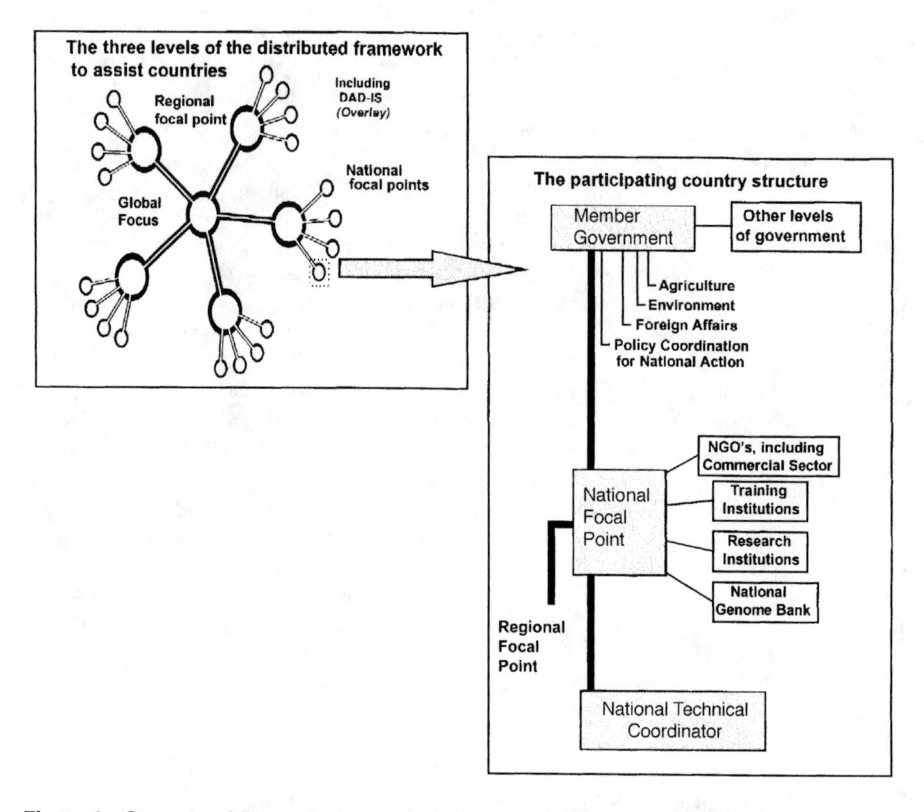

Figure 1 Structure of Domestic Animal Diversity Information System (DAD-IS).

The technical program of activity is structured to facilitate implementation and reporting and is being developed to align with the CBD (see Figure 2). Each technical element provides for a set of connected activities, for example,

1. **Characterization**. Indicators and assessment of AnGR are essential to mount a program of management. These are required at four levels:
   a. *Baseline surveys* — national inventories of AnGR and of primary production environments form the essential starting points for the development of Action Plans, including the Early Warning System for AnGR.
   b. *Monitoring* — review of the population status of particular AnGR at risk for effective and efficient conservation and as the ongoing basic need for Early Warning.
   c. *Comparative genetic evaluation* — increased knowledge of the unique qualities of breeds under each production environment, as the basis for making best use of these resources in the short and longer term.
   d. *Genetic distancing* — a comparative molecular description strategy is being developed to establish which breeds harbor significant unique genetic diversity, allowing better-targeted conservation actions and facilitating baseline surveying.
2. ***In situ* utilization and conservation**. These activities primarily involve the active breeding of animal populations such that genetic diversity is best utilized in the short term and maintained for the longer term. *In situ* operations include establishing

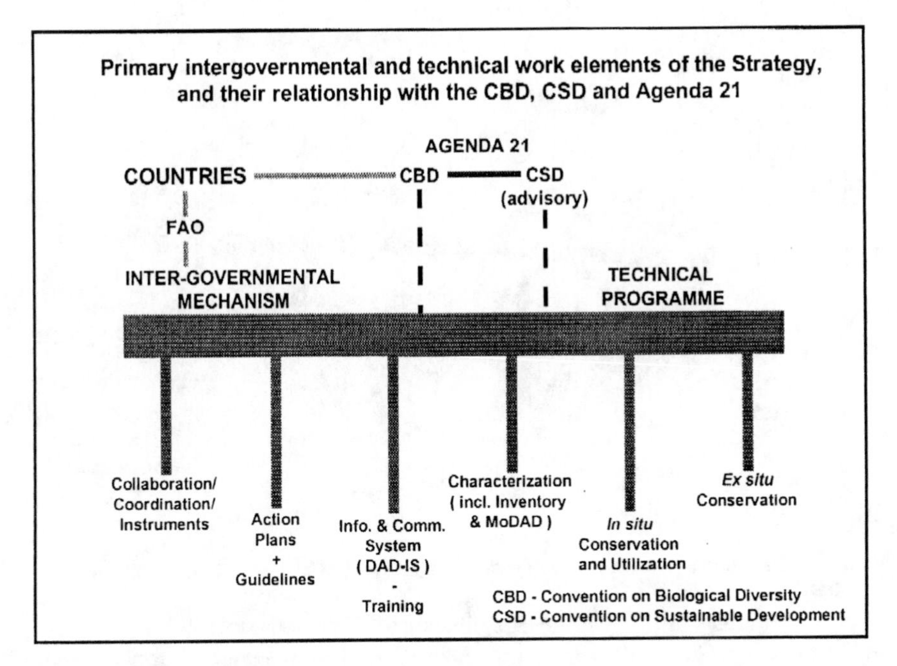

Figure 2 Primary intergovernmental and technical work elements of the Strategy and their relationship with the CBD, CSD, and Agenda 21.

breeding goals for sustainable production systems and performance recording and genetic development and dissemination of improved germplasm. *In situ* conservation covers the maintenance of live populations of animals in their natural environment for possible future use. Incentive systems will generally need to be developed to enable farmers to develop and reliably maintain such conservation activities.

3. ***Ex situ* conservation**. This activity includes cryogenic preservation and the maintenance of live animals in farm parks or zoos, beyond their development environment. The Strategy focuses on the use of live animals backed up by cryopreservation where technology exists, combining within-country genome banks with likely regionally distributed global genome repositories of last resort. For the latter in particular, policy development is required to assist countries to secure diversity associated with resources currently at high risk.
4. **Guidelines and Action Plans**. Cost-effective management of AnGR is complex technically and operationally. The Strategy provides assistance to countries through the development and provision of comprehensive guidelines for use as decision aids in the planning and implementation of national action strategies. These guidelines will form much of the detailing at the center of the Strategy; consequently, their rapid development and field testing will best assist countries. The Strategy will then be further developed by integrating all national action strategies.

FROM STRATEGY TO A GLOBAL INITIATIVE

The components of the Strategy are interdependent, and require concurrent implementation to maximize cost-effectiveness, generate momentum, and guarantee

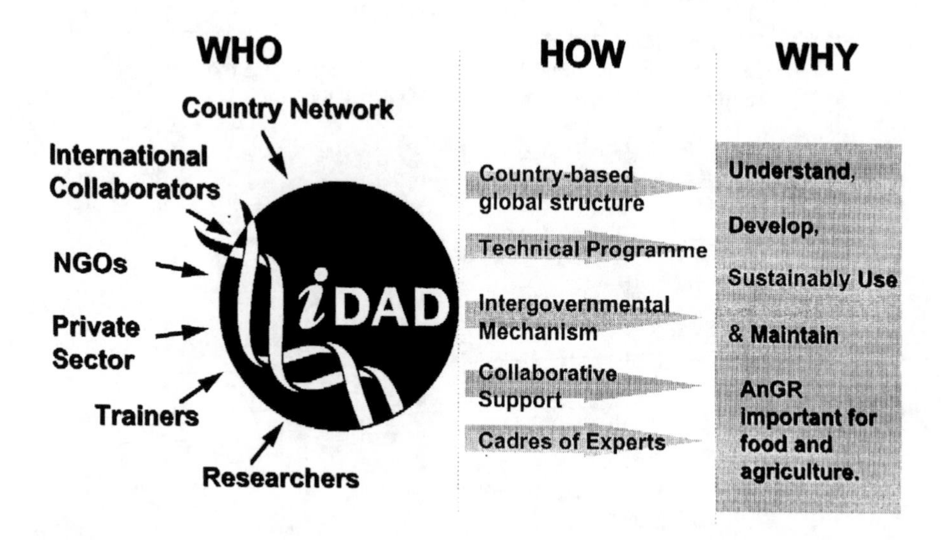

Figure 3 The Global Initiative for Domestic Animal Diversity involves all stakeholders.

success over time. Still, once the guidelines and basic infrastructure are in place, involvement must be developed over time, as human and financial resources become available.

The collective collaborative effort being developed to implement the Strategy is termed the Initiative for Domestic Animal Diversity (iDAD). The Initiative comprises all stakeholders: the countries themselves — their farmers, governments, and focal points; financial institutions; international agencies; and nongovernmental organizations, including the commercial sector. Figure 3 summarizes the Initiative. An informal mechanism for stakeholder involvement is being developed utilizing both real and virtual conferencing to achieve the range and continuity of involvement required. Building upon the Strategy, the key objectives of the Initiative are summarized in Table 4.

Table 4 Objectives of the Initiative for Domestic Animal Diversity

- Support and develop a global, country-based structure to assist countries;
- Help countries design and implement action plans and management strategies;
- Promote a better understanding of farm AnGR and the threats to them;
- Facilitate the use and improvement of locally adapted genetic resources;
- Establish the comparative uniqueness of resources in order to maximize the cost-effectiveness of their management;
- Enable the maintenance of unique AnGR by assisting development of national and global genome banks;
- Ensure the Strategy conforms to the CBD;
- Stimulate and coordinate the involvement of the range of stakeholders;
- Promote and enable intergovernmental discussion and policy development;
- Promote development and use of better technologies for cost-effective strategies;
- Upgrade training and facilitate capacity building;
- Be aware of the status of the world domestic animal diversity; and
- Facilitate access to irreplaceable animal genetic resources.

Table 5 Species of Livestock Receiving Initial Focus in the Global Strategy

Ass	Domestic goose
Buffalo	Horse
Camel	Lamoids
Cattle	Pig
Chicken	Rabbit
Duck	Sheep
Goat	Turkey

The Strategy will be implemented over 15 years, progressing through the phases of implementation from awareness and development of a global structure for coordination, to assisting countries implement national action strategies. The FAO governing body has requested the focus initially be on the 14 or so most important livestock species which account for over 90% of agriculture and food production (Table 5).

As the UN technical secretariat for agriculture, FAO is uniquely placed to facilitate, coordinate, and report on AnGR. FAO is providing some essential core resources for the operation of the Global Focus for the iDAD. However, it is neither possible nor appropriate for FAO to provide all the human, financial, and other resources required for effective implementation of the Strategy.

The Strategy aims to increase awareness and understanding of AnGR, markedly improve their use and sustainable development, upgrade the maintenance of unique resources, and enable improved access and benefit sharing. It will promote international collaboration and markedly increase the cost-effectiveness of AnGR-related initiatives worldwide, avoiding duplication of effort. The Strategy provides for a range of bilateral and multilateral assistance and for public and private sector support. It is important to note that such support need not necessarily be through FAO. Importantly, the Strategy provides for countries to discuss and develop directly the broad range of policy issues concerned with the critical upgrading now required for the management of AnGR of interest to food and agriculture, with their improved characterization, use, development, maintenance, and access.

BENEFITS OF THE GLOBAL STRATEGY

Based on compelling evidence and experiences, specific benefits associated with the Strategy include the following:

1. Framework for country assistance and action planning:
   a. A structural framework to assist countries manage their AnGR;
   b. A mechanism for country decision making and international policy development;
   c. An integrated and comprehensive technical framework to provide direction for national effort, promote expert comment and advice prior to expensive implementation of action plans, and give sound guidance for the maintenance of action strategies; and

   d. Increased country and regional institutional, managerial, scientific, and technological capacity to improve productivity of AnGR and maintain domestic animal diversity.
2. Development, sustainable use, and maintenance of Animal Genetic Resources:
   a. Coordinated surveying, monitoring, and description of AnGR;
   b. Establishment of the comparative genetic uniqueness of AnGR to minimize management costs over time;
   c. Improved development, use, and maintenance of AnGR according to a strategic combination of *in situ* and *ex situ* conservation and utilization;
   d. Better matching AnGR to feed resources and other inputs, and to required outputs, to increase productivity, reduce environmental degradation, and maintain sustainable production systems; and
   e. Conservation of domestic animal diversity of national and global importance, retaining for future generations the necessary broad flexibility required to respond quickly and efficiently to unpredictable change in production environments.
3. International collaboration and policy development:
   a. A mechanism to assure continuity and coordination of AnGR-related activities over time;
   b. A strong project identification mechanism to propose costed projects and activities for financing and to minimize duplication;
   c. Detailed country national action strategies to enable donors to recognize investment priorities; and
   d. Through the CGRFA, a mechanism for intergovernmental deliberation and policy development.
4. A communication and information tool, DAD-IS
   a. Has been developed as the key communications and information tool of the Strategy for country use, to link and involve all other stakeholders with the Strategy;
   b. Incorporates security features to enable countries to use the system as their own, with complete control over national data and other information;
   c. Provides databases for cataloging domestic animal breeds, their characteristics and risk status, production environments, and the range of activities underway for each breed, thereby acting as a comprehensive information system for AnGR use, development, access, and early warning;
   d. Enables countries and international collaborators to set conservation priorities;
   e. Supports utilization and conservation by communicating the values and roles of AnGR;
   f. Enables global networking by putting countries and their farmers in contact with other farmers; research, training, and education institutions; and scientists from around the world;
   g. Provides guidelines for development of national action strategies and supports action planning for the sustainable development, use, and maintenance of AnGR;
   h. Supports cost-effective training and research; and
   i. Provides customized project management software to support AnGR activity planning, reporting, and monitoring.

A comprehensive communications element for the Global Strategy will support all stakeholders and provide a least-cost system for awareness generation at all levels.

5. CBD objectives:
   The components and elements of the Strategy are being developed to align with the objectives of the CBD: conservation, sustainable use, and sharing of the benefits of biological diversity.
6. World Food Summit Plan of Action:
   For the farm animal biodiversity important to developing "Food for All" on a sustainable basis, the Strategy provides an integrated framework for action, particularly with respect to Commitments Three and Seven of the Plan, which refer to signatory governments pursuing participatory and sustainable development policies in high and low potential areas, and implementing, monitoring, and following up the World Food Summit Plan of Action at all levels.

The preceding benefits represent a powerful incentive for implementing a strong Strategy for the Management of Farm AnGR.

STRENGTHS AND INNOVATIVE FEATURES OF THE GLOBAL STRATEGY

The benefits are secured through a number of strengths and innovative features of the Strategy, which

1. Facilitates assessment of AnGR and incorporates development of a comprehensive early warning system, providing the basis for establishment of conservation priorities and action plans;
2. Provides the necessary framework for a participatory approach to building country consensus around complex issues, for resolving conflicting situations, and developing effective policy;
3. Incorporates a specific mechanism to facilitate country, subregional, regional, and other international cooperation in AnGR management;
4. Offers a unique mechanism to identify, link, and strengthen the network of regional and national institutions, based in the main on existing structures and their involvement in the development of the Strategy;
5. Provides a framework for assistance that promotes clear identification of priorities and links country, regional, and global priorities and concerns to maximize the impact of funded activities;
6. Makes available a unique moderated, comprehensive information and communications tool — DAD-IS — which can be used by countries as their own secure system while also enabling all stakeholders to be informed and involved; this new system also provides the primary clearinghouse mechanism for the farm animal sector of agrobiodiversity;
7. Provides the capability for ongoing project and activity documentation and monitoring by countries and donors, and for least-cost formulation of action plans at the species, country, regional, and global levels; through the DAD-IS Action Planner;
8. Promotes, for this sector of biological diversity, the development of a management approach which is in harmony with the CBD and in pursuance of Agenda 21;
9. Comprehensively provides for the involvement of the spectrum of donors and other stakeholders;
10. Allows for coordination of effort while respecting stakeholders' need for control of their own activity; and

11. Incorporates guidance from the necessary disciplinary array of technical specialists convened regularly to develop the Strategy further and advise on its implementation.

While the Strategy has been supported and early implementation has commenced,[1] success depends on its ability to mobilize the commitment, energies, and resources of the many different groups which have a stake in conserving and utilizing agrobiodiversity — farmers and pastoralists throughout the world, those involved in training and research into improved farming systems and technologies, environmentalists, agroindustries, and ultimately the consumers. While countries will be formally responsible for developing and implementing national policy, supported by the international collaborative effort, implementation of the Strategy must involve these additional stakeholders and be responsive to their varied concerns, creating mechanisms for interaction.

REFERENCES

FAO/UNEP, 1995. *World Watch List for Domestic Animal Diversity,* 2nd ed., FAO, Rome.

Hammond, K. and Leitch, H., 1996. The FAO Global Programme for the Management of Farm Animal Genetic Resources, in *Beltsville Symposia in Agricultural Research XX, Biotechnology's Role in the Genetic Improvement of Farm Animals,* R. H. Miller, V. G. Pursel, and H. D. Norman, Eds., American Society of Animal Science, Savoy, IL, 24.

IUCN/UNEP/WWF and FAO/UNESCO. 1980 World Conservation Strategy, Living Resources Conservation for Sustainable Development, IUCN, Switzerland.

[1] A listing of activities underway and progress to date implementing the Strategy is available on the Internet at http://www.fao.org/dad-is.

CHAPTER 16

Agroecosystem Quality: Policy and Management Challenges for New Technologies and Diversity

Joel I. Cohen

CONTENTS

1-56670-290-9/99/$0.00+$.50

INTRODUCTION

Providing a meaningful contribution to the topic of agroecosystems, new technology, and diversity poses many challenges. First, it is difficult to obtain agreed-on definitions or standards for "agroecosystem quality." The second difficulty occurs when considering how new technologies affect agroecosystem quality, including issues related to biodiversity. These difficulties, and the management and policy issues which they raise, are illustrated by examples of technical and adaptive challenges facing agricultural policy makers, managers, and end users concerned with maintaining levels of biodiversity or enhancing agroecosystem quality.

The objectives of this chapter are to first consider the differences between these technical and adaptive problems, the nature of the situations they each address, and the learning required when facing an adaptive challenge. Second, agroecosystem complexities and the difficulties in determining quality indicators are presented. Applications of biotechnology are presented as derived from international collaborative research using examples compiled by the Intermediary Biotechnology Service (IBS), executed by the International Service for National Agricultural Research (ISNAR). Some of these examples, as used in IBS policy seminars, highlight emerging policy and management needs which were identified and discussed. It is hoped that this chapter clarifies adaptive challenges regarding agroecosystem diversity and quality, and prepares stakeholders for the challenges and opportunities of new technologies.

CONFRONTING THE DIAGNOSTIC CHALLENGE: TECHNICAL VS. ADAPTIVE PROBLEMS

When confronting "technical problems," difficulties are faced which can be clearly defined and understood, and for which solutions are readily available. They have become problems of a technical nature by virtue of lessons learned through experiences confronted over time. The benefits derived from these accumulated experiences let us know both *what to do,* through the use of knowledge (organizational procedures for guiding our actions), and *who should do it,* by identifying whoever is authorized to perform such work (Heifetz, 1996).

When facing an "adaptive problem," however, ready organizational responses are absent, the problem is difficult to define, and expertise and/or established procedures are lacking. Technical responses to the problem are at best only part of the solution. When facing such difficulties, time is required for learning, as this is a central task of the adaptive process. Learning occurs before solutions and implementation modalities become apparent. Those holding competing values with regard to the problem are identified, questions are posed to define the issues, and stakeholders are given time to adjust values to accommodate the nature of the problem. The learning phase of adaptive work diminishes the gap between the original stakeholder values, the realities they now face, and the adjustments that may be necessary to adapt their values to the new realities (Heifetz, 1996).

Differences between technical and adaptive problems are used to diagnose issues presented in this chapter as related to agricultural productivity (see Table 1). Agricultural problems of a technical nature are often remedied by choosing among appropriate technologies, whether they are from conventional or nonconventional sources. One chooses between or combines various cultural, crop, or livestock options to address problems, needs, or deficiencies in productivity of agricultural ecosystems. However, when technologies are considered beyond their technical dimensions, in the broader sense of affecting agroecosystem quality, then adaptive problems may be encountered for the following reasons. First, no universal definition of quality exists, especially for the variable nature of agricultural ecosystems in the tropical climates of developing countries. Second, stakeholder opinions may vary as to utility vs. risk of new inputs or technologies. Third, values (whether cultural, economic, or health) create perceptions which must be addressed in relation to the realities of the proposed inputs and the changes they may cause. It is in this context that new technologies can raise adaptive challenges to farmers, system managers, and policy makers.

Consequently, questions regarding agroecosystem quality are "adaptive challenges." In this paper, two indicators of agroecosystem quality are proposed, one based on biodiversity and the second on the use of chemical inputs. These indicators can be affected by the introduction of new technologies, using biotechnology products as examples. Biological differences among agroecosystems and stakeholder values and perceptions will be critical to defining specific quality indicators. Policy and management challenges posed by new technologies and considerations of biodiversity and use of chemical inputs are then analyzed in relation to agroecosystem quality.

INTRODUCING AGROECOSYSTEMS AND INDICATORS OF QUALITY

Defining Agroecosystems

Agroecosystems include highly managed, productivity-oriented systems which vary widely in their dependence on chemical, energy, and management inputs, and are one conservation tactic identified to protect extant diversity (Soule, 1993). Defining "quality indicators" associated with agroecosystems relies on concepts *not* inherent in the system itself, just as do efforts to define sustainability. Rather, concepts such as sustainability or "quality" imply values derived from a human or cultural perspective for a particular management system (J. Tait, personal communication). These perspectives help determine whether a particular agricultural input enhances agroecosystem quality or not.

Four major components of agricultural systems have been proposed by Antle (1994) in studies on pollution and agriculture. His work highlighted relations among (1) agricultural production, (2) the broader agroecosystem, (3) human health considerations, and (4) valuation and social welfare, with each possessing characteristics

Table 1 Summarizing the Technical and Adaptive Problems, Solutions, and Questions Related to Agroecosystem Quality, Biodiversity, and New Technologies

I.A	Technical problems characterized by:	• Clear problem definition • Clear problem solution • Able to identify relevant authority/developer for solution	
I.B	Technical problems and solutions posed:	Problem 1:	Is durable resistance available for rice blast in farmer's fields?
		Technical Solution:	Improved varieties, with new sources of genetic resistance
		Problem 2:	Is insect resistance using B.t. available in tropical maize?
		Technical Solution:	Improved varieties, with new sources of genetic resistance
II.A	Adaptive problems characterized by:	• Organizational responses are absent, • The problem is difficult to define, • Expertise and/or established procedures are lacking • Technical responses are at best only part of the solution • Time required for learning	
II.B	Adaptive problem posed in this paper:	Does the introduction and use of described products require changes in stakeholder values, perceptions, or attitudes with regard to agroecosystem quality?	Two indicators of quality selected in this paper: • Biodiversity, conservation and use • Minimize use of chemical inputs
III	Answers depend on ability to address questions, such as:	In the view of the stakeholders: • Have new sources of resistance affected the composition of extant biodiversity, including possibility for horizontal gene transfer? • Have the new varieties diminished the need for chemical insecticides or fungicides? • Have new varieties included management packages for gene deployment, and extending or guarding the length of time available for resistance? • Are clear understandings available for current chemical input levels? • Are measures of productivity or other economic gains available? • Was the technical problem solved?	

valued by society. By using the divisions presented by Antle, the introduction of novel sources of genetic diversity would occur in the agricultural production. Coupling the introduction of biotechnology with the management of biodiversity and agroecosystem quality would influence a range of perspectives regarding overall quality of the agroecosystem component (2) and, often, values of human health and welfare (3 and 4).

Factors Affecting Quality Indicators

Determining practices to enhance the sustainability of a given agricultural system, as presented by Tait (personal communication), and the components used by Antle (1994) in his pollution study are also useful for this discussion. Here, these two concepts (dependence on human values and four components depicting introductions to agricultural systems) are used in the context of managing agroecosystems in developing countries. They provide a foundation for understanding the interrelations between quality indicators, inputs derived from biotechnology, and agroecosystem biodiversity. Examples of inputs are given, using cultivars as technical solutions to specific environmental and productivity problems, but which can also be valued in the context of the ecosystem.

QUALITY INDICATORS — LINKING BIODIVERSITY WITH NEW TECHNOLOGIES

Relevant agroecosystem quality indicators, which could be applied to products derived from new technologies, now need to be selected. Examples of products, like virus resistance and applications of B.t. (see section on Examples from IBS Seminars, later), illustrate both technical and adaptive challenges when considered in relation to agroecosystem quality. With such examples in mind, two indicators were selected which would relate them to agroecosystems: (1) biodiversity and (2) diminishing use of chemical inputs.

Conserving, Maintaining, and Using Biodiversity

Many traditional agroecosystems are undergoing some process of modernization (Altieri and Merrick, 1988). This process of modernization and its relation to the use of high-yielding varieties can threaten indigenous diversity or other repositories of crop germplasm. Pressures to modernize can have a drastic effect on the conservation of diversity, and indicators of quality will depend on our knowledge of natural populations in each ecosystem. In many agroecosystems, premiums are placed on maintaining and conserving sources of biodiversity. Different and often competing values exist for what constitutes an ecologically correct mix or use of diversity within a given agroecosystem. Whether this diversity can be increased or decreased reflects values attributed to ecosystem quality. Placing premiums on maintaining diversity recognizes the importance of multiple-crop agroecosystems which make use of indigenous as well as introduced sources of diversity (Gliessman, 1993). Complex crop mixtures, rotations, and practices developed by local farmers can protect the environment under tropical conditions and provide an array of products for harvest.

Several case study examples illustrate the importance of using and conserving extant biodiversity within managed agricultural and forest ecosystems (Potter et al., 1993). An important, if not essential, element of these systems is the involvement of native peoples in these managed areas, and their application of the knowledge gained over time for the care and management of such areas (Padoch and Peters, 1993). In addition, it has been argued that maintaining traditional agroecosystems is an important strategy for preserving *in situ* repositories of crop germplasm (Altieri and Merrick, 1988). For example, Latin American farming systems studied demonstrate a high degree of plant diversity (Altieri and Montecinos, 1993). The authors also recognize the importance of small farmer holdings in these ecologically diverse systems.

Minimizing Chemical Inputs

Biotechnology and sustainable agricultural systems are often portrayed as antagonistic ends of a continuum. However, this portrayal lacks evidence, especially given that the use of biotechnology-derived agricultural products within either production systems or agroecosystems is still largely an unknown factor. In fact, there are many applications of biotechnology which seek to minimize the use of chemical inputs as pest, weed, or disease control strategies in developing country agriculture. The relation between these applications and broader concerns of sustainability have been recognized (Hauptli et al., 1990). In this regard, technical solutions to pressing pest or weed management problems are becoming available from biotechnology. For this reason, minimizing chemical inputs to agroecosystems was selected as the second potential quality factor to be presented.

Both of these indicators will rely on mobilizing, understanding, and taking into account stakeholder values and perceptions. Management of agricultural systems will be complicated by the fact that indicators of quality are difficult to measure, highly location specific, and reflect "value judgments." Such indicators will by necessity incorporate values held or determined by the stakeholders of each system, and will reflect values that are not part of the biological system being considered (J. Tait, personal communication). Solutions to stakeholder problems, such as the need to combat pests or minimize chemical applications, can take the form of technical solutions by using new inputs. However, adaptive problems may also occur after interventions are identified and new technical solutions are employed. Here, stakeholder opinions may differ with the claims made by or for technical solutions, such as can occur with new products from agricultural biotechnology, or when levels of extant diversity are threatened.

It is necessary to identify the real stakeholders, to learn their expectations regarding the issue, and to gain an understanding of their opinions regarding these options to the problem at hand. Mobilizing stakeholder response is a key facet of adaptive problems, and a major task for those managing such situations (Heifetz, 1996). Constituents of specific agroecosystems will help determine quality indicators and work with those advocating new inputs, or cultural options which may affect levels of diversity. Introducing new sources of diversity raises further complications in agreeing whether such additions reflect an improvement in overall quality. These complications are expected, based on the increases in stakeholder involvement regarding the question of genetically engineered crops and introductions to areas rich in extant or indigenous biodiversity.

INTERNATIONAL COLLABORATION IN BIOTECHNOLOGY RESEARCH

With the two indicators of agroecosystem quality determined, attention is now placed on examples of new technologies. Examples have been selected that take into account the emerging needs of developing countries regarding biotechnology and their ability to collaborate with international research programs. These examples are taken from information collected from IBS policy seminars and its Registry of Expertise. IBS began to collect, analyze, and discuss with client countries its information on international collaboration in biotechnology by organizing a meeting held at ISNAR in 1993 (Cohen and Komen, 1994).

Information was collected through survey forms from some 40 international biotechnology programs. Taken together, this material clearly demonstrated that international collaboration in agricultural biotechnology offers developing countries access to a range of specific technologies, and unique opportunities for developing improved crop plants, livestock, vaccines, and diagnostic probes. An aggregate analysis of this information was made, as described below, for which specific conclusions are most relevant for a discussion on new technologies and agroecosystem quality.

Findings

Among the international programs studied by IBS, most research is undertaken on essential commodities, or foods on which significant numbers of people depend, often with regional significance (Brenner and Komen, 1994; Cohen and Komen, 1994; IBS, 1994). Analysis of the 22 international crop biotechnology research programs indicates that they address five broad research objectives, containing 126 separate activities. These primary objectives, crops, and research activities are shown in Table 2. As such, they represent solutions to many technical situations facing farmers and growers in developing countries.

With regard to crop transformation, research supported by the international programs concentrates primarily on resistance to viruses and insects, and improving quality factors (IBS, 1994). In Table 3, general categories and specific examples of transformation are shown for agriculture in industrialized countries, using examples from Day (1993). The third column summarizes research being conducted specifically for developing country agriculture with illustrations of specific applications.

These data indicate a strong commitment to improving crop plants through biotechnology by addressing agricultural needs and objectives for developing countries. Approximately 50% of the expenditures in these international biotechnology programs are devoted to research needed to develop these modified crops (Cohen,

Table 2 Number of Research Activities Undertaken by International Biotechnology Projects as Shown for Five General Research Objectives and for Crops of Major Importance to Developing Countries

	Objectives					
Crops	Disease Resistance	Insect Resistance	Virus Resistance	Quality Traits	Micropropagation	All
Cereals	**9**	**13**	**8**	**12**		**42**
Rice	5	4	6	6		21
Maize	1	6	2	3		12
Sorghum	1	3		2		6
Other	2			1		3
Root Crops	**4**	**5**	**7**	**2**	**1**	**19**
Potato	1	3	2			6
Cassava	1		3	2		6
Yam	2		1		1	4
Sweet potato		2	1			3
Legumes	**4**	**6**	**4**	**6**		**20**
Bean	1	2	1	2		6
Groundnut	1	1	3	1		6
Chickpea	1	1		2		4
Other	1	2		1		4
Horticulture	**2**		**3**		**1**	**6**
Perennial	**2**	**1**	**2**	**2**	**15**	**22**
Banana/plantain	2		1		5	8
Industrial crops				1	4	5
Coffee			1		4	5
Sugarcane			1	1	1	3
Cocoa					1	1
Forestry Species				**2**	**5**	**7**
Miscellaneous	**3**	**3**		**2**	**2**	**10**
All	**24**	**28**	**24**	**26**	**24**	**126**

Note: Figures are based on information gathered from 22 international research programs that include activities in crop research. For this table, we used those research activities with a specific applied objective, excluding research activities aimed toward general technology development.

From IBS *BioServe* Database, 1997.

1994). This percentage of available resources increases their ability to solve technical problems, as defined in this chapter, and as shown in the examples below. However, this also means that a much smaller amount of resources is available to address questions of a more adaptive nature arising as their products move from research into agricultural production, and then enter the broader agroecosystem, confronting human health or valuation considerations (Antle, 1994).

Anticipating Adaptive Challenges for Developing Countries

Over the past 4 years, IBS has organized a series of Agricultural Biotechnology Policy Seminars, held regionally for collaborating countries. In these seminars, attention is given to examples of biotechnology providing solutions to technical problems faced by farmers in developing countries. These same examples are

Table 3 Cloned Genes of Interest for Crop Plant Improvement and Related Applications of the International Biotechnology Programs

General Category[a]	Specific Examples[a]	International Biotechnology Program Applications[b]
Disease resistance: viruses	Virus coat protein subunits (TMV, cucumber mosaic, potato virus X) Potato leaf roll virus Potato virus S Soilborne wheat mosaic virus Plum pox virus Tomato spotted wilt virus Viral replicase gene (PVX)	African cassava mosaic virus, common cassava mosaic virus Bean gemini viruses Rice stripe virus, yellow mottle virus, tungo virus, ragged stunt Potato virus X and Y Tomato yellow leaf curl virus Sweet potato feathery mottle virus Groundnut stripe virus, Rosette virus, and clump virus
Fungal diseases	Chitinase gene, H1 gene for resistance to *H. carbonum* from maize, systemin gene — a peptide signal molecule which controls wound response in plants, infectious viral CDNA	Potato late blight Rice blast
Insect resistance	B.t. genes, cowpea trypsin inhibitor, wheat agglutinin gene for resistance to European corn borer	B.t. toxin genes applied to borers in maize, rice, sugarcane, potato, coffee Potato glandular trichomes Sweet potato weevil Pigonpea: *Helicoverpa* and podfly
Storage protein genes	Wheat low-molecular-weight glutenin gene, maize storage protein	No applications reported
Carbohydrate products	Polyhydroxybutyrate as an alternative to starch for the production of biodegradable plastics	No applications reported
Ripening	Antisense polygalaturonase in tomato, regulation of ACC synthase gene	No applications reported
Breeding systems	Self-incompatibility genes from *Brassica,* anther specific genes used for male sterility with a ribonuclease gene	Male sterility in rice
Flower color	Petunia, *Antirrhinum*	No applications reported
Herbicide resistance	Glyphosate, bialaphos, and imidazolinone resistance	No applications reported

[a] General categories and specific examples from Day, 1993.
[b] Examples from IBS (1994) *BioServe* database of international agricultural biotechnology programs.

explored with regard to the adaptive challenges posed when new technologies enter agricultural systems. As in many complex social situations, agricultural managers and policy makers can face substantially more complex adaptive challenges from situations originally perceived as technical in nature. Often, the problem itself is unclear because of divergent opinions regarding the nature of the problem and its possible solutions (Heifetz, 1996). One stakeholder's technical solution is another stakeholder's adaptive challenge. In these cases, there is also often disagreement among scientific experts, particularly at early stages of problem definition, hence the time needed for learning.

In the seminars, technical examples are explored from the perspective of multi-disciplinary and diverse national delegations. In facilitating these delegations, IBS ensures involvement of individuals with responsibility for, or vested interest in, the design, implementation, and use of agricultural biotechnology. This range of stake-holder interests enriches the debates which occur within each delegation as the delegates identify needs for services to help with the learning phase of adaptive work, often taking the form of policy dialogues, management recommendations, or responses needed for various international agreements. As such, IBS builds on scientific data and available understanding to expand discussions to address the broader needs of stakeholders, including policy makers, managers, and researchers, and farmers, end users or non-governmental organizations (Komen et al., 1996).

Seminar Findings

Participant action planning methodology, carried out by the 17 attending countries, identified needs and/or constraints. In total, 227 needs were identified from the delegations. These needs were systematically analyzed, identifying nine general policy issues, their relative degree of emphasis, and whether or not there was a convergence of these needs (Table 4). In addition, seven implementation issues and three issues related to priority setting have been summarized. Most relevant to a discussion on new technologies and agroecosystem diversity are the needs identified for biosafety, socioeconomics, and priority setting. Here, the specific needs related very clearly to the adaptive policy challenges facing developing countries, particularly those located in centers of diversity. These issues will be presented later, in the section on Quality Indicators and New Technologies.

EXAMPLES FROM IBS SEMINARS: THE TECHNICAL AND ADAPTIVE CHALLENGES

In the most recent policy seminar for selected countries of Latin America, three case studies were presented on issues related to biotechnology, productivity, and the environment. These case examples are most relevant to the discussion above. They illustrate solutions to agricultural problems having, to a greater or lesser extent, an adaptive and technical component (Roca et al., 1998; Serratos, 1998; Whalon and Norris, 1998).

Table 4 Number of Policy Needs Identified by Members of 17 National Delegations Attending Policy Seminars for Africa, Asia, and Latin America

General Policy Issues	No. of Countries Responding	No. of Needs Identified	No. of Convergent Needs
1 Biosafety	14	19	4
2 Socioeconomic assessment	12	19	3
3 Integration	9	11	2
4 Policy development/coordination	9	9	2
5 End user/beneficiary linkages	9	10	2
6 Technology transfer system	8	8	2
7 Intellectual Property Rights (IPR)	7	8	3
8 Biodiversity	6	7	3
9 Public awareness	5	5	1

The first example uses the introduction of improved rice varieties with the potential to curtail use of toxic and expensive fungicides. This case is primarily technical, as the products and techniques used have not posed adaptive challenges. In this case, the new varieties are not products of transgenic technologies. Rather, biotechnology tools have been used after varietal development to understand sources of resistance and to type resistance against lineages of the pathogen. For the second case, the introduction of maize containing novel sources of resistance to insect pests is considered. In the case of maize, insect resistance is derived from transgenic technologies allowing for the insertion of genes encoding a pesticide from bacteria. In the third case, broader implications of managing and deploying transgenic crops using *Bacillus thuringiensis* (B.t.) technologies are considered. As can be seen from the maize and the B.t. examples, complex situations can be anticipated when introducing new inputs into traditional agroecosystems.

The Case of Durable Resistance to Rice Blast Fungus

Blast is the most widespread and damaging disease of rice. When control is needed, and is not present in the form of cultivar resistance, then fungicide treatments are applied which may not be effective, economically sound, or desirable from an environmental perspective. Conventional resistance has been made available genetically, but it has traditionally been weakened or lost after 3 years. However, durable resistance has been achieved in rice cultivars using conventional breeding, resulting in Oryzica Llanos 5, developed as a resistant variety by Centro International de Agricultura Tropical (CIAT), the National Federation of Rice Growers, and the National Research Institute of Colombia (F. Correa-Victoria, personal communication).

The variety was introduced to tropical agroecosystems in Colombia and represented a technical solution to the problem of blast, as well as the potential to improve system quality by reducing the unwise or ineffective use of fungicides. The cultivar was adopted across Colombia in the season following its release, and has been planted in at least 50,000 ha/year until 1996. Since then, newer high-yielding cultivars were released and widely adopted by farmers (F. Correa-Victoria, personal communication).

More recently, techniques derived from biotechnology have been coupled to these applied breeding strategies (Tohme et al., 1992; Roca et al., 1998). These molecular tools are helping to understand the mechanisms controlling durable resistance in Oryzica Llanos 5 by typing resistance genes to different genetic families of blast, identifying molecular markers associated with resistance genes in other highly resistant cultivars, and guiding rice breeders in selection of potential parents leading to lines with durable blast resistance. Genes are being identified that express resistance to six pathotype lineages of the blast pathogen. This analysis depended on the use of DNA probes containing cloned fragments of the blast fungus genome which could then be used to construct DNA fingerprints of the fungus. Molecular markers were then used by breeders to confirm the manipulation and selection of various sources of resistance to these six lineages of the blast fungus. This resistance will bring a reduction in the use of fungicides by farmers as in the case of the cultivar Oryzica Llanos 5 (Tohme et al., 1992).

Decreases in the use of fungicides as a result of farmers growing these new varieties have been reported. Unfortunately, it has not been possible to review these data at this time. Measures of declining use of fungicides in agroecosystems of Colombia can be estimated, in that farmers' expenditures on these chemicals range from 6 to 50% of total crop protection costs. Actual estimates of how much farmers have saved over this period of time and how much the use of fungicides has been reduced as a result of resistance will be obtained later (F. Correa-Victoria, personal communication).

The Case of *Bacillus thuringiensis* and Transgenic Crops

By using genetic engineering, it is possible to introduce novel sources of insect resistance to crop plants. In this case, resistance comes from genes encoding the production of various endotoxins, which is being done by some of the international programs as shown in Tables 2 and 3, including work on maize. Engineered varieties would eventually be suitably adapted for growth in areas of Latin America, some areas of which are associated with centers of diversity for maize. It is essential to prepare Latin American countries for the advent of transgenic maize containing genes for insect resistance, for which it is claimed that dependence on pesticides would be eliminated, thereby enhancing the quality of the agroecosystem.

Studies on the introduction of transgenic maize in Mexico were one of the cases selected by IBS for the Latin American seminar. Serratos (1998) stated that research criteria for transgenic corn to be introduced in Mexico should be based on characterization of agroecological, social, and economic aspects of the area where it is to be grown. The introduction of transgenic cultivars seems inevitable to developing countries. Thus, it is important to consider the impact of transgenic cultivars on the agroecosystems of countries with extensive diversity of native germplasm.

Research on the use of endotoxins in maize is also being done on tropical maize at CIMMYT's (Centro Internacional de Mejoramiento de Maiz y Trigo) Applied Biotechnology Center. These activities include screening of cloned B.t. genes for toxicity against *Heliothis zea* and other tropical maize pests. They are also working on the transformation of tropical maize inbreds containing *cry* gene constructs and greenhouse evaluations of acquired transgenic germplasm containing *cry* gene(s) and introgression of *cry* gene(s) into tropical germplasm (IBS, 1994).

Research at CIMMYT and by commercial companies on hybrid maize suitable for growth in tropical climates suggests the need for further study of their potential effects on these complex agroecosystems. Thus, it is important to study, as a multi-institutional task, gene flow and biological risks which may be associated with transgenic maize in Mexico (Serratos, 1998). This could include genetic flux, hybridization, and introgression among the transgenic cultivators, native cultivators, and wild parents. Addressing factors such as these would contribute to an analysis of benefits from the transgenic maize in relation to potential environmental concerns.

In the final case (Whalon and Norris, 1998), the role of resistance management when deploying transgenic B.t. plants is discussed within a resistance management framework. Here, it was noted first that transgenic technology will help reduce reliance on chemicals, reduce environmental contamination, and reduce human health impacts by conventional pesticides. Second, this technology appeals to developing countries lacking effective pesticide safety regulations because transgenic plants do not carry the human and environmental risks that conventional pesticides do.

However, it was argued that some type of management is needed to sustain the effectiveness of these pest control tactics by managing the factors that may contribute to resistance development. This may require commitment and participation by farmers, pesticide or seed suppliers, and regulators to help prevent insect resistance through detection and proactive management. The preservation and management of genetic resources, i.e., susceptible genes, is the key goal of resistance management (Whalon and Norris, 1998).

The authors recommend that, as regards a specific group of technologies, the decision to deploy transgenic crop plants should be based on an assessment of indigenous ecological, environmental, socioeconomic, and agricultural conditions. Criteria to consider include the risk of gene transfer from transgenic plants to related species, availability of refugia to counteract resistance development, economic importance of the target pests, and the level of cooperation among growers and industry in the management of transgenic resources. The assessment should include input from scientists, policy makers, agricultural specialists, public and private institutions, and local farmers.

QUALITY INDICATORS AND NEW TECHNOLOGIES — SYNTHESIS OF ABOVE DISCUSSION

Concerns regarding the use of crops modified by new technologies vary, as shown by the case of rice and for B.t. technologies. Clearly, more issues are expected for the use of products containing B.t.-derived genes. These differences point to the need for some of the international crop biotechnology programs (see Table 2) to consider their research, testing, and use of products in the context of integrated pest or resistance management can be anticipated. It may also require more-detailed consideration of the two indicators of agroecosystem quality presented in the section on Quality Indicators — Linking Biodiversity with New Technologies, above.

The need for such approaches is often discussed in reports and workshops enumerating biosafety considerations for the introduction of transgenes into tropical agroecosystems. By summarizing these reports (see World Bank, 1993; Frederick et al., 1995; Frederiksen et al., 1995; Beachy et al., 1997; Hruska and Pavon, 1997; Serratos 1998; Whalon and Norris, 1998), the more specific considerations regarding biosafety can be covered by the following categories:

- Transgene flow in centers of diversity: crops becoming weeds, transgenes moving to wild plants, or erosion of genetic diversity
- Development of new viruses
- Resistance developed rapidly to the transgenes
- Affects on unintended targets
- Other ecosystem damage

Addressing these concerns begins with technical solutions, including data collection and experimentation. However, there is also a more adaptive component found in biosafety considerations, indicating agroecosystem complexities, the stakeholders involved, and the need for information addressing the two quality indicators selected. Generally, the more adaptive components of these concerns are voiced in terms of educating policy makers and public regarding consequences of use and deregulation, developing educational materials, and providing cost/benefit analysis reflecting the needs or priorities of each country. These points are often raised by participating countries during IBS policy seminars, and at biosafety meetings where this topic is stretched to accommodate other debates. These more adaptive challenges relate directly to the policy and management challenges facing leaders in developing countries seeking to employ the products of new agricultural technologies.

Initiating programs to address some of the above considerations often exceeds the funding base provided for the international programs. However, some of the international programs have begun this experimentation and data collection, as is being done for rice (Gould, 1997). There is an equally great need to build such understanding among those responsible for agricultural research in the developing countries. Unfortunately, developing countries cannot derive much information from analysis by regulatory agencies in developed countries for permits or notification for small-scale field testing of transgenic products, because the trial is conducted within parameters taking into account isolation, pollen flow, and avoiding persistence of crops at field sites.

These criteria and parameters enable those conducting tests to demonstrate that transgenic plants are as safe as other plant varieties. However, such isolation practices established for the needs of trials in the U.S. and Europe do little to satisfy the concerns (as listed above) anticipated for tropical ecosystems or centers of diversity. Of course, this is not the purpose of trials carried out in developed countries. The questions are: who will determine and how, whether the new plants are of no greater danger to tropical ecosystems than plants produced traditionally, and how will technical estimates for the two quality indicators be prepared in this regard?

AGROECOSYSTEM QUALITY AND CHALLENGES AHEAD — ADAPTIVE PROBLEMS REVISITED

The various points to be covered in this chapter are now complete, as summarized in Table 1. While it is not common to pose agricultural questions in the context of technical and adaptive problems, this distinction has much to offer discussions concerning biotechnology, especially when considering the range of questions that may be asked by various stakeholders regarding agronomic inputs and biodiversity. For biotechnology-derived improvements to have acceptability, clear demonstrations of utility with regard to stakeholder concerns for environmental and productivity considerations are needed.

As mentioned above, agroecosystem quality may be improved by eliminating or minimizing dependence on chemical inputs (quality indicator 2), although clear data on this is lacking at the present time. They may also affect perceptions regarding biodiversity (quality indicator 1) leading to widespread use of a variety or, in the case of transgenic maize, have implications for gene transfer in a center of diversity, or on horizontal gene transfer (Harding, 1996). The examples used (durable blast resistance and B.t. technologies) indicate potential suitability for farmers lacking access or money for chemical inputs, where it is desired to reduce chemical inputs in traditional systems or where minimal disruption of biological populations is desired. In the case of tropical maize with insect resistance, since the technologies have not yet been used or tested in the field, it was not possible to obtain estimates on expected decreases in the use of pesticides, as related to the second quality indicator.

As seen in the policy seminars, new products often focus attention on acceptance issues, which can be related to indications of agroecosystem quality. Consequently, in each seminar, socioeconomic methodologies are explored in regard to how stakeholders benefit from investments in biotechnology, and how such analysis can contribute to the learning required to address environmental and productivity questions. Follow-up to the seminars gives attention to identified needs, providing the opportunity to approach them as adaptive problems, often requiring changes in stakeholder values, attitudes, or behavior.

This supports points emphasized by Whalon and Norris (1998), as much remains to be learned regarding the wise management and deployment of genes introduced through biotechnology. Thus, findings point to where future work can be anticipated that it is hoped will diminish the learning required for adaptive situations. In many cases, these situations will weigh productivity issues of profitability, market acceptability, and overall agronomic performance with effects on agroecosystem quality. Neither dimension (environment or productivity) can be ignored. At the present time, adaptive problems arising from international biotechnology efforts are encountered not in the context of agroecosystem quality, but under the heading of biosafety considerations. The relation among biosafety, solutions offered by biotechnology, and more complex considerations of ecosystem effects is seen at many workshops.

New biotechnologies used by farmers will raise adaptive problems, of which biosafety deliberations may be one part. Stakeholder involvement will be essential in considering these cases, especially given that local land-use knowledge continues to be essential to food production in the tropics and in traditional agroecosystems (Gliessman, 1993). Such knowledge reflects experience gained over many generations, and can contribute much toward sound management practices. Using local knowledge when determining quality indicators could be done in conjunction with efforts to determine natural resource or ecologically sound management practices. However, as already stated, such measurements have human biases or judgments attached to them and reflect the stakeholders involved.

REFERENCES

Altieri, M. A. and Merrick, L. C., 1988. Agroecology and *in situ* conservation of native crop diversity in the third world, in *Biodiversity,* E. O. Wilson, Ed., National Academy Press, Washington, D.C., 361–369.

Altieri, M. A. and Montecinos, C., 1993. Conserving crop genetic resources in Latin America through farmers' participation, in *Perspectives on Biodiversity: Case Studies of Genetic Resource Conservation and Development,* C. S. Potter, J. I. Cohen, and D. Janczewski, Eds., AAAS Press, Washington, D.C., 45–64.

Antle, J. M., 1994. Health, environment and agricultural research, in *Agricultural Technology: Policy Issues for the International Community,* J. R. Anderson, Ed., CAB International, Wallingford, Oxon, U.K., 517–531.

Beachy, R., Eisner, T., Gould, F., Herdt, R., Kendall, H. W., Raven, P. H., Swaminathan, M. S., and Schell, J. S., 1997. Bioengineering of Crop Plants, Report of The World Bank Panel on Transgenic Crops, World Bank, Washington, D.C.

Brenner, C. and Komen, J., 1994. International Initiatives in Biotechnology for Developing Country Agriculture: Promises and Problems, Technical Paper No. 100, OECD, Paris.

Cohen, J. I., 1994. *Biotechnology Priorities, Planning, and Policies: A Framework for Decision Making,* International Service for National Agricultural Research, The Hague.

Cohen, J. I. and Komen, J., 1994. International agricultural biotechnology programmes: providing opportunities for national participation, *AgBiotech N. Inf.,* 6(11):257N–267N.

Day, P., 1993. Integrating plant breeding and molecular biology: accomplishments and future promise, in *International Crop Science I,* D. R. Buxton, R. Shibles, and R. A. Forsberg, Eds., Crop Science Society of America, Inc., Madison, WI, 517–523.

Frederick, R. J., Virgin, I., and Lindarte, E., 1995. Environmental concerns with transgenic plants in centers of diversity: potato as a model, in *Proceedings from a Regional Workshop,* Parque National Iguazu, Argentina, 2–3 June, 1995.

Frederiksen, R., Shantharam, R., and Raman, K. V., 1995. Environmental impact and biosafety: issues of genetically engineered sorghum, *Afr. Crop Sci. J.,* 3:131–244.

Gliessman, S. R., 1993. Managing diversity in traditional agroecosystems of tropical Mexico, in *Perspectives on Biodiversity: Case Studies of Genetic Resource Conservation and Development,* C. S. Potter, J. I. Cohen, and D. Janczewski, Eds., AAAS Press, Washington, D.C., 65–74.

Gould, F., 1997. Integrating pesticidal engineered crops into Mesoamerican agriculture, in *Transgenic Plants in Mesoamerican Agriculture,* A. J. Hruska and M. L. Pavon, Eds., Zamorano Academic Press, Tegucigalpa, Honduras, 6–36.

Harding, K., 1996. The potential for horizontal gene transfer within the environment, *Agro-Food-Industry-Hi-Tech,* 31–35.

Hauptli, H., Katz, D., Thomas, B. R., and Goodman, R. M., 1990. Biotechnology and crop breeding for sustainable agriculture, in *Sustainable Agricultural Systems,* C. A. Edwards, R. Lal, and P. Madden, Eds., Soil and Water Conservation Society, Ankeny, IA, 141–156.

Heifetz, R. A., 1996. *Leadership without Easy Answers,* Belknap Press of Harvard University, Cambridge, MA.

Hruska, A. J. and Pavon, M. L., 1997. *Transgenic Plants in Mesoamerican Agriculture,* Zamorano Academic Press, Tegucigalpa, Honduras.

Komen, J., Cohen, J. I., and Ofir, Z., 1996. *Turning Priorities into Feasible Programs, Proceedings of a Policy Seminar on Agricultural Biotechnology for East and Southern Africa,* Intermediary Biotechnology Service and Foundation for Research Development, The Hague.

IBS, 1994. *International Initiatives in Agricultural Biotechnology: A Directory of Expertise,* Intermediary Biotechnology Service, The Hague.

Padoch, C. and Peters, C., 1993. Managed forest gardens in West Kalimantan, Indonesia, in *Perspectives on Biodiversity: Case Studies of Genetic Resource Conservation and Development,* C. S. Potter, J. I. Cohen, and D. Janczewski, Eds., AAAS Press, Washington, D.C., 167–176.

Potter, C. S., Cohen, J. I., and Janczewski, D., 1993. *Perspectives on Biodiversity: Case Studies of Genetic Resource Conservation and Development,* AAAS Press, Washington, D.C.

Roca, W., Correa-Victoria, F., Martinez, C., Tohme, J., Lentini, Z., and Levy, M., 1998. Desarrollo de resistencia durable al anublo del arroz: consideraciones de productividad y ambientales, in *Proceedings of IBS–CamBioTec Regional Policy Seminar on Planning, Priorities and Policies for Agricultural Biotechnology,* Peru, October 5–10, 1996. ISNAR, The Hague, (in press).

Serratos, J. A., 1998. Evaluación de variedades novedosas de cultivos agarícolas en su centro de origen y diversidad. El caso del maíz en México, in *Proceedings of IBS–CamBioTec Regional Policy Seminar on Planning, Priorities and Policies for Agricultural Biotechnology,* Peru, October 5–10, 1996, ISNAR, The Hague, (in press).

Soule, M. E., 1993. Conservation: tactics for a constant crisis, in *Perspectives on Biodiversity: Case Studies of Genetic Resource Conservation and Development,* C. S. Potter, J. I. Cohen, and D. Janczewski, Eds., AAAS Press, Washington, D.C., 3–17.

Tohme, J., Correa-Victoria, F., and Levy, M., 1992. *Know Your Enemy: A Novel Strategy to Develop Durable Resistance to Rice Blast Fungus through Understanding the Genetic Structure of the Pathogen Population,* CIAT/Purdue University, Cali, Colombia and West Lafayette, IN, 140.

Whalon, M. E. and Norris, D. L., 1998. Pest resistance management and transgenic plant deployment: perspectives and policy recommendations for the developing world, in *Proceedings of IBS–CamBioTec Regional Policy Seminar on Planning, Priorities and Policies for Agricultural Biotechnology,* Peru, October 5–10, 1996, ISNAR, The Hague, (in press).

World Bank, 1993. Rice Biosafety, World Bank Technical Paper, Biotechnology Series No. 1, World Bank, Washington, D.C.

Index

A

B

C

D

E

F

G

H

I

J

K

L

M

N

Q

R

S

T

U

V

W

Y